TOWARDS A COMPARATIVE
HISTORY OF COALFIELD SOCIETIES

Dedicated to Klaus Tenfelde

Towards a Comparative History of Coalfield Societies

Edited by

STEFAN BERGER
University of Manchester, UK

ANDY CROLL
NORMAN LaPORTE
University of Glamorgan, UK

LONDON AND NEW YORK

First published 2005 by Ashgate Publishing

2 Park Square, Milton Park, Abingdon, Oxon OX14 4RN
711 Third Avenue, New York, NY 10017, USA

Routledge is an imprint of the Taylor & Francis Group, an informa business

First issued in paperback 2017

British Library Cataloguing in Publication Data
Towards a comparative history of coalfield societies. –
(Studies in labour history)
1. Coal miners – Social conditions – 20th century – Cross-cultural studies 2. Coal miners – Social conditions – 19th century – Cross-cultural studies 3. Coal mines and mining – History – 20th century – Cross-cultural studies 4. Coal mines and mining – History – 19th century – Cross-cultural studies
I. Berger, Stefan, 1964- II. Croll, Andy, 1948- III. LaPorte, Norman
331.7'622334

Library of Congress Cataloging-in-Publication Data
Towards a comparative history of coalfield societies/edited by Stefan Berger, Andy Croll, Norman LaPorte.
    p. cm. – (Studies in labour history)
    Includes index.
    ISBN 0-7546-3777-8 (alk. paper)
1. Coal miners – Cross-cultural studies. 2. Coal miners – Labor unions. I. Berger, Stefan. II. Croll, Andy, 1948- III. LaPorte, Norman. IV. Series: Studies in labour history (Ashgate (Firm))

    HD8039.M615T69 2005
    331.7'622334–dc22

                                                                    2004012653
ISBN 978-0-7546-3777-6 (hbk)
ISBN 978-1-138-26641-4 (pbk)

Typeset by Tradespools, Frome, Somerset

Contents

Studies in Labour History
General Editor's Preface

Labour history has often been a fertile area of history. Since the Second World War its best practioners – such as E.P. Thompson and E.J. Hobsbawm, both Presidents of the British Society for the Study of Labour History – have written works which have provoked fruitful and wide-ranging debates and further research, and which have influenced not only social history but history generally. These historians, and many others, have helped to widen labour history beyond the study of organised labour to labour generally, sometimes to industrial relations in particular, and most frequently to society and culture in national and comparative dimensions.

The assumptions and ideologies underpining much of the older labour history have been challenged by feminist and later by postmodernist and anti-Marxist thinking. These challanges have often led to thoughtful reappraisals, perhaps intellectual equivalents of coming to terms with a new post-Cold War political landscape.

By the end of the twentieth century, labour history had emerged reinvigorated and positive from much introspection and external criticism. Very few would wish to confine its scope to the study of organised labour. Yet, equally, few would wish now to write the existence and influence of organised labour out of nations' histories, any more than they would wish to ignore working-class lives and focus only on the upper echelons.

This series of books provides reassessments of broad themes of labour history as well as some more detailed studies arising from recent research. Most books are single-authored but there are also volumes of essays centred on important themes or periods, some arising from major conferences organised by the Society for the Study of Labour History. The series also includes studies of labour organisations, including international ones, as many of these are much in need of a modern reassessment.

Chris Wrigley
British Society for the Study of Labour History
University of Nottingham

List of Figures

List of Tables

Introduction

Stefan Berger

Much is familiar about the history of miners, their unions and their industrial struggles. In Britain many of the major debates on working-class and labour history involved discussion of the coal industry.[1] In many other countries too, coal and the people who dug it up attracted considerable attention from labour historians. However, the emphasis was traditionally on the organizational history of miners and their employers as well as on the diverse ideologies of trade unions and employers' federations. That kind of traditional labour movement history is still buoyant, if one considers the number of quality publications. In particular the quality of local studies employing microhistorical approaches is high. They tend to complicate and differentiate the grand heroic narratives which have often built up around labour movement history. The move away from homogeneous national narratives and towards the diverse local and regional experiences is also noticeable in this collection, with several contributions focusing on specific towns or regions. As economic history in particular has demonstrated, nations can indeed be too big for comparison.[2] A regional approach is often more appropriate.

If, as I have argued elsewhere, it is inappropriate to speak of a general crisis of labour history, the subject has been characterized by some serious rethinking of approaches.[3] Many of the cutting-edge histories of labour, which have been appearing over the last two decades, made use of approaches, broadly informed by postmodernist theories,[4] by gender theory[5] and by ideas about ethnicity and race.[6] These are also the theories and themes which are prominent in this collection. Many of the chapters are concerned with issues of ethnicity and gender, and the discursive construction of class identities receives considerable attention. Labour historians have increasingly questioned the old wisdom about politics being rooted in 'objective' class relationships. A new political history of labour has stressed the relative autonomy of politics from social history. Its practitioners have been primarily interested in the precise relationship between working-class organizations and their constituencies. This has resulted in studies which ask how links were forged, solidarity produced and memberships consolidated.[7]

Class, by contrast, has been carrying much less weight as a blanket explanatory factor for the development of organizations such as trade unions and political parties. Gone are the straightforward assumptions that social conditions produce class positions which inevitably result in the formation of strong labour movements. Instead it is now widely accepted that the latter constantly had to construct and reconstruct notions of class unity and solidarity, and they were, as Leighton James points out in his comparison of the Ruhr and South Wales, successful to very different degrees. Most trade unionism was, as Peter Alexander confirms for the Transvaal, defensive in character. The idea of miners as the avant-garde of the class-conscious and Marxist/socialist proletariat has been in need of reassessment for some time now, and Dick Geary provides us with a superb dismantling of the 'myth of the radical miner'.

If the more recent emphasis in labour history generally and in coalfield history specifically has been on the local, on gender, ethnicity and the discursive construction of class, the growth of comparative and transnational approaches to labour history has also been among the more noticeable innovations for the subject.[8] One of the leading coalfield historians who has consistently championed comparative approaches to the study of mining and miners is Klaus Tenfelde. As early as the mid-1980s he edited a path-breaking collection of essays on comparative labour history which sought to explore the innovative potential of the comparative method.[9] In 1989, he was the main organizer of a massive attempt to write the social history of mining as a genuinely transnational and comparative undertaking.[10] He established a comparative framework for the study of coalfield societies, which is based on the assumption of a great commonality of social conditions. Everywhere, he argued, mining was dependent on geology – a 'primeval industry' which needed large investments and often was the subject of early state legislation. Although technological advances were increasingly different, the work process itself was quite similar everywhere. Finally, there was a strong connection between work environment and the wider lifeworld: the former determined much of the latter, with miners building up tightly-knit communities inside and outside the mine.[11]

Tenfelde has repeatedly reflected on the need for comparative approaches to shed further light on the many similarities but also the significant differences of coalfield histories: 'the history of the miners' community is better suited for international comparison than any other branch of the industrial working class'.[12] This volume sets out to contribute towards the further comparativization of coalfield histories. It originates in a joint conference of the British Labour History Society and the Welsh People's History Society at the University of Glamorgan in 2002, which received further generous support from the British Academy, the Royal Historical Society and the School of Humanities and Social Science at the University of Glamorgan. The editors and organizers of the conference would like to record their thanks to all of these institutions. The location of the University of

Glamorgan in the heart of the former coalfield of South Wales and its origins as the School of Mining in the early twentieth century made it an appropriate location for such a conference. We would like to dedicate this volume to Klaus Tenfelde who has been such an inspiration to comparative coalfield historians worldwide and was good enough to accept our invitation to participate in the conference from which this volume derives.

The comparative history of 'coalfield societies' (for long a favourite term of description, but much less commonly a rigorously defined category of analysis) has been taking shape slowly but surely over the last two decades.[13] Comparative history is, of course, inevitable, unless history is understood as a mere list of dates and events. But conscious comparisons remain the exception rather than the rule, not the least because they are time-consuming, difficult and fraught with a range of methodological problems. The units of comparison have to be carefully chosen. The purpose of the comparison has to be rigorously defined. What specific questions are to be answered by the comparison? The comparativist needs to be familiar with diverse social and cultural contexts and often needs considerable linguistic competence. However, the rewards of comparative history, if it is mastered, are rich, and not for nothing has it been frequently described as the pinnacle of historical writing, the icing on the cake.[14] Comparisons have successfully relativized explanations which seemed perfectly good when given for one particular social context. They have tested and challenged a wide range of conceptual ideas in history and established typologies which have brought about a better understanding of the differences and similarities between related historical phenomena.

As Andrew Taylor underlines in his chapter, coalfield historians have, over the last decades, presented us with increasingly complex stories of coalfield societies based on local and micro-studies which often use methods of 'thick description' (Clifford Geertz). These have enriched our understanding of the workings of mining communities immeasurably. But they have also left us with the problem of how to integrate these many stories into an overarching story of coalfield society development. If the historiography of coalfield societies is not to sink into the abyss of local parochialisms, comparisons are needed which shed light on what coalfield societies had in common and explain their differences. If we want to know what was specific about particular mining communities, we need to compare them to others. Taylor's chapter discusses a variety of different comparative frameworks and concepts. He looks in particular to comparative politics and sociology for conceptual frameworks which could help establish some kind of order in the multitude of different coalfield experiences. But he also, rightly in my view, warns about the dangers of scientificity: *pace* Durkheim, comparison in history is not the equivalent of the experiment in the natural sciences. For all the right reasons historians will continue to be mistrustful of certainties and laws in history; positive proof remains evasive; falsification, in the Popperian sense of the word, seems all that we are able to establish.

Another dangerous pitfall for comparative historians is discussed by Stefan Berger and Neil Evans. Taking the Ruhr and South Wales as case studies, they argue that differences revealed by comparisons might well be rooted, not in the actual history, but in the different historiographical traditions of the cases to be compared. A richer and more diverse historiography of the Ruhr has produced very different research questions and perspectives than a more linear historiographical tradition in South Wales. Hence any comparison of coalfield societies needs to be preceded by a comparison of the historiographical traditions present in these societies.

More recently transnational historians have pointed to another major problem with comparative studies: by constituting units of comparison they tend to isolate cases artificially and ignore the history of cultural transfer between the units of comparison.[15] Societies never existed in a vacuum and on their own. A constant interchange of ideas led to the selective appropriation of 'foreign' models in indigeneous societies. Cultural appropriation and mediation has also been important for the functioning of coalfield societies. As the following chapters demonstrate, Welsh miners in the USA, Korean miners in Japan and Polish miners in the Ruhr and in Pennsylvania all brought their own ideas with them and adapted them in a dialogue with their respective host societies. The latter, however, were also deeply influenced by such migratory movements and the diverse knowledge that they transported from one part of the world to another. The discursive and agential connections between the units of comparison need to be explored fully before any comparison can be carried out.

The chapters below emphasize both the singularity and the plurality of the coalfield experience worldwide. Some of them, like Dick Geary's and Bert Hogenkamp's chapters, consider more than two cases, but the majority of chapters compare two specific coalfields. In the latter category, some compare transnationally while others compare specific locations within one and the same nation. Overall the range of chapters in this collection reflects coalfield experiences across all five continents. Many chapters deal with three separate, although interlocking, strands: 'identities', 'communities' and 'organizations'. The different spheres of workplace, neighbourhood, family and working-class organizations have for too long been treated as separate entities. This volume presents chapters which bring together these different realms and emphasize their connectedness. It sheds light on the interface between community, identity and organization and problematizes the solidarities and fissures within and between them.

Coalfields and coalfield societies have for long been represented in distinctive ways. The nature of the work process, the character of the social structure of the communities, the topography of coalfields and the often turbulent histories that unfolded within them are all factors that have contributed to strong public images of the world of mining. 'The miner', 'the miner's wife', 'the mining community', not to mention 'the mine' itself, are all charged with meanings. Dick Geary argues strongly against any occupational determinism in coalfield histories. Debunking the idea of the heroic, militant and radical miner as 'archetypal proletarian', Geary begins to explore the wide variety of factors that contributed to the prevalence of diversity and fragmentation among and within coalfield societies. The nature of the state, the prevalence of company paternalism, the workings of the industrial relations system, geological conditions, ethnic and confessional divisions, specific economic conjunctures, the introduction of technology and differential pay systems (often related), generational experience and age, migration and gender relationships all played a major role in determining levels of solidarity and economic as well as political militancy. Even where such levels were high, Geary concludes, the solidarity of miners was occupational rather than class-based.

If Geary reminds us that it is about time that we changed some of our more entrenched images of miners, Janet Wells Greene demonstrates in her chapter that photography has been of crucial importance when it comes to producing many of the stereotypical images of miners. Photography was used both by those who wanted to further and those who wanted to prevent social reform. Employers, who recognized the power of the cameras, attempted in vain to keep them out of the coalfields. Trade unionists could, after all, use the power of photographic images to highlight bad housing conditions, inadequate water supplies or sub-standard medical care. Prolonged disputes over the quality of life of American miners finally produced an extensive federal government investigation in 1946 involving a wide range of different groups with different political agendas. Greene shows convincingly that the more than 4000 photographic images produced by this investigation have dominated the public image of the American miner ever since.

Bert Hogenkamp's contribution also deals with images of miners, but his chapter is concerned with moving images. As a genre the miners' film has been very popular from the beginnings of feature films to today. Hogenkamp explores various reasons for such popularity and comments on how miners were represented in diverse miners' films from Germany, France, Belgium and the USSR. Pit disasters and strikes were popular themes, as were the importance of the family unit and strong mother figures. Miners' leisure time activities and the process of becoming a miner were extensively documented in the movies. Finally, we frequently encounter the theme of getting out of the coalfield, through social mobility or illness, or through the decline of the industry manifesting itself in increasing pit closures.

Meredith Fletcher, in her comparison of two Australian coalmines in the first half of the twentieth century, provides a fascinating example of how state interventionism can act to prevent the development of a more active sense of citizenship and civil society in mining communities. Where, as in Wonthaggi, the state relinquished its control over town and community, the coalfield society developed strong forms of solidarity and self-help. A strong civil society developed institutions and networks which in turn caused the state-owned mining company no end of problems and conflicts. That was precisely the reason why the state, when it opened a new mine at Yallourn a few years later, sought to keep everything in the mining town under its control, thereby preventing the emergence of a strong and independent civil society. Fletcher's chapter provides much food for thought for the large number of comparativists who have been interested in the role and influence of the state in community and society building.

The same can be said for Sean Patrick Adams's comparison of the Richmond coal basin and the Pennsylvania anthracite fields. Political institutional support, or the lack of it, was crucial in explaining the success of the Pennsylvania coalfield development and the failure of its Richmond rival. Whereas the Virginia legislature prioritized agricultural interests over those of coal, Pennsylvania's political institutions protected and promoted anthracite coal. Of course, other reasons were also important. There were negative economic factors for Virginia, such as the high cost and relative scarcity of slave labour, the lack of technological expertise in the local community, and the ineffective transportation system. Conversely a number of economic factors proved important for the success of Pennsylvania. These included the scientific re-evaluation of anthracite's heating capacity, a successful marketing campaign to change consumer preferences, a more effective transportation system and the rapid expansion of a free (and cheap) labour force. But overall, Adams argues, the political and institutional factors were more important than the economic ones.

Historians have become increasingly aware of the importance of 'place' in the development of political, cultural and social patterns. The 'community' and the 'neighbourhood' have become categories of analysis in their own right, independent variables in the historical process. In her analysis of the development of mining communities in the Nigerian Iva Valley between the First World War and the aftermath of the Second World War, Carolyn Brown points to the crucial role of the colonial state in developing a working-class consciousness among those who lived in what rapidly became known as the 'Red Enugu'. Brown explores the interesting topic of a transfer of strategies of working-class radicalism from the metropolitan centre (in this case the UK) to the Nigerian periphery. As metropolitan management practices constructed 'the African worker' as feminine, childlike and immature, the workforce itself re-enforced its masculine identities and began to demand, in line with its European comrades, a 'family wage'. As images of a tribal and rural Africa were used by the colonial administration to withhold from African workers the wages and rights enjoyed by metropolitan white workers, the miners established their own organizations to give weight and focus to their demands and, increasingly, to reject the colonial administators' desire to intervene in and regulate the everyday lives of African workers.

Inter-ethnic tension is also the theme of Leen Beyers's chapter on the relationship between ethnicity, class and gender in the town of Zwartberg, which is located in the Dutch mining region of Limburg. A micro-historical study relying on oral history, it demonstrates the strength of ethnic identities of subsequent waves of immigrants, who settled in the small coalfield on the western side of The Netherlands. Where ethnic groups were numerous enough to provide their own networks of support, very little inter-ethnic bonding took place. Strong gender identities did not necessarily produce inter-ethnic solidarity but rather served to bolster ethnic identities. The same can be said for class identities which were no bulwark against ethnic tensions. Beyers also produces intriguing comparative glimpses with a London-based working-class neighbourhood, demonstrating that comparisons of micro-historical studies can produce very stimulating results.

It makes particularly interesting reading to put Beyers's chapter on Limburg next to Stephen Catterall's and Keith Gildart's comparison of miners' responses in North Wales and Lancashire to immigration from Poland and Italy after the Second World War. Once again ethnic tension characterized the relations between British miners and immigrants from other parts of Europe. The union leadership attempted to convince its members that the migrants were crucial in raising the overall efficiency of the industry. They were keen to maintain their largely consensual relationship with the National Coal Board (NCB). By contrast, the miners were more concerned about the impact of 'foreign' labour on pay, working conditions and the long-term future of employment in the mines. According to Catteral and Gildart, their resentment of 'foreign' labour was not an expression of working-class racism. In fact there were cases of solidarity with 'foreign' miners at the workplace. Socioeconomic rather than ethnic motives dominated the British miners' rejection of migrant labour in the 1950s.

Polish miners must belong to the most studied species of workers worldwide. We encounter them again in Brian McCook's comparison of Polish mining communities in the Ruhr and in Northeastern Pennsylvania. McCook underlines the crucial role of religion as an anchor of Polish identity, especially in areas and at times of hostility from the host community. He investigates the conflicts of Catholic Polish miners with the indigenous Catholic hierarchy which took very different turns in the two coalfields. In the Ruhr they were dealt with inside the Catholic church, whereas in Pennsylvania they led to the breakaway foundation of a separate Catholic Polish national church. Ironically, strong religious–ethnic communities defending (through different strategies) their autonomy from the host society found themselves eventually adapting to and integrating into these host societies. By dealing with the representatives of that host society on a day-to-day basis, they eventually acquired a share in that society.

Ethnic networking is also the theme of Ron Lewis's chapter, which deals with Welsh migrants in diverse American coalfields. Lewis describes the various forms of ethnic networking and discusses the emergence of trade unionism among the Welsh in the USA. He highlights the role played by individual Welsh managers who deliberately hired Welsh miners, thereby building up ethnic Welsh communities in the coalfield. In the north, managers like Benjamin Hughes practised extensive ethnic networking, based on the support for the Republican Party, craft unionism and the rejection of slavery in the south. In the south, a manager like Llewellyn Jones did not feel less Welsh than his counterpart Hughes, but Jones rejected the practices of ethnic networking. Cheap black (often black convict) labour was readily available to him, which gave him few incentives to hire Welsh miners. Ultimately, Lewis argues, the different social environments in the northern and southern states of the USA produced diverse employment strategies.

Ethnic and gender identities remain to the fore in Donald Smith's chapter on Japan's Chikuho coalfield. As in Limburg, solidarity across ethnic lines was rare in Chikuho. Korean workers gained a reputation for industrial militancy which was often unorganized and forced Japanese unions to take a stance. In an international context the Chikuho mines were also unusual in that they allowed for the employment of women underground until well into the twentieth century. Women miners were economically exploited and sexually harassed, but their experience, Smith suggests, was not entirely negative. They earned more money than elsewhere, often took pride in their work, enjoyed the workplace camaraderie and also had greater control over their husbands, with whom they would often work side by side. When women's work was finally banned in 1933, economic rather than moral arguments were far more important.

Ethnic identities of miners and their impact on the integration of immigrants into their host societies is a theme which unites the contributions by Beyers, Gildart, Catterall, McCook, Lewis and Smith. From the evidence presented here it would seem as though such integration is far more difficult in societies with strong dominant ethnic identities, although factors such as employment practices, unionization and the law on 'foreign' labour were all crucially important in determining the eventual success or failure of integrationist policies.

Labour migration in the European coal industry was of such importance that the European Union played an increasing role in regulating such migratory movements after 1945. René Leboutte demonstrates the importance of such EU regulation by examining in some detail the labour migration into the Belgian coalmines of the Borinage – the greatest recipient of 'foreign' workers in the European coal industry. Initially most of these 'foreigners' came from Italy. As they were largely unskilled, their hiring was the subject of bilateral agreements between Belgium and Italy. The EU or European Coal and Steel Community (ECSC), as it was known at the time, only regulated the movement of skilled labour across European borders at that time. But, given the poor safety record of Belgian mines, the bilateral agreement with Italy soon came to an end. Throughout the 1950s the ECSC took an increasingly greater role in regulating and controlling mine safety, ultimately forcing employers to take the issue seriously. In the 1960s and 1970s, the EU also played a crucial role in stabilizing the coalfield societies in Europe and helping them in their transition to post-industrial societies. The transnational organization of the EU thus had an important impact on the development of coalfield societies across Western Europe and contributed to the homogenization of workplace legislation (especially regarding work safety).

Perhaps most is known about the organizations which miners formed and participated in. Trade unions and their struggles (in particular, strikes) have attracted the attention of generations of labour historians. However, the comparative approach to trade union history is still in its infancy, and there are many other organizations (cultural and political) about which relatively little is known. Peter Alexander looks at the strike activities in the Transvaal from the 1920s to the 1940s. Because of the deep racial divisions among the workforce he questions the validity of concepts of mining communities in South Africa. Both South Africanization and Christianity were of limited importance in integrating miners into local society. Low levels of trade unionization among black miners in the Transvaal meant that black miners' militancy was largely unorganized. Following Cronin in understanding militancy as a 'means of communication', Alexander explores the issues behind the various complaints raised by the striking miners. He concludes that in most instances the strikes had a defensive character. They were rooted in the black miners' understanding of a 'moral economy'. Exploring the tactical repertoire of the strikers and discussing their methods of dispute resolution, Alexander highlights the importance of ethnic networking which provides intriguing comparative perspectives with Lewis's chapter on the Welsh in the USA.

The last two chapters in the collection focus squarely on trade unions and trade union politics. Leighton James's comparison of the trade union development in the Ruhr and South Wales stresses the importance of discursive constructions of solidarity and community in South Wales and asks why similar constructions failed in the Ruhr. According to James, the lifeworld of the miners of South Wales was more homogeneous and less rigged by conflicting ethnic, religious and political conflicts than was the case in the Ruhr. Discursive ideological identities in the Ruhr, in particular Marxism and Christianity, were far more divisive than they were in South Wales.

John McIlroy and Alan Campbell are also concerned with ideological divisions: they compare the careers of two leading trade union officials in Scotland and South Wales, respectively. Abe Moffat and Arthur Horner were both Communists who preached 'trade union loyalism' beyond party politics and maintained trade union unity in the face of strong antagonisms between Social Democratic and Communist politics in the inter-war period. Their eventual acceptance of gradualism and reformism went hand-in-hand with a strong belief in Stalinism and absolute loyalty to Soviet-style Communism. McIlroy and Campbell take up a long-running debate with other distinguished scholars of British communism, such as Nina Fishman and Kevin Morgan, about the character of the Communist Party. Whereas Fishman and Morgan have argued that the CPGB's actions should be analysed primarily in the context of domestic politics, McIlroy and Campbell have insisted that domestic politics were always heavily influenced by decisions taken in Moscow.[16]

The contributions in this volume all underline the strong connections between politics, community and identity. The construction of miners' identities, class-oriented or not, had a major impact on the shape of miners' communities and politics. Conversely, miners' politics, socialist or not, influenced the outlook of coalfield communities and (partially) informed miners' identities. Last, but not least, the specific characteristics of coalfield communities made possible the emergence of particular identities and prevented the formation of others while they also influenced to a considerable degree the kind of miners' politics emerging in different coalfields.

Notes

1 Peter Ackers and Jonathan Payne, 'Before the storm: the experience of nationalization and the prospects for industrial relations partnership in the British coal industry, 1947–1972: rethinking the militant narrative', *Social History*, **27** (2002), 184–209.

2 Pat Hudson, 'The regional perspective', in P. Hudson (ed.), *Regions and Industries: A Perspective on the Industrial Revolution in Britain* (Cambridge, 1989), pp.5–32.

3 Stefan Berger (ed.), *Labour and Social History in Great Britain: Historiographical Reviews and Agendas*, special issue of the *Mitteilungsblatt des Instituts für soziale Bewegungen*, 27 (Essen, 2002).

4 Andy Croll, 'The impact of postmodernism on modern British social history', *Mitteilungsblatt*, 137–52.

5 Karen Hunt, 'Gender and labour history in the 1990s', *Mitteilungsblatt*, 185–200; Kathleen Canning, *Languages of Labor and Gender: Female Factory Work in Germany 1850–1914* (Ithaca, 1996); Maggie Walsh (ed.), *Working Out Gender: Perspectives from Labour History* (Aldershot, 1999).

6 See Kenneth Lunn (ed.), *Race and Labour in Twentieth Century Britain* (London, 1985).

7 Steven Fielding, ' "New" Labour and the "new" labour history', *Mitteilungsblatt*, 35–50.

8 Marcel van der Linden, *Transnational Labour History* (Aldershot, 2003).

9 Klaus Tenfelde (ed.), *Arbeiter und Arbeiterbewegung im Vergleich* (Munich, 1986).

10 Klaus Tenfelde (ed.), *Towards a Social History of Mining in the Nineteenth and Twentieth Centuries* (Munich, 1992).

11 Klaus Tenfelde, 'Comparative research in the history of mining workers: some problems and perspectives', in Gustav Schmidt (ed.), *Bergbau in Großbritannien und im Ruhrgebiet. Studien zur vergleichenden Geschichte des Bergbaus 1850–1930* (Bochum, 1985), pp.18–35. On the concept of 'lifeworld' see Jürgen Habermas, *The Theory of Communicative Action*, vol. 2, *Lifeworld and System: A Critique of Functionalist Reason* (New York, 1989). According to Habermas, the lifeworld comprises all everyday communicative practicies and consists of three structural components: culture, society and personality.

12 Klaus Tenfelde, 'The miners' community and the community of mining historians', in idem (ed.), *Towards a Social History*, p.1209.

13 An arbitrary selection of book-length comparative studies on coalfield histories that have appeared include Werner Berg, *Wirtschaft und Gesellschaft in Deutschland und Großbritannien im Übergang zum 'organisierten Kapitalismus'. Unternehmer, Angestellte, Arbeiter und Staat im Steinkohlenbergbau des Ruhrgebietes und von Südwales, 1850–1914* (Berlin, 1984); Joel Michel, *Le Mouvement Ourvrier chez les Mineurs d'Europe Occidentale (Grand-Bretagne, Belgique, France, Allemagne). Etude Comparative des Années 1880 à 1914*, 6 vols, Thèse d'Etat, Université de Lyon (Lyon, 1987), Roger Fagge, *Power, Culture and Conflict in the Coalfields: West Virginia and South Wales 1900–1922* (Manchester, 1996); David Gilbert, *Class, Community and Collective Action: Social Change in Two British Coalfields 1850–1926* (Oxford, 1992); John H.M. Laslett, *Colliers Across the Sea. A Comparative Study of Class Formation in Scotland and the American Midwest, 1830–1924* (Urbana/Champaign, 2000).

14 Stefan Berger, 'Comparative history', in Stefan Berger, Heiko Feldner and Kevin Passmore (eds), *Writing History: Theory and Practice* (London, 2003), pp.161–79.

15 Methodologically innovative was, in particular, Michel Espagne and Michael Werner (ed.), *Transferts. Les Relations Interculturelles dans L'Espace Franco-Allemand* (Paris, 1988).

16 John McIlroy and Alan Campbell, 'Histories of the British Communist Party: a user's guide', *Labour History Review*, **68**(1) (2003), 33–59; see also a whole range of replies and rejoinders in *Labour History Review*, **69**(3) (2003).

CHAPTER ONE

So Many Cases but So Little Comparison: Problems of Comparing Mineworkers

Andrew Taylor

A common sentiment is that miners are the same all over the world, a sentiment often articulated by miners themselves. How do we know? It is ironic that a group of workers often identified as at the forefront of working-class organization and politics in so many national contexts have generated so few comparative studies. It is this paradox I wish to explore in this chapter.

Why are there so few comparative studies of mineworkers? Part of the answer is the national focus of research and methodologies which produce 'thick' description. There is little commitment to comparative analysis as a primary specialization amongst students of mineworkers and mining.[1] A more important explanation is that mining communities have created a particularly opulent store of material which permits academics to penetrate mining's complexities at a range of levels. The richness of this legacy encourages a focus on the specific and the micro 'and the form in which the concern for comparative analysis comes out is in the obsession with the uniqueness of each nation's labor history'.[2]

Might the lack of comparative analysis be due to the lack of a comparative framework? Again, this is not quite accurate. For example, it is now *de rigueur* for any mining scholar to include a critique of the Kerr and Siegel isolated mass hypothesis.[3] Mining is one of the few areas in the study of the working class which generated a comparative framework that was employed in various contexts.[4] The theoretical models are sociological in origin and methodological in inspiration and are well represented by Bulmer's 1975 article.[5] They have provided scholars with a common set of perceptions and images which have often been used to emphasize, not similarities in the mineworker's experiences, but differences. Criticism of existing comparative frameworks has been used to undermine and weaken comparison. Cronin comments on labour history's 'curious ambivalence' towards comparison, noting that, 'In method and style, labor historians tend toward the unique and the local, if not the microscopic; in interpretation and conceptualization, however, they routinely work with models that are highly general and at least implicitly comparative.'[6] This applies forcibly to mining historiography.

The 'new' labour history of the 1960s and 1970s moved the focus away from the study of organizations towards the examination of the day-to-day 'experiences' of the working class. More recent work has drawn on the cultural and linguistic turns. Both developments have served to emphasize further complexity, uniqueness and the importance of meanings. Such complexity is frequently seen as the enemy of comparison. Contemporary methodology is dominated by the requirement that the mineworker's organizations and experiences be grounded in the locality, coupled with a determination to emphasize exceptions to any generalization. Neither are objectionable in themselves as long as any attempt to compare experiences and behaviour is not greeted with the cry, 'Ah, but it's different in Borsetshire.' Difference should not be used as a triumphalist trump card. It *will* be different in Borsetshire, an important comparative statement, but is that sufficient reason for not trying to make general statements about mineworkers? Mineworkers are commonly studied with respect to reference points such as the workplace, the union and community, all of which provide the basis for comparative studies. Methodology has increased in sophistication but, as a recent collection of essays noted, 'these developments have located mining unionism more sensitively within the distinctiveness of individual coalfields and local communities. The dominant sense is of diversity – of experience and of response – and also of the contingent'.[7]

Examining politics through the lens of diversity would seem to militate against any comparison. However one of the best known studies of mining politics, Gregory's study of the politics of the Miners' Federation of Great Britain between 1906 and 1914, is comparative. He showed how regional diversity was crucial in explaining the timing and nature of the mineworkers' affiliation to the Labour Party.[8] Campbell *et al.* are correct to caution that 'generalization is hazardous', but this should be seen as a warning, not a prohibition. Sensitivity to complexity and the fine grain of everyday life encourages the avoidance of structural reductionism but the emphasis on the local and regional has been used to support the view that fragmentation and a lack of solidarity was 'normal' in working-class politics.[9] Studies of mining in a wide variety of contexts showed that solidarity can be constructed.

'Archetypal Proletarians' Reconsidered

The 'archetypal proletarian' was perhaps the first comparative framework available to students of mining. This historiographical tradition has been subject to challenge by labour historians.

Laslett's 1974 paper was an 'exploratory and analytic' study of the differences in the political behaviour of British and US coal miners focusing on the former's abandoning of the Liberals and the latter's continued involvement with the two-party system. He focused on the miners for the usual reasons: 'large numbers, their strong tradition of political activism, and their special ability to influence the course of labour politics because of their concentration in specific, relatively autonomous, geographical areas'.[10] Laslett intended to undertake four regional case studies in the UK and the USA, analysed via seven variables: *geology* (accessibility and character of the coal), *the market* (export, import, stable, unstable), *the work system* (longwall, degree of independence at the coalface), *the employers* (small companies, large combines), *the ethnic/religious character of the workforce, the degree of unionization* (their political character, types of leaders) and *opportunities for advancement*. A comparative agenda was central:

> it was important to try and systematize areas chosen for comparison both to lend greater precision to judgements about the sources of radicalization in both countries, and ascertain how far these sources derived from the nature of the coalmining industry and how far they came from the differences in the broader context of American and English society as such.[11]

The discussion of Laslett's paper focused on local peculiarities and exceptionalism. Methodologically, the paper is suspicious – but not dismissive – of the Kerr and Siegel hypothesis and addresses this by using a framework which integrates workplace and community to explain similarities and differences.

The most significant assault on the 'archetypal proletarian' thesis came with Harrison's 1978 collection, *Independent Collier*.[12] Harrison's explicit intention was to challenge the miner as the quintessential proletarian because it had become 'impossible to think of him [*sic*] outside the context of a relatively large-scale enterprise and a highly socialized process'. This is why many of the essays in the volume examined proto-industrial situations.[13] For most mineworkers in most contexts over the past two centuries, the *absence* of large-scale enterprise and a highly socialized production process was the dominant (but not exclusive) experience. Harrison's volume, located in nineteenth-century Britain, presented mineworkers as occupying a complex class position.[14] The emphasis was on the complexity and diversity of the mineworkers' experience; militancy sat easily alongside moderation, and conflict alongside cooperation. This dualism is a characteristic feature of mining unionism wherever it is found. Harrison made a powerful (and ignored) case for the comparative study of mineworkers.[15]

Independent Collier attacked the use of models in general (and the isolated mass in particular) as a means of understanding mineworkers. Many of the most influential studies of the past twenty years have focused on variety of experience and differentiation, challenging the image of the mineworkers as the 'archetypal proletarians'.[16] Yet the authors' common motivation is to discover the bases of collective action and solidarity. An emphasis on differentiation and fragmentation highlights the variability of mineworkers' experience and *Independent Collier* emphasized the choices open to mineworkers and the factors which influenced those choices 'born of the conviction that we need more historical, micro-comparative, studies of coal-mining communities if we are ever to return, with profit, to histories of coal-mining trade unionism'.[17]

The work of Church and Outram flows from this. Clearly influenced by *Independent Collier*, *Strikes and Solidarity* is methodologically complex – a mixture of quantitative and qualitative methods – and is an ambitious attempt to advance the study of mineworkers to a more sophisticated level. Central is a desire 'to offer generalizations of comparable interest to those offered by scholars who have chosen wider fields to investigate'. The methodology is 'both historical and firmly rooted in concepts and methodologies drawn from the social sciences' and it 'throws light on some of those issues which hitherto historians have assumed or implied were not susceptible to generalization because they were beyond quantitative analysis'.[18] But, as Cronin has noted, 'even the most sophisticated approaches to comparative labor history have yet to provide a model for how actually to conduct effective comparison'.[19] Can other disciplines which have striven to develop a comparative method help?

Has Comparative Politics Anything to Offer?

Comparison is the political scientist's laboratory and they aspire to use comparison to verify or falsify whether a generalization holds.[20] Excessive optimism is characteristic of many comparativists and 'Most comparative political scientists could be said to behave very much like magpies. These birds are famous for amassing collections of shiny objects. These have no real use and have little or no connection with each other, but they please the magpie.'[21]

Comparative politics is based on trade-offs. Complexity and the variety of cause and effect limits means the comparativist's task is to isolate those factors which *appear* to have the most direct effect on that which we are trying to explain. This can then be tested and further relationships can be explored. The problem is understanding the interrelationship between the multitude of factors which might explain the phenomena which interest us.[22] A 'most similar' approach selects cases with the most characteristics in common so that differences can be more easily identified and explained. The obvious problem is that differences might be explained by more than one factor, or combination of factors, and we can never be sure we have identified all the relevant factors. The 'most different' formulates a statement which seems to hold regardless of the context (for example, 'mineworkers support left-of-centre political parties').[23] It is irrelevant whether we are discussing Chile or France, we are concerned with the factors underpinning left-wing mineworker voting generally. If a relationship holds in differing contexts then we can have confidence that we have identified a 'true' relationship.

Comparativists deal with probabilities (not laws). Central problems of comparison are definition (there is no agreed definition of militancy, for example) and measurement (for instance, of strike statistics). The dominant form of comparison is the case study, employed in an effort to create generalizations. Theory provides guidance, helps manage complexity, underpins explanation and influences case selection. The search for 'grand' theory in comparative politics has 'produced much thinking, writing and disputing about frameworks' which 'often have seemed to amount to little more than elaborate preparations for small journeys or journeys never taken', and as Guy Peters points out, 'Theory can be a strait-jacket as much as a life-jacket'.[24] Most comparative politics is meso-level (or middle-range), which permits generalizations by using concepts applied to a variety of cases. Eclecticism – 'the strategy of combining intermediate levels of analysis of a limited population of localities and collieries with a smaller number of case studies'[25] – is a response to the shortcomings of historical and social science method. Hence:

> In the rush to generalize, major differences are being canceled on the thin basis of secondary, trivial similarities. It would hardly make sense to say that men [sic] and fishes are alike in that both classes share a 'swimming capability'. Yet, much of what we are saying in global comparative politics may not make much more sense.[26]

Comparativists strive to use transparent criteria derived from existing scholarship. Concepts with unique or idiosyncratic definitions are useless; scholars might dispute the definitions used but they know their derivation, which is the beginning of a common language of comparison.

Concepts, Satori argues, suffer from 'travelling' and 'stretching': 'we can cover more only by saying less, and by saying less in a far less precise manner', but concepts facilitate comparison because without concepts we cannot relate our data to other contexts, so in comparison concepts come first.[27] Concepts precede theory and analysis because evidence must be organized and concepts provide an organizing framework. Research should be based on generic concepts which can 'travel' without too much 'stretching' to hold together disparate cases. 'Extension' refers to the class to which something belongs and the cases to which it is applied. Can a concept developed to explain something in one context be readily transferred ('travel') to another without losing rigour and meaning? The more instances to which a concept is applied, the vaguer becomes its meaning. This is 'intension'. What number of characteristics are needed to define a concept adequately? Combined, these produce the 'ladder of abstraction' and is the comparativist's core problem: how to make extensional gains without suffering drastic losses in precision and testability.[28]

To avoid becoming bogged down in an endless cycle of definition, one answer is to use 'radial categories'.[29] Based on 'family resemblances', because in the real world we seldom (if ever) find a case which has all the attributes of a concept, these assume a central core of meaning with variants and sub-variants which, while not sharing characteristics with each other, do have characteristics in common with the 'best case' which should be based on historically derived cases.

A comparative study has the following characteristics:[30]

1 a focus explicitly or implicitly on one or more country,
2 use of concepts which are applicable in more than one country,
3 rejection of both universalism and particularism,
4 it highlights differences and similarities between countries,
5 it explains (or suggests explanations for) these differences and similarities.

Church and Outram combine 'traditional historical research methods and quantitative analysis with concepts and medium range theory borrowed from the social sciences ... we have attempted to combine the statistical approach with narratives which are illustrative rather than descriptive, interpretative rather than synoptic'. Their work is a 'test case for an innovative methodology applied in a historical and comparative analysis'.[31] Comparative analysis is neither easy nor neat; any theory will be challenged by new research but this is no reason for abandoning comparison.

Back at the Coalface

Confronted by a vast literature emphasizing uniqueness and complexity, it is tempting to conclude that comparison can generate only the most trite generalizations. Carew argued that 'significant generalization about the social history of mining communities can still be made but only if we are prepared to live with considerable ambiguity and contradiction'. He continued, 'comparison is still possible, but we must decide exactly where the line should be drawn. There are indeed oranges and apples in the history of mining communities'.[32] Oranges and apples *are* comparable: both are fruit and should therefore enjoy some common characteristics. Are we trying to explain similarities or differences? There would seem to be little point in comparing apples and oranges if we were interested in commonalities as there would be few of a highly general nature, but if we were interested in identifying sources of variation, such an analysis would make more sense. Deciding 'exactly where the line should be drawn' conflicts with 'ambiguity and contradiction', so is there an unresolvable tension in any comparative study? A characteristic of the study of mineworkers is geographical range and multidisciplinarity, a combination which cries out for synthesis using miners and mining to explore general issues. There is, however, no evidence of sustained and coherent cross-disciplinary cooperation as 'the tendency continues for each discipline to delimit the purview of its practitioners, but there has been a growing and very salutary *leakage* across disciplinary boundaries'.[33] Cross-fertilization 'is most likely to work on similar topics. To increase the chances for such fertilization, to make more historians aware of the potential gains from comparative studies, we need to transcend the existing fragmentation'.[34] Mineworkers provide the ideal ground for such work.

Carew's solution was 'to put historical specificity more firmly back into mining history' (a surprising aspiration, as historical specificity never left) accompanied by 'removing many of the more abstract sociological concepts that have invaded the field in recent years'.[35] Examples of such concepts are 'community', 'collective action' and 'protest', but these concepts are too deeply embedded to be removed, and what would replace them? Carew was objecting to sociologically derived concepts (historically derived concepts were a very different matter) but railing against concepts generally can only weaken analysis and nullify comparison as concepts are the basis of knowledge and comparison. No concepts, no comparison. The correct methodological response is not to take concepts as given, or as hopelessly contested and unusable, but to specify clearly which concepts we are using and why. Carew hints at this: 'close examination of concrete historical contexts may often contradict the claims of rather more abstract generalizations'.[36]

Concepts must come first and they can be deployed in specific historical contexts. One advocate of the comparative method concluded:

> Contact with the comparative work by sociologists and political scientists can help historians cultivate methodological awareness and rigor. At a bare minimum, historians' efforts to define what they are comparing will require theoretical attention to the meaning of such analytical categories as slavery, race, class, gender, urbanization, government, nationalism, and social movement.[37]

Other concepts might include community, collective action and protest. Unanimity about a concept's meaning is not essential, but meanings can be clearly specified. Far from sounding the death-knell of comparative studies, the detailed work of labour historians makes possible a renaissance in comparison because it offers 'the historian the opportunity to escape the constraints of national boundaries, and consider working-class formation and behaviour at a different level; one which again can allow broader analysis of variables in, and causes of, contrasts in the nature of communities and their patterns of protest'.[38] Fagge sees the potential for the development of a 'new Labour History' reconstructed around the analysis of broad patterns of national experiences. The problem remains of the bases of generalization. Fagge's observation is based on a preference for generalization based on historical analysis rather than a priori sociological ideal-types.

Labour historians are drawn to broad theories which assume a common process is at work, resulting in similar outcomes across nations. Universalizing theories clash with the complexity of working-class experience and encourages the explanation of variation from a (presumed) norm.[39] When comparison occurs it tends to homogenize experience, overemphasize small variations and magnify differences. Miners and mining communities have some common properties, so comparison is based implicitly on strategies which emphasize similarities and, by so doing, identify dissimilarities. The comparative study of mineworkers does have a major advantage: the brute facts of geology. No-one would support a geological determinism but the commonalities in the organization and conduct of mining do provide a common basis to studies of mineworkers. As Fagge notes of the experiences of coalfields in South Wales and West Virginia:

> For all the differences, however, it is important that we do not forget the shared base of the coal industry. Although the act of mining may have differed in these two regions, the instabilities within the industry helped a particularly aggressive attitude toward cost cutting on the part of the coal owners. This laid the base for the fierce relationship between owner and miners, and the consequent high levels of protest in both these regions.[40]

These are significant points of comparison which we cannot ignore.

Towards a Comparative Strategy?

Fredrickson argues, 'cross-national comparative work exists primarily as a vehicle for the exploration of a particular problem or topic. Historians normally start with concerns arising from a particular national history and then seek to gain insights by examining an analogous phenomenon elsewhere'. A social scientist compares to explore 'a transnational process or recurring condition, such as a class formation, ethnicity, state building or revolution'. Both see comparison as 'a way of isolating the critical factors or dependent variables that account for national variables' but historians 'value the discovery and explanation of differences primarily as a stimulus to the reinterpretation of national histories', social scientists 'value their contribution to the development of a better model or theoretical understanding of the process, structure or condition being studied'.[41]

Culver and Greaves make a persuasive case that the study of mining must be multidisciplinary, but multidisciplinarity is not necessarily comparative. They claimed to detect after 1970 a move to 'a self-developing, multidisciplinary scholarly field' in mining studies, but the explosion in research and publication did not break down disciplinary boundaries or increase comparative work.[42] The contest between historically derived concepts and sociologically derived ideal-types appears to have been resolved in favour of the former, but the latter remain important either as targets or as Aunt Sallies and their legacy cannot be removed from the existing corpus of work. In a mining historiography marked by complexity and fragmentation, comparison can offer an important perspective on (but not substitute for) detailed studies. Comparison can fill the gap left by the retreat from meta-theory and renew our appreciation of broad patterns. Complexity and variety provided a valuable corrective to static and deterministic theories but scholars 'should not allow the subject to collapse into an increasingly dislocated empiricism'. The solution 'must entail mining historiography escaping its own academic isolated mass and locating itself within the more general approaches of working-class history, particularly in relation to questions of structure, culture, power and politics'.[43]

Culver and Greaves reflect this concern and identify common research areas such as class and class consciousness; the strategies whereby mineworkers adapt to various production settings and political contexts; the impact of export-dependent mineral economies on local and national development or its retardation; and the conjunction of politics and mine communities orchestrated by national ideological and class-based interests.[44] This agenda is influenced by geography and non-carbon mining in Latin America, but these four factors could be used as the basis for a comparative study. Fagge's study of South Wales and West Virginia focuses on the point of production and the 'huge contrasts in community, culture and power, which explains why protest took differing and complex forms'. Nevertheless, protest did occur in West Virginia and South Wales. The results were differing 'levels of coherence within political identities, and of organizational forms both reflecting and helping to solidify these'.[45] Solidaristic consciousness exists, but manifests itself differently in different contexts, but contexts can be compared to illustrate how the tensions between the general and the particular generate (or not) solidaristic class consciousness.

Shubert's study of the Asturian miners illustrates the most extreme form of working-class behaviour: an attempt to create a revolutionary *foco* via armed insurrection against the state 'of such epic proportions as to shape our perceptions of the history of those workers as a whole'. The Asturian Rebellion is undeniably attractive but it is attractive because it is not typical and Shubert argues that the shift from 'homogeneity and solidarity to diversity and divisiveness has been accompanied by a reconsideration of the idea that miners are unusually radical in their behaviour'. Classic theories were static and universalistic and Shubert's work is clearly hostile to sociological models as the basis for comparison, but he is not hostile to comparison per se.[46] Shubert's work prompts an obvious comparative question: why were there not more Asturian-type rebellions? The key variable is not 'sociology' but 'history' (or perhaps 'political culture') and the two are perhaps reconcilable, he argues, by reference to Rimlinger's 1959 article. Rimlinger acknowledges the importance of history 'but does not recognize the contradiction between his acceptance of the role of specific historical circumstances and his insistence upon the universal characteristics of the mining world'.[47]

This shifts our attention to the central problem of comparison, namely, the relationship between the general and the specific. Pushed to an extreme this would entail the rejection of any meaningful comparison or generalization, but Shubert does not do this:

> when historians have looked at individual mining communities they have found that these bear little resemblance to the static, homogenous world prescribed in the theories. After analysing the actual experience of actual miners, they have found few generalizations that retain their validity. Perhaps the only one is that mining is always a hard and dangerous profession.

So great is the deviation that ideal-types are no use other than to point out that they do not work and that there are significant differences even between apparently similar cases, but rejecting ideal-types is not to reject comparison. Shubert does not develop this point but does indicate how mineworkers might be compared:

> there is not, nor has there been, a single experience shared by miners at all times and in all places. The work and social environments vary substantially from place to place and over time, and it is here, *in these differences,* and not in any presumed universal characteristics that the *sources of the behaviour of different groups of miners may be found.*[48]

The comparative method's value lies in explaining variation and exceptions: 'We must approach the study of the mining world accepting each community as a unique historical process in which a number of variables combine in different ways.'[49] Which variables? Shubert, following Royden Harrison, suggests a comparative analysis which identifies common patterns through an exploration of the reasons for difference.[50]

Mining in the developing world begs a whole set of comparative questions. There is, for example, a substantial amount of work on African mineworkers which could provide the basis for comparative analysis.[51] A wide range of products, geographical and cultural complexity, an extensive time frame, varieties of imperialisms and so on, would seem to comprise a list which vitiates comparative analysis. But breadth and complexity can also be an advantage because commonalities stand out clearly on a broad canvas. Richardson argues that goldmining in India, West Africa and Papua New Guinea shows that, despite different geographical, social, cultural and political settings, all reproduced enduring themes in the history of mining: the 'making' of a working class, the role of the state and, finally, class relations and how they are similar to, or different from, those of mature industrial economies.[52]

Class formation provides an interesting conceptual lens through which various national experiences can be viewed. Some scholars have gone so far as to identify a common industrial trajectory:

> Mining has always entailed the systematic application of labour, but as technology and capital have moved the most productive mines beyond the small-scale diggings of the past, the entrepreneurial role shifted from those who toiled to those who capitalized technology, organized workers and profited from their sweat. Wherever this transition has occurred the skills of the miners have evolved less rapidly than those of management. Mining originally called for personal acumen and the intangibles of long experience to evaluate the ore while doing a complicated series of manual tasks. Over the years, mining for most has become a machinery operation.[53]

Phimister's study illustrates how a detailed case study can inform a series of research questions of relevance to mining in southern and central Africa. Phimister cites, inter alia, the development of gendered histories of mining, the political sociology of mining compounds, responses to imperialism and 'the myriad ways in which labour, especially but not exclusively black labour, was mobilized and controlled, and of course made its own history'.[54]

'The making of an industrial working class' is a process recognizable to all students of mineworkers. It would seem feasible to research comparatively the nature and extent of this process in various contexts, and thereby to identify variations and exceptions. Becker's study of Peruvian mineworkers, for example, shows them subject to a conventional process of 'proletarianization', but the result was a sectional consciousness which sustained political activism and industrial militancy.[55] Studies of mineworkers show that conflict consciousness is the *normal* form of consciousness. What is of interest is not whether mineworkers fulfil a 'vanguardist' role but how (and if) they engage in collective action, and the methods used in differing circumstances and contexts against the power of state and capital.[56]

Conclusions

Accepting that a phenomenon is the product of unique and particular history does not mean that phenomenon is not part of a general pattern. Moreover, one needs to ask why one case should be regarded as exceptional and another as the norm. Lipset and Marks, for example, entertain few illusions about the ability of comparison to deliver 'answers' but insist the effort should be made:

> All that one can do is use evidence that is available to develop and test hypotheses about causal processes. We compare in several ways, across and within countries, in order to isolate as far as possible the causal effects of particular factors. We also pay close attention to strategic choices of relevant actors – socialists and unionists – in trying to evaluate the consequences of their choices and whether they could have acted differently.

They describe their work as applied social science whose purpose is not testing generalizations but understanding particular outcomes. Thick description is important but understanding *why* requires a comparative dimension, there is a plethora of hypotheses but a 'deficit of systematic testing of those hypotheses cross-nationally'.[57]

The dominant methodologies in mining historiography predispose studies towards particularist/exceptionalist conclusions. There is no point in forcing cases into a theoretical straitjacket, so 'the comparative historian needs to begin with the assumption that each of her cases may be equally distinctive, equally likely to embody a transnational pattern or to depart from it'. Fredrickson discusses a number of studies which do this 'without privileging one case or establishing a normative framework', a non-exceptionalist comparative method which produces variations on a common theme rather than a contrast between one case and the rest.[58] Similarities and dissimilarities coexist and are interdependent. For example, mining's existence

> was extremely dependent on mining's highly sensitive business cycles which oscillated usually more than those of other industries. For another, paternalistic forms of management and control were evidently to be found in all European and American mining districts, although to characteristically varying degree. The influence of national political culture is undeniable...[59]

Church and Outram describe their attempt at international comparison as 'indeterminate' not just because of data weaknesses but because there has been virtually no significant theoretical development since the 1950s and because of weaknesses in social science methodologies 'which hitherto have been more sociological than historical, which have sought certainties rather than ambiguities'.[60] Campbell condemns *Strikes and Solidarity* for 'an unsubtle positivism which seeks to reduce the complexities of miner's history to a "manageable" number of quantifiable variables in order that they can in turn produce a series of generalizations which possess little explanatory power'.[61] Central to Campbell's argument is a detailed critique based on his unrivalled knowledge of the Scottish coalfields. Campbell sees Church's and Outram's constant desire to generalize as a flaw, but calls for 'an analytical framework which can distinguish structural and cultural differences between regions as well as localities within them, and also encompass political agency and the construction (or the absence) of solidarity upon these local and regional terrains'. Though critical of generalization via quantification Campbell is anxious to avoid exceptionalism and advocates 'a multi-level comparative study of mines and mining communities, employing the full range of qualitative and quantitative techniques in the toolbox of the modern social historian'.[62]

Exceptionalist analyses privilege national experiences but often fail to explain national variations over time in the experience of mineworkers. If we want to explain exceptionalism and particularism in a wider context we need to engage with similar processes of change and other national experiences. Every society is exceptional and we must avoid seeing national exceptionalisms as defective, partial or malformed examples of an 'ideal' mineworker consciousness; neither can we ignore the fact that the mineworkers' response to mining capital and state power varies despite common patterns. It would, for example, be instructive to compare the reactions of British and Russian mineworkers to the impact of neoliberal ideologies and policies on their industry.[63]

Perhaps we should be concerned with explaining, not differences or similarities per se, but 'variability' instead. The dominant theme in mineworker historiography is diversity, yet my feeling is that many scholars are uncomfortable with this, suggesting there is a subconscious feeling that there *ought* to be commonalities.[64] There is no methodological reason why comparison cannot focus on diversity and variability in the mineworkers' experience and then synthesize it into an analytical comparative framework. Language is imprecise and value-laden. When we compare we use words like 'more', 'less', 'stronger', 'weaker', 'many', 'few'; the search for a perfect framework is pointless, but, as Sartori elegantly puts it:

> I do not wish to encourage in the least the overconscious thinker, the man [sic] who refuses to discuss heat unless he is given a thermometer. My sympathy goes, instead, to the 'conscious thinker', the man [sic] who realizes the limitations of not having a thermometer and still manages to say a great deal by saying hot and cold, warmer and cooler.[65]

If we refrained from analysis until there was agreement on meaning then no analysis would, or could, take place.

Notes

1   G.M. Fredrickson, 'From exceptionalism to variability: recent developments in cross-national history', *Journal of American History*, **82**(2) (1995), 604.
2   J.E. Cronin, 'Neither exceptional nor peculiar: towards the comparative study of labor in advanced society', *International Review of Social History*, **38**(1) (1993), 59.
3   C. Kerr and A. Siegel, 'The interindustry propensity to strike: an international comparison', in A. Kornhauser, R. Dubin and A.M. Ross (eds), *Industrial Conflict* (New York, 1954), pp.189–212. Critiques are P.K. Edwards, 'A critique of the Kerr–Siegel hypothesis of strikes and the isolated mass: a study of the falsification of sociological knowledge', *Sociological Review*, new series, **23**(3) (1975), 551–74, and R. Church, Q. Outram and D.N. Smith, 'The "isolated mass" revisited: strikes in British coal mining', *Sociological Review*, new series, **39**(1) (1991), 55–87.
4   Apart from Kerr and Siegel and Bulmer, the classic article is G. Rimlinger, 'International differences in the strike propensity of coal miners: experience in four countries', *Industrial and Labour Relations Review*, **12**(3) (1959), 389–405.
5   M. Bulmer, 'Sociological models of the mining community', *Sociological Review*, new series, **23**(1) (1975), 61–92.
6   Cronin, 'Neither exceptional nor peculiar', 59.
7   A. Campbell, N. Fishman and D. Howell, 'Editors' introduction', in A. Campbell, N. Fishman and D. Howell (eds), *Miners, Unions and Politics, 1910–1947* (Aldershot, 1996), p.4.
8   R. Gregory, *The Miners and British Politics, 1906–1914* (Oxford, 1969). L. Whitehead, 'Miners as voters. The electoral process in Bolivia's mining camps', *Journal of Latin American Studies*, **14**(1) (1981), 1–32 examines Bolivia's miners between 1900 and 1952 to explore the development of collective action to influence candidates and parties.
9   Campbell, Fishman and Howell (eds), *Miners, Unions and Politics*, p. 6.
10 J.H.M. Laslett, 'Why some do and some don't: some determinants of radicalism among British and American mineworkers, 1872–1924', *Society for the Study of Labour History Bulletin*, **20** (Spring, 1974), 6.
11 Laslett, 'Why some do and some don't', 7.
12 R. Harrison (ed.), *Independent Collier: The Coal Miners as Archetypal Proletarian Reconsidered* (Brighton, 1978). See also J.H.M. Laslett, *Nature's Noblemen: The Fortunes of the Independent Collier in Scotland and the American Mid West, 1855–1889* (Los Angeles, 1983).
13 Harrison (ed.), *Independent Collier*, p.2.
14 An important element in the historiography of Chilean copper mineworkers is whether or not their privileged position in the Chilean working class and political economy makes them a 'vanguard' or 'labour aristocracy'. See F.S. Zapata, *Los Mineros de Chuquicamata: Productores o Proletarios?*, Cuadernos del Centro de Estudios Sociologicos 13 (Mexico City, 1975).

15   Harrison (ed.), *Independent Collier*, p.12.

16   For example, A. Campbell, *The Lanarkshire Miners: A Social History of their Trade Unions* (Edinburgh, 1979); D.F. Crew, *Town in the Ruhr: A Social History of Bochum* (New York, 1979); J. Gaventa, *Power and Powerlessness: Quiescence and Rebellion in an Appalachian Valley* (Oxford, 1980); S.H.F. Hickey, *Workers in Imperial Germany: The Miners of the Ruhr* (Oxford, 1985); A. Shubert, *The Road to Revolution: The Coal Miners of Asturias, 1860–1934* (Urbana, IL, 1987); and R.L Lewis, *Black Coal Miners in America: Race, Class and Community Conflict 1780–1980* (Lexington, KY, 1987).

17   Harrison (ed.), *Independent Collier*, p.14.

18   R. Church and Q. Outram, *Strikes and Solidarity: Coalfield Conflict in Britain, 1889–1966* (Cambridge, 1998), p.xvi.

19   Cronin, 'Neither exceptional nor peculiar', p.65.

20   G. Sartori, 'Comparing and miscomparing', *Journal of Theoretical Politics*, **3**(3) (1991), 243–57, at 245.

21   B. Guy-Peters, *Comparative Politics. Theory and Method* (Basingstoke, 2000), p.221.

22   A. Przeworski and H. Teune, *The Logic of Comparative Social Inquiry* (New York, 1982), pp.31–9.

23   L.M. Wolfe, 'Socialist voting among the coal miners, 1900–1940', *Sociological Forum*, **16**(1) (1983), 37–47, uses economic factors to explain socialist voting amongst mineworkers and test various models, concluding that wages levels and propensity to vote socialist are significantly related and that support for mass socialist parties fluctuates with the economic cycle.

24   H. Eckstein, 'Unfinished business: reflections on the scope of comparative political studies', *Comparative Political Studies*, **31**(4) (1998), 505–34, at 513; Guy-Peters, *Comparative Politics: Theory and Method*, p.218.

25   Church and Outram, *Strikes and Solidarity*, pp.9 and pp.11ff.

26   G. Sartori, 'Concept misformation and comparative politics', *American Political Science Review*, **64**(4) (1970), 1033–53.

27   Sartori, 'Concept misformation', 1035; Rose, 'Comparing forms of comparative analysis', *Political Studies*, **39** (1991), 446–62.

28   Sartori, 'Concept misformation', 1041.

29   D. Collier and J.E. Mahan, 'Conceptual stretching revisited: adapting categories in comparative analysis', *American Political Science Review*, **87**(4) (1993), 845–55.

30   Rose, 'Comparing forms of comparative analysis', 447.

31   Church and Outram, *Strikes and Solidarity*, pp.xv, xvi.

32   D. Carew, 'Rapport/Berich', in K. Tenfelde (ed.), *Towards a Social History of Mining in the 19th and 20th Centuries*; papers presented to the International Mining History Congress Bochum, Federal Republic of Germany, 3–7 September 1989 (Munich, 1992), p.54.

33   W.C. Culver and T.C. Greaves, 'Miners and mining in the Americas', 1, original emphasis, in T.C. Greaves and W.C. Culver (eds), *Miners and Mining in the Americas* (Manchester, 1985). An interesting example of 'leakage' and comparison is J.L. Finn, *Tracing the Veins: Of Copper, Culture, and Community from Butte to Chuquicamata* (Berkeley, Los Angeles, 1998). Flinn's methodology is anthropological. An earlier anthropological study of African copper miners which, despite its age, remains useful is A.L Epstein, *Politics in an Urban African*

Community (Manchester, 1958) and it is interesting to compare Epstein with the contemporary N. Dennis, F. Henriques and C. Slaughter, *Coal Is Our Life: An Analysis of a Yorkshire Mining Community* (first published 1956, 2nd edn, London: 1969).

34 Fredrickson, 'From exceptionalism to variability', 588.

35 Carew, 'Rapport/Berich', p.54.

36 Ibid.

37 Fredrickson, 'From exceptionalism to variability', 607.

38 R. Fagge, *Culture and Conflict in the Coalfields: West Virginia and South Wales, 1900–1922* (Manchester, 1993), p.3.

39 Cronin, 'Neither exceptional nor peculiar', 63.

40 Fagge, *Culture and Conflict in the Coalfields*, p.262.

41 Fredrickson, 'From exceptionalism to variability', p.587.

42 W.C. Culver and T.C. Greaves, 'Miners and mining in the Americas: an introduction' p.2.

43 R. Fagge, 'A comparison of the miners of South Wales and West Virginia, 1900–1922', p.118.

44 Culver and Greaves, 'Miners and mining in the Americas', p.2. See also E.D. Langer, 'The barriers to proletarianization: Bolivian mine labour, 1826–1918', *International Review of Social History*, **41**, Supplement 4(1996), 27–51.

45 Fagge, *Culture and Conflict in the Coalfields*, p.263.

46 A. Shubert, 'A divided community: the social development of the Asturian coalfields to 1934', in Tenfelde (ed.), *Towards a Social History of Mining in the 19th and 20th Centuries*, pp.284–5; idem, *The Road to Revolution*, p.18.

47 Shubert, *The Road to Revolution*, p.19.

48 Ibid., my emphasis.

49 Ibid., pp.19–20.

50 Harrison (ed.), *Independent Collier*, p.13.

51 For example, C. van Onselen, 'The 1912 Wankie colliery strike', *Journal of African History* **15**(2) (1974); C. van Onselen, *Chibaro: African Mine Labour in Southern Rhodesia 1900–1933* (London, 1976); J. Crisp, *The Story of an African Working Class: Ghanaian Miners' Struggles 1870–1980* (London, 1984); M. Burawoy, 'The hidden abode of underdevelopment: labour process and the state in Zambia', *Politics and Society*, **11**(2) (1982); J. Parpart, *Labour and Capital on the African Copperbelt* (Philadelphia, 1983); E. Berger, *Labour, Race and Colonial Rule: The Copperbelt from 1924 to Independence* (Oxford, 1974); C. Perrings, *Black Mineworkers in Central Africa* (New York, 1979); J. Higginson, *A Working Class in the Making: Belgian Colonial Labour Policy, Private Enterprise and African Mineworkers 1907–1951* (Madison, WI, 1989). Africa (like Latin America) is more complex because mining involves ores as well as coal.

52 P. Richardson, 'Berich/Report', pp.833–4.

53 Culver and Greaves, 'Miners and mining in the Americas', p.3. See also, D. Simeon, 'Coal and colonialism: production relations in an Indian coalfield, c.1895–1947', *International Review of Social History*, **41**, Supplement **4** (1996), pp.83–108.

54 I. Phimister, *Wangi Kolia: Coal, Capital and Labour in Colonial Zimbabwe, 1894–1954* (Harare, 1994), p.155. It is instructive to compare *Wangia Kolia* with Domitila Barrios de Chungara (with Moema Viezzer), *Let Me Speak! Testimony*

of Domitila, a Woman of the Bolivian Mines, translated by Victoria Ortiz (London, 1978).

55   D.G. Becker, 'The workers of the modern mines in Southern Peru: socio-economic change, and trade union militancy in the rise of a labour elite', in Culver and Greaves (eds), *Miners and Mining in the Americas*, pp.226–56. A study of Chilean copper miners which covers similar concerns is T. Miller Klubock, *Contested Communities: Class, Gender, and Politics in Chile's El Teniente Copper Mine, 1904–1951* (Durham, NC, 1998).

56   For example, S. Cohn, *When Strikes Make Sense – And Why: Lessons from Third Republic French Coal Miners* (New York, 1993).

57   S.M. Lipset and G. Marks, *It Didn't Happen Here: Why Socialism Failed in the United States* (New York, 2000), p.11.

58   Fredrickson, 'From exceptionalism to variability', p.595.

59   Tenfelde, 'The miners' community and the community of mining historians', in Tenfelde (ed.), *Towards a Social History of Mining in the 19th and 20th Centuries*, p.1208.

60   Church and Outram, *Strikes and Solidarity*, p.268.

61   A. Campbell, 'Exploring miners' militancy 1889–1966: I', *Historical Studies in Industrial Relations*, **7** (1999), 147–63 at 155.

62   Ibid., 161. This clearly harks back to Harrison.

63   There is a substantial literature on mineworkers under glasnost, perestroika and 'shock therapy'. W.D. Connor, *The Accidental Proletariat: Workers, Politics, and Crisis in Gorbachev's Russia* (Princeton, 1991); S. Clarke, P. Fairbrother, M. Burawoy and P. Krotov, *What About the Workers? Workers and the Transition to Capitalism in Russia* (London, 1993); L.J. Cook, *The Soviet Social Contract and Why It Failed: Welfare Policy and Workers' Politics from Brezhnev to Yeltsin* (Cambridge, MA, 1993); D. Filtzer, *Soviet Workers and the Collapse of Perestroika: The Soviet Labour Process and Gorbachev's Reforms 1985–1991* (Cambridge, 1994); S. Clarke, P. Fairbrother and V. Borisov, *The Workers' Movement in Russia* (Dartmouth, 1995); S. Crowley, *Hot Coal, Cold Steel: Russian and Ukrainian Workers from the End of the Soviet Union to the Post-Communist Transformations* (Ann Arbor, 1997); S. Ashwin, *Russian Workers: The Anatomy of Patience* (Manchester, 1999); P.T. Christiansen, *Russia's Workers in Transition: Labor Management, and the State under Gorbachev and Yeltsin* (De Kalb, IL, 1999).

64   Cronin, 'Neither exceptional nor peculiar', 59.

65   Sartori, 'Concept misformation', 1033.

Two Faces of King Coal: the Impact of Historiographical Traditions on Comparative History in the Ruhr and South Wales

Stefan Berger and Neil Evans

Dependence on existing monographic scholarship is particularly noticeable in comparative work done by historical sociologists and political scientists. By relying on national historiographies for the basic building blocks of comparative history, scholars ... reflect the disciplinary traditions of the academic world they have inherited.[1]

Ian Tyrrell's warning about the dangers of relying on secondary literature for generalized analyses is one which has been sounded intermittently in the past decade or so but has not been tested in detail. John Goldthorpe's critique of 'grand historical sociology' is grounded in similar concerns: the comparativists may simply choose from a range of interpretations those which suit their particular model.[2] Are comparativists comparing historiographies rather than histories?

We have chosen to explore this issue through a systematic comparison of the historiography of two of the nineteenth and early twentieth centuries' major coalfields. We start with the work of historians on the Ruhr and then shift our attention to South Wales, before returning to the problems posed in this introduction. The most striking facts about the historiography of the Ruhr are its complexity and deep-rootedness.[3] It can be traced back to the works produced by Prussian civil servants, often in retirement, in the nineteenth century. Many of these had taught at mining academies in addition to their day jobs, reflecting the premium they placed on education. Much of the work they produced was critical of the capitalist practices introduced after the mines passed out of state control and was nostalgic about former times. Their responsibilities had included social welfare and their histories took on a paternalist tone.[4] In their pages, miners were given the virtues of industriousness, thrift and loyalty to the monarchy. This analysis was supplemented in the nineteenth century by urban histories which gave much attention to the development of industrial conurbations. Local historical associations were established throughout the Ruhr and often continue to publish their weighty transactions, such as *Beiträge zur Geschichte von Stadt und Stift Essen*, *Beiträge zur Geschichte Dortmunds und der Grafschaft Mark* and *Vestische Zeitschrift*.

Research into the history of the Ruhr was also conducted under the auspices of the Verein für Socialpolitik, the so-called 'socialists of the chair', from the late nineteenth century onwards.[5] This had a formative effect on the questions asked over a long period. These included problems of economic growth and development, biographies of industrialists, the history of technical change and the pattern of connection between companies in the region. They speculated over the relationship between migration and the propensity to strike, debated wage statistics and related their history to contemporary concerns for social reform. Such scholars had little sympathy with the labour movement but their commitment to social reform often brought them into conflict with the employers. But there could also be praise for the patriotic role of the employers in turning the Ruhr into the 'economic heart' of Germany.

Employers did not put forward or even finance a particular view of the history of mining in the Ruhr before 1914. The Association for Mining Interests (*Verein für die bergbaulichen Interessen*) published Festschriften and financed the journal *Glückauf* which contained some historical material. It also had considerable influence over the local press and invested in publishing anti-union and anti-strike literature, especially in the 1860s and 1870s.[6] It was only in the inter-war period and under the impact of the impression made by the Ruhr struggle of 1923 that the Association for Mining Interests commissioned a multi-volumed series 'Twelve Years of Mining at the Ruhr', written by Hans Spethmann, a scholar whose right-wing views chimed with those of the industrialists. His work showed a good deal of regional patriotism, anti-French nationalism, concern for great men, especially great captains of industry, anti-socialism, anti-unionism and hatred of the Weimar Republic. In the 1930s, he strongly endorsed Nazi ideas. He then wrote a three-volume history of the Ruhr from ancient to contemporary times. The economic chaos of the industry in the 1920s and early 1930s was held to have been resolved by the 'wise planning' of the Nazis from 1933 onwards. His overt support for National Socialism proved to be no impediment to his postwar career, writing about more remote periods rather than the immediate past. But even his later work is not totally free of the taint of Nazi 'blood and soil' ideology.[7]

Much academic writing of the inter-war period shared this partiality to the right and the Nazis in particular. *Volksgeschichte* and *Volkstumssoziologie* supported the idea of the workers forming a homogeneous estate as an ideological bridge leading to the unity of employers and workers. This harmonized with Nazi ideas of *Volksgemeinschaft*. Other writers went even further in their embrace of fascist ideas. The major example is the work of

Wilhelm Brepohl, who had started his career as a journalist in the 1920s.[8] He published studies of the people of the Ruhr (*Ruhrvolk*) which drew on an eclectic mix of biology, anthropology, psychology and sociology. In 1935, he became head of a small research institute at Gelsenkirchen which developed an approach based on empirical research combined with pseudoscientific racist ideas. Observation of everyday life was combined with speculation about the mixture of blood flowing through the veins of the workers of the region. He continued to publish in the postwar period in a similar manner, although with considerable toning down of the racism. Even so he classified behaviour in the Ruhr according to the distribution of Poles and 'lower' types of people. Much of this chimed with the social conservatism of the early Federal Republic of Germany (FRG) and he gained a chair at Münster in 1957 and an honorary doctorate from Bochum in 1968. He was celebrated as the 'father of the Ruhr'. Anti-Marxism and anti-materialism had served him well under the Nazis and they were far from being a disadvantage in the Bonn Republic either.

Much of the historical writing on the Ruhr in the first decade and a half after the Second World War retained this partiality towards the employers. They were depicted as heroic creators of German and European prosperity and much more successful than either Nazi or Allied attempts to manage the Ruhr. Their freedom to manage was seen as the foundation of the growth and dynamism of the region. There was little consideration given to the immiseration of the workers or the employers' tactics in denying workers the ability to organize. Social partnership was endorsed, but within a firm capitalist framework.

Apart from Nazi and employer-friendly histories of the Ruhr, there existed a strong tradition of histories written by workers' representatives. Both the Christian and the Social Democratic labour movements had recourse to history, because it could be used as a political weapon in its modern struggle.[9] Thus, the labour press, for example, regularly carried historical features and made historical references, in order to justify some of its key demands, including nationalization of the mines. As Otto Hue argued: 'The past has to teach the present what needs to be done for the creation of a better future.'[10] But while such labour activists cum historians were sympathetic to the workers, they also felt that they had a limited understanding of their 'real' needs and were not sufficiently trade union conscious. They would need to appreciate the benefits of strong organization. The main enemy, however, was the employer rather than the 'ignorant' worker. The strikes of 1899 and 1905 were justified responses to the 'industrial feudalism' of the employers. The Imperial state was also criticized for supporting the employers rather than backing workers' rights. This tradition of workers' history continued into the inter-war period. Spethmann, the voice of the employers, was balanced by Lothar Erdmann the articulator of trade union concerns. He perceived trade unions to be protectors of the true national interest rather than defenders of sectional interests.[11] This approach was supplemented by heroic biographies of workers' leaders like Otto Hue. He was depicted as having sprung from the people but exhibiting a superior understanding and enlightenment; this 'man of steel' therefore faced a massive struggle to educate 'obstinate human material'. Future generations were encouraged to follow his example.[12]

After 1945, workers' history was still written but was usually associated with commemorations and merely local in scope. It was nostalgic rather than scholarly. A more substantial pro-worker historiography was developed in the German Democratic Republic (GDR), though relatively little of this concerned the Ruhr.[13] What was written stressed the conflict between workers and employers and the support which the employers gained from the militaristic state. The revolutionary credentials of the working class were stressed and quotations from the thoughts of Marx, Engels, Lenin, Stalin and Ulbricht (clearly not in ascending order) maintained doctrinal purity. This work continued the old tradition of workers' history linked with the contemporary struggle. Continuities between Nazism and the FRG were implied. The German Communist Party (KPD) was the only safe leadership for the working class but it was banned in the FRG: social democracy remained a key enemy. Inspirational stories were told of struggles against employers and the Nazis and these played their role in the creation of 'the first socialist state on German soil'. The past was searched for anticipations of the two German states which were duly found in differing vectors of development.

It was not difficult to criticize such crude methodologies, weak theoretical foundations and naked service of immediate political masters. Yet, by the 1960s and 1970s, some younger West German historians found a challenge in the work emanating from the GDR. It raised issues of social class, interest formation and class conflict which, they thought, might be fruitfully explored in a more flexible theoretical framework. There were several valuable sources of these: *Annales* history from France, the work of those historians exiled by the Nazis and the modernization theory emanating from the United States. In addition certain proponents of Nazi *Volksgeschichte* had a significant influence, especially Theodor Schieder and Werner Conze. The most important figure to emerge from this ferment of ideas was Klaus Tenfelde, who published a magisterial social history of the Ruhr miners in 1977.[14] He rejected the approach of Brepohl but not of Conze, who was acknowledged as a major inspiration. Conze had been enthusiastic about *Volkgeschichte* under the Nazi's and emerged as one of the doyens of social history in 1950s West Germany. But Tenfelde also drew on French, British and North American historians, some of whom were Marxists. He embraced a rigorous structural approach rooted in macroeconomic analysis. Labour markets, technological change, employers' and workers' interests, trade unions, working-class politics, social conditions, urbanization, workplace organization and legal–material conditions are the key themes addressed. This kind of social history was closer to the West German labour movement than the conservative employer-friendly historiography of the 1950s. The Institute for the History of Social Movements, now headed by Tenfelde, at the University of Bochum has enjoyed close links with the trade unions. The radical mood of the 1960s also influenced historical writing and the student movement provided the inspiration to rediscover the revolution of 1918–20 and to interpret it as a democratic revolution supported by many rank-and-file workers.[15] The willingness of structural social history to borrow methodologies and theories from the social sciences also contributed to new fruitful interdisciplinary projects, such as Karl Rohe's studies of elections in the Ruhr.[16] From the 1970s onwards, the focus of much West German research on the Ruhr has been on the dynamic of capitalist modernization and the slow integration of the working class into the democratic structure of the modern welfare state.

Tenfelde's work already included a concern with the everyday life of the miners and their families, but by the 1980s some historians wanted to place these matters at the centre of their work. Lutz Niethammer and Detlev Peukert launched a project on 'Life History and Social Culture in the Ruhr, 1930–1960'.[17] It rejected the idea that massive political changes were necessarily caesuras in the memory of the people. For them changes in work or private life might be more significant. National Socialism was shown to have mobilized both women and youth, drawing them out of traditional milieus. Heroic stories of resistance to fascism were countered by the evidence of a fascist consensus in working-class families. The project also put forward a powerful argument for the importance of local Social Democratic leaders who adopted a multitude of functions in their communities and thus paved the way for the social democratization of the Ruhr after 1945. Everyday life history used different sources and methodologies from structural social history and rejected its focus on structures, systems and the dynamism of progress. Institutions were subjected to searching questions: how well had they served people in their efforts to improve their lives? The new work was related to the attempts to democratize historical studies through the history workshop movement. Proponents of this approach found their way to university chairs in the 1990s and became senior figures in the discipline. An umbrella organization for the multitude of grassroots research projects emanating from history workshops, the 'Forum Geschichtskultur an der Ruhr und Emscher', was created in the mid-1990s and it boasts its own journal and the financial support of the Land government.

It should be clear from this brief pen sketch that the body of work on the history of the Ruhr is rich and complex. It defies easy summary but contains work emanating from civil servants, the employers, the labour movement and academics. These strands run side-by-side for over a century. They are fractured by the impact of National Socialism and state socialism. The spectrum of opinions expressed ranges from the biological determinism and concern for racial purity of the Nazis to the simplistic Marxism of the GDR, to the populist celebration of difference in the history workshop movement.

South Wales had a very different history from the Ruhr and perhaps it is not surprising that the historical work produced was very different as well. At the state level there was no experience of Nazism or state socialism. The industry developed late and with none of the traditions of guilds and state control which generated early historical research in the Ruhr. Universities were late developers in Wales – in the nineteenth century educationalists claimed that Wales was the only European nation to lack a university – and when they did emerge in the late nineteenth century they did not establish an easy connection with the industrial history of the country.[18] The University College of South Wales and Monmouthshire was opened in Cardiff in 1883, partly to foster scientific and technical education in the area. But it was underfunded and much of its work in mining engineering was absorbed by the rival Glamorgan School of Mines (now the University of Glamorgan) in 1913.[19] The first incumbent of the Chair of Mining at Cardiff, in 1891, was the noted mining engineer William Galloway, who wrote his classic account of mining history in Britain, *Annals of Coalmining and the Coal Trade*, before he resigned his chair owing to pressure of other work in 1902.[20] The University College also provided other sources for the study of the coalfield. Stanley Jevons taught political and economic science at the university from 1905 to 1911 and drew on his knowledge of South Wales to write his encyclopaedic *The British Coal Trade* (1915).[21] But this survey of the current state of the industry contained only a limited amount of history. The history department was rooted in the Stubbsian verities of the English state and its remote origins and Welsh history did not emerge as a subject until 1930. Even then it had little connection with the industrial world. There was, as a result, a limited pool of talent on which to draw for historical writing. Compared with the complex traditions of historical writing we find in the Ruhr, South Wales had essentially a single tradition which went through a series of important modulations.

The new industrial areas of South Wales in the nineteenth century attracted only a relatively small amount of attention from employers in historiographical terms. These were entirely contemporary history, linked with major industrial disputes with the workers. A year after the major strike of 1871 and the emergence of the Amalgamated Association of Miners in the coalfield, William Dalziel produced *The Colliers' Strike in South Wales*. It is a valuable source of documentation about industrial relations, but mainly an attempt to justify the employers' stand in the recent dispute. The position was repeated forty years on, when David Evans, another employee of the coalowners' association, produced his *Labour Strife in the South Wales Coalfield, 1910–11*. This was an account of the Cambrian Combine dispute of 1910–11, and the

Tonypandy riots firmly from the viewpoint of the employers. It has achieved some status in the historiography of South Wales because the South Wales Area of the National Union of Minerworkers chose to reproduce it in 1963, at around the same time that they reprinted the classic industrial unionist pamphlet *The Miners' Next Step*, which arose from the same era. Though a communist-led union was clear about the relevance of the pamphlet, it is hard to believe that they had understood – or even read – Evans's book.

These writings were the limits of coalowner interest in history. Nor did technical bodies, especially the South Wales Institute of Engineers, show much interest in the history of the coalfield. The interest in the past of these rapidly growing areas was filled by works written for local eisteddfodau.[22] It was a regular feature of these gatherings to offer prizes for works of local history, and out of these emerged many local studies, most of which probably remained unpublished, but some found their way into print. They vary enormously in quality. Many are simply compilations of statistics of growth which are used to frame and unite otherwise disparate material, usually organized topic by topic. They lack an underlying idea of society or community which might help give some more meaningful shape to their stories. An extreme example is the journalist, Morien's, *History of Pontypridd and the Rhondda Valleys*, published in 1903.[23] Admittedly this is the work of a journalist and derives from newspaper articles but it is typical of the genre of writing and throws its problems into relief. Its brief chapters follow no clear plan and seem to be largely random pieces of information. To illustrate the point we may take the case of the Welsh national anthem, *Hen Wlad fy Nhadau* (The ancient land of my fathers) written in Pontypridd in 1856. Clearly this belongs in a history of the area, but Morien follows this chapter with one on the writing of *La Marseillaise*! Occasionally this tradition produced works of distinction. The main example is the *History of Tredegar* by Evan Powell, which won a local eisteddfodic competition in the 1880s and was then revised and enlarged by his brother and son in 1902. Perhaps the experience of a company town like Tredegar was easier to organize into a coherent story, but what distinguishes the book is an overarching sense of the local economy and society.

Out of this background came the major historical writer on the South Wales coalfield in the nineteenth century. Charles Wilkins (1831–1913)[24] was born in Stonehouse, Gloucester, the son of a postmaster who came to Merthyr as a bookseller. His father became postmaster of Merthyr in 1851 and Charles was his clerk until he succeeded him in 1871. He wrote successively *The History of Merthyr Tydfil* (1867; enlarged edn 1908); *The History of the Coal Trade of South Wales* (1888) and *The History of the Iron, Steel, Tinplate and other Trades of South Wales* (1903). Wilkins had the virtues and the defects of the autodidact. His wide interests were not confined to a single academic discipline. He had exhaustive knowledge and gathered an immense amount of documentary material, yet this was only loosely organized. If he did not confine himself to the artificial divisions of a discipline, neither did he subject his material to the organizing principles of one. He explained the nature of his work well in his preface to his book on the coal trade: 'the description of our coalfields, the annals of our coal trade, and the biography of the many worthy men whose lives have been interwoven with the history of our industries'.

His interests embraced geology and he set his account of the trade in this setting and he used archaeological evidence to trace its earliest days. The basic organizing principle was geographical; he sought the places where the trade first emerged and then moved on with the flow of the industry. His book was dedicated to the coalowner W.T. Lewis, the foremost of his gallery of heroes who were the dynamic of the process. His theory of history, as far as he had one, derived from Samuel Smiles. But the book contains much listing and many rather undigested facts, and its structure is shaky. Within this indispensable quarry the working class does not appear even as an angel in the marble. They first emerge within the biography of his hero W.T. Lewis when strikes in the 1850s are mentioned and there is a flash forward to the Sliding Scale which would dominate the trade from 1875 to 1902 (the brainchild of Lewis). This is later amplified into two chapters on strikes and the coalowners' association. The major strike of 1873 escapes his generally encyclopaedic approach and he portrays the coalowners' association as an industrial parliament rather than as an agency of the class war. There is little reference at all to trade unions, though a few mentions of trade union leaders like Thomas Halliday.[25]

In his book on Merthyr there is some discussion of the great Merthyr rising of 1831 and of other episodes in the dramatic working-class history of the town. Even Wilkins could not airbrush these out of the picture, though he did (as was characteristic of the time) demean the events of 1831 as a riot rather than the insurrection that it was. These events were also mentioned in civic histories of Merthyr, showing that a tradition of industrial history had emerged in the chaotic process of industrialization: but unlike the north-east of England, which had deeply-rooted trade unions, there were no sustained histories of working-class movements and few accounts of the general history of the coalfield.

The task of rectifying this omission fell to the independent working-class education movement which emerged out of the militant confrontations of 1910–26. It laid foundations upon which most subsequent historical writing in South Wales has been erected. Independent working-class education (IWCE) took a stream of local leaders to London in search of a grounding in the principles of Marxism.[26] One of these was Mark Starr, who produced the Central Labour College's primer in industrial history, a remarkable book, more theoretically aware than anything academics were producing in Wales at the time, though it suffered from the mechanical nature of its theory. It is an indication of the parlous state of workers' history in South Wales that the book makes no real reference to the specifics of the place in which it was written. It is *English* industrial history.[27] Within a decade another pilgrim to London would transform this position.

Ness Edwards was born in Abertillery in 1897 and worked in local collieries before going to the CLC in 1919–21. Presumably it was then that he began the work on the Home Office Papers in the PRO which is one of the bases of his historical research on industrial history in South Wales. The other was the nineteenth-century press which would have been available to him in South Wales. By the eve of the Second World War he had produced three volumes which laid scholarly foundations for the history of the miners in South Wales.[28] They were politically engaged works and one consequence of them was the awareness of history which is present in the 1930s in the politics of South Wales.[29]

Edwards's analysis started in tribal times, a clear indication of the evolutionary framework into which his research was poured. The complex moves from subsistence peasant production to industrial capitalism were outlined and the miners were placed within an overarching economic system. Material conditions determined his outlook. His account of the harshness of early industrial conditions is bleak and heart-rending:

> The men, women and children whose conditions we have attempted to describe are the stock from which the population of these days is descended. The wonder is that they left any offspring at all.[30]

These conditions, he thought, produced only a limited class consciousness and hence only sporadic and violent movements. This was history which was synchronized with the vision of the authors of *The Miners' Next Step* and the end result of historical evolution was industrial unionism; the experience of struggle provided the necessary lessons and demonstrated the need for trade unionism. As he approached the present, Edwards could write a much more conventional institutional history and effectively stagger from resolution to resolution.

In 1939, Edwards entered parliament and moved progressively to the right. But independent working-class education had produced strong foundations upon which an academic history of the South Wales miners could be built. It had also helped create a widespread historical consciousness amongst the workers of South Wales: a remarkable achievement given the paucity of historical work in the nineteenth century. Others could build on this in the inter-war period and shortly afterwards. By the 1930s, the Communist Party was trying to establish the main outlines of the history of the class struggle in South Wales, and others tried to write localized histories in the same manner.[31] A rather belated expression of this tradition was the official histories of the South Wales miners written by Robin Page Arnot in his declining years.[32] They come from the same intellectual tradition as Edwards and Starr but they are closest to the 'resolutionary history' practised by Edwards in his more modern accounts. Modern historians find them useful as sources of narratives and as quarries of information rather than as analyses. Essentially they are works of left-wing antiquarianism.

By the time that Page Arnot's works appeared, academics were beginning to engage with the history of the coalfield. Academics only really entered the scene after 1945, with the exception of David Williams, who produced a fine biography of the Chartist leader, John Frost, on the eve of the war.[33] The postwar trend in historiography issued from the Department of Economics at the University of Wales Aberystwyth. Eric Wyn Evans produced both a biography of the late nineteenth-century moderate trade union leader, William Abraham (known by his bardic name, 'Mabon') and a general history of the South Wales miners down to 1914.[34] These were scholarly works with a heavy economic emphasis, though Evans frequently resorted to ethnic explanations for the slow unionization of the South Wales miners and stressed the need for English leadership. Some aspects of his work were the subject of a critique from within the same department, in the work of L.J. Williams. His starting point was the coalowners' association, on whose records he had worked, but his published work was mainly in the form of articles which offered an alternative view of industrial relations in the coalfield, with a stronger economic foundation than Evans and emphasizing the efforts of the leadership to overcome structural impediments. In the same period A.H. John provided the first full-length scholarly analysis of industrialization in South Wales, while L.J. Williams joined with John Morris to produce a fine study of the coal industry in the mid-Victorian years.[35] At around the same time, political historians with a wider remit in the history of Wales began to discuss the politics of the coalfield.[36]

Yet these vital developments did not remain confined to the academy. The foundation of Llafur: The Society for the Study of Welsh Labour History in 1970–71 provided a forum for young scholars who emerged through the political turmoil of the 1960s and established close links with the South Wales Area of the NUM. John Williams, Ieuan Gwynedd Jones and Gwyn A. Williams in particular were closely involved in this, so that much of the scholarly advance remained intimately connected with adult education and the miners' union.[37] Most of it was loosely Marxist in orientation and almost all left of centre. Major works on the coalfield were produced in the 1970s and 1980s, including Hywel Francis's and Dai Smith's path-breaking history of the miners in the twentieth century, which was rooted in oral history and community studies rather than the high politics of industrial relations, and

Dai Smith's fine cultural studies which became the basis for a reinterpretation of the history of Wales.[38]

In subsequent years, Llafur extended its remit into ethnic relations, women's history, popular culture and many other aspects of the history of Wales. It has just reinvented itself as 'The Welsh People's History Society' in 2001, but retains its commitment to reaching out into the wider community, beyond the confines of the university. Nor has this been much challenged by postmodernism, revisionism, theory or 'the crisis of labour history'. Serious debate in the subject is a recent phenomenon, though it shows signs of continuing. In some ways it avoided these things by never having been an 'orthodox' labour history organization and constantly widening the scope of its interests.[39] Though the scope of its work is very much wider than that of Ness Edwards, it is possible to detect lines of continuity back to his day. Adult education has been a central thread in Welsh labour/people's history and in the history of the coalfield.

So it is clear that the traditions of historical writing in the two coalfields are markedly different. It would be an unwise historian who wrote comparatively about them without taking account of these differences. South Wales has been portrayed as a left-wing, militant coalfield: but to what extent is this because its history has been written by left-wing historians? The existing comparative studies of the two coalfields do not raise this issue. Nor, in the extensive literature on the problems of comparison, is this often broached. It is possible to argue that a comparativist could simply extract 'facts' from such a literature and leave the interpretations behind, but it would be only a hard-boiled positivist who could do this.[40] A slightly different tack would be to argue (more reasonably) that comparativists should focus on what is agreed between historians, rather than what is in dispute. There is always a substantial area of consensus, even behind the most heated controversies: indeed some historical controversies are fought out over details rather than general principles. But this may not take us much further as it risks rooting comparative discussions in the bland rather than the exciting and illuminating areas of research and interpretation. And the point of comparison is surely that it opens up exciting new views and offers new perspectives. Engaging in archival work in both locations may offer some controls over the problem but it is unlikely that many scholars will have the resources to be able to do this on an equal basis and even then they will have to ensure that different sources encountered in the archives do not highlight differences between the chosen units of comparison.

In a short account we cannot offer an easy solution to this problem. We can, however, offer three related observations. Firstly, our survey of the historiography of the two coalfields shows that the differing historiographical traditions are not arbitrary but closely related to the experiences of the two areas. Perhaps we should not be surprised to find that different areas have different historiographical traditions. There are clear reasons which explain the richness and variety of the Ruhr tradition compared with the slender, more or less continuous, thread of South Wales historiography. In short, we are suggesting, tentatively, that the historiographical tradition is related to the history. Secondly, analysing historiographical traditions can generate insights into the conditions of intellectual production in coalfields and their relations with the wider society which might not easily be produced in other ways. It follows from these points that, thirdly, historiographical traditions need to be part of what we compare when we make comparisons between coalfields, or any other units of study. They need to become part of the working equipment of the comparative historian. This is not a comforting conclusion in a field of study which is already acknowledged to be difficult and all too rarely practised. We are adding another warning to all the difficulties which exist in the field. But this is not meant as a deterrent. The approach is one to which we are firmly committed.[41] Comparative history is a high-risk activity but it can bring correspondingly high gains. In drawing attention to the problems of historiographical traditions we are attempting to insure against one of the risks.

Notes

1   Ian Tyrrell, 'American exceptionalism in an age of international history', *American Historical Review*, **96**(4) (1991), 1037.

2   John H. Goldthorpe, 'The uses of history in sociology: reflections on some recent tendencies', *British Journal of Sociology* [hereafter *BJS*], **42** (1991), 211–30.

3   Klaus Tenfelde, '"Klassische" und "moderne" Themen in der Bergbau-geschichte', *Landwirtschaft und Bergbau: Zur Überlieferung der Quellen in Rheinischen Archiven* (Cologne, 1996), pp.127–42.

4   See, for example, H. Brassert, *Die Bergrechtsreform in Preußen: Historischer Überblick* (Berlin, 1862).

5   See, for example, Karl Oldenberg, *Studien zur Rheinisch-Westfälischen Bergar-beiterbewegung* (Leipzig, 1890).

6   Klaus Tenfelden, *Sozialgeschichte der Bergarbeiterschaft an der Ruhr im 19. Jahrhundert* (Bonn, 1977), pp.288, 481.

7   Hans Spethmann, *Die Großwirtschaft an der Ruhr* (Bochum, 1924); *Zwölf Jahre Ruhrbergbau*, vol. 1: *Aufstand und Ausstand bis zum zweiten Generalstreik April 1919* (Berlin, 1928); *Fünfzig Jahre technischer Grubenbeamter Oberhausen 1885–1935. Eine Festgabe* (Gelsenkirchen, 1935); *Der Verband technischer Grubenbeamter 1886–1936: Eine Festschrift zu seinem 50jährigen Bestehen, verfaßt im Auftrage des Vorstands* (Gelsenkirchen, 1936); *Die ersten Mergelzechen im Ruhrgebiet* (Essen, 1947), as well as several contributions in *Essener Beiträge* between 1947 and 1956; see also *Franz Haniel. Sein Leben und seine Werke* (Duisburg, 1956).

8   Stefan Goch, 'Wege und Abwege der Sozialwissenschaft: Wilhelm Brepohls industrielle Volkskunde', *Mitteilungsblatt des Instituts für soziale Bewegungen*, **26** (2001), 139–76.

9   The classic accounts are Heinrich Imbusch, *Arbeitsverhältnis und Arbeiterorganisation im deutschen Bergbau* [1908] (Bonn, 1980); Otto Hue, *Die Bergarbeiter*, 2 vols (Berlin, 1910–13).

10 Hue, *Bergarbeiter*, vol. 1, p.vii.

11 Lothar Erdmann, *Die Gewerkschaften im Ruhrkampf* (Berlin, 1924), especially pp.23f. On Erdmann see also Ilse Fischer, *Versöhnung von Nation und Sozialismus?* Lothar Erdmann (1883–1939). Biographie und Auszüge aus den Tagebüchern (Bonn, 2004).

12 Nikolaus Osterroth, *Otto Hue. Sein Leben und Wirken* (Bochum, 1922), especially pp.24, 28.

13 Exceptions to the rule are Dieter Fricke, *Der Ruhrbergarbeiterstreik von 1905* (Berlin, 1955); Henri Walther and Dieter Engelmann, *Zur Linksentwicklung der Arbeiterbewegung im Rhein-Ruhr-Gebiet*, 3 vols (Leipzig, 1965).

14 Tenfelde, *Sozialgeschichte*.

15 Erhard Lucas, *Märzrevolution im Ruhrgebiet*, 3 vols (Frankfurt am Main, 1971–8).

16 Karl Rohe, 'Vom alten Revier zum heutigen Ruhrgebiet', in Karl Rohe and H. Kühr (eds), *Politik und Gesellschaft im Ruhrgebiet* (Königstein im Taunus, 1979).

17 Lutz Niethammer (ed.), *'Die Zeit weiss man nicht, wo man die hinsetzen soll'. Faschismuserfahrungen im Ruhrgebiet* (Bonn, 1983); *'Hinterher merkt man, daß es richtig war, daß es schiefgegangen ist': Nachkriegserfahrungen im Ruhrgebiet* (Bonn, 1983); Lutz Niethammer and Alexander von Plato (eds), *Wir kriegen jetzt andere Zeiten. Auf der Suche nach der Erfahrung des Volkes in nachfaschistischen Ländern* (Bonn, 1985).

18 J. Gwynn Williams, *The University Movement in Wales* (Cardiff, 1993), p.15.

19 G.W. Roderick, 'Education, culture and industry in Wales in the nineteenth century', *Welsh History Review*, **13**(4) (1987), 438–52; 'The Institute of South Wales Engineers and the South Wales economy in the late nineteenth century', *Welsh History Review*, **14**(4) (1989), 596–609.

20 Stanley B. Chrimes (ed.), 'University College Cardiff: a centenary history, 1883–1983' (unpublished ts in Cardiff Central Library and other major libraries in Wales), 296; *Annals* was published in the *Colliery Guardian* in 1896–7 and as a book in 1898.

21 Chrimes, 'University College', 295, 303.

22 'Eisteddfod' (pl. eisteddfodau: lit. sitting place) is the Welsh word for a cultural festival, which in the nineteenth century involved competitions of a literary and musical nature. The institution can be traced back to the Middle Ages, but was much transformed in the process of nation building in the nineteenth century. A network of local competitions was established which was capped by a National Eisteddfod from 1858.

23 His non-bardic name was Owen Morgan (?1836–1921)

24 R.T. Jenkins (ed.), *The Dictionary of Welsh Biography* (London, 1959), p.1019.

25 Wilkins, *Coal Trade*, chs xv, xvi.

26 See Richard Lewis, *Leaders and Teachers: Adult Education and the Challenge of Labour in South Wales, 1906–1940* (Cardiff, 1993)

27 Mark Starr, *A Worker Looks at History* (London, 1917; 3rd edn, 1919).

28 Ness Edwards, *The Industrial Revolution in South Wales* (London, 1924); *The History of the South Wales Miners* (London, 1926); *History of the South Wales Miners' Federation, Vol. 1* (London, 1938). The material for the projected second volume of this work exists in a proof copy in Nuffield College Library.

29 Neil Evans, ' "South Wales has been roused as never before": marching against the means test, 1934–1936', in David Howell and Kenneth O. Morgan (eds), *Crime, Protest and Police in Modern British History: Essays in Memory of David J.V. Jones* (Cardiff, 1999).

30 Edwards, *Industrial Revolution*, p.38.

31 J.E. Morgan, *A Village Workers' Council: Being a Short Account of the Lady Windsor Lodge* (Pontypridd, n.d., c.1950); Edmund Stonelake, *The Aberdare Trades and Labour Council, 1900–1950* (Aberdare, 1950).

32 R. Page Arnot, *The South Wales Miners/Glowyr de Cymru: A History of the South Wales Miners' Federation, 1898–1914* (London, 1967). The second volume, covering the years 1914–1926, was published in Cardiff in 1975.

33 David Williams, *John Frost: A Study in Chartism* (Cardiff, 1939).

34 E.W. Evans, *Mabon: A Study in Trade Union Leadership* (Cardiff, 1959); *The Miners of South Wales* (Cardiff, 1961).

35 John Williams, *Was Wales Industrialised?* (Llandysul, 1995), collects most of these articles along with some unpublished papers; A.H John, *The Industrial Development of South Wales, 1750–1850* (Cardiff, 1950); J.H. Morris and L.J. Williams, *The South Wales Coal Industry, 1841–1875* (Cardiff, 1958).

36 For a superb sample of this, including essays by Gwyn A. Williams, Ieuan Gwynedd Jones and Kenneth O. Morgan, see Glanmor Williams (ed.), *Merthyr Politics: The Making of a Working Class Tradition* (Cardiff, 1966).

37 For a more substantial account of these developments, see Neil Evans, 'Writing the social history of modern Wales: approaches, achievements and problems', *Social History*, **17**(3) (1992), 479–92.

38 Hywel Francis and David Smith, *The Fed: The South Wales Miners in the Twentieth Century* (London, 1980); Dai Smith, *Aneurin Bevan and the World of South Wales* (Cardiff, 1993); *Wales! Wales?* (London, 1984).

39 See the editorial in *Llafur*, **8**(3) (2002), which issue has a substantial section of debate and the perceptive critique of the journal: Andy Croll, ' "People's remembrancers" in a post-modern age: contemplating the non-crisis of Welsh labour history', *Llafur*, **8**(1) (2000), 5–17, along with a short response in the editorial.

40 Joseph M. Bryant, 'Evidence and explanation in history and sociology: critical reflections on Goldthorpe's critique of historical sociology', *BJS*, **45** (1994), 13f.

41 Neil Evans, 'Two paths to economic development: Wales and the north-east of England', in Pat Hudson (ed.), *Regions and Industries: A Perspective on the Industrial Revolution in Britain* (Cambridge, 1989); 'Patterns of protest and regional labour implantation in South Wales and the north-east of England, 1780–1950', *Tijdschrift voor Sociale Geschiedenis*, **80** (1992), 212–30; Stefan Berger, 'Working-class culture and the labour movement in the South Wales and the Ruhr coalfields, 1850–2000: a comparison', *Llafur*, **8**(2) (2001), 5–40; 'And what should they know of Wales? Why Welsh history needs comparison', *Llafur* **8**(3) (2002), 131–9.

CHAPTER THREE

The Myth of the Radical Miner

Dick Geary

Introduction: the Radical Miner

The miner has often been depicted as the archetypal proletarian. This is scarcely surprising in industrial nations, which were, for a relatively long time, heavily dependent on coal and possessed a large and often truculent mining community. In a way, which is perhaps more unusual, other sections of society have recognized something almost heroic, even romantic, in the physical exertion of the miner in frightful and dangerous conditions underground. Great mining disasters and the high incidence of disease, sickness and invalidity have in the past generated some general sympathy. So, too, has the intransigent behaviour of autocratic coal barons, whose local economic *and* political power some American miners sought to denounce as 'un-American' after the turn of the century in an attempt to win the support of both the Federal state and the American public. In this they were sometimes successful, as in the case of the Lehigh anthracite miners of Pennsylvania, who were supported by their local business community against the mine owners during the 1887 dispute. Miners in the French town of Decazeville sought to mobilize the communal politics of radical republicanism against the dominant local coal company in the second half of the nineteenth century; and even in the British coal disputes of the 1970s Sheffield newspapers organized kitchens for striking pitmen. They did not do so in 1984–5; and the death of sympathy can be listed as one of many reasons for the failure of this last dispute.[1]

The miner has been portrayed as 'heroic' in another sense by those who have taken labour's side in the class war. The miner has been seen as a prime vehicle of industrial militancy and proletarian radicalism. In Britain and the USA, miners have stood at the forefront of strike activity, as Kerr and Siegel pointed out long ago. Knowles's statistical analysis of strikes between 1911 and 1947 confirms that miners were more strike-prone than any other group of British workers. Between 1889 and 1921, for example, miners in the UK struck between two and three times more frequently than any comparable group of workers. Though the gap between the strike rate of other groups of workers and of miners has narrowed since 1945 and especially since 1959, it remains true that, from the end of the Second World War to the early 1960s, over half of all British strikes were in mining. The miners' strikes of 1972 and 1974 entailed confrontation with the Tory government of the day, as did, most famously, the stoppage of 1984–5. The largest industrial disputes in Imperial Germany were the coal strikes of 1889, 1905 and 1912, whilst in France between 1890 and 1914 miners downed tools much more frequently than the national strike average. Miners have been the most likely group of workers to strike in Peru, and in Japan in the 1870s eight of the 11 industrial disputes recorded in the country took place in the mines. Miners have also been at the forefront of union mobilization in Chile, Peru and Bolivia since the 1940s.[2]

Miners' strikes have often been characterized by violent confrontation, which arguably confirms the bitterness of class struggle in the coalfields. This was so in the Japanese case, mentioned above, where company officials and foremen were beaten up and office buildings, warehouses and the homes of managers burnt down. The strikes of Lanarkshire miners before the First World War were on occasion accompanied by sabotage and violent disorder. The Lattimer Massacre of 1897 in the USA saw at least 19 striking Pennsylvanian miners murdered and 32 wounded. In West Virginia, the Paint/Cabin Creek strike of 1912–13 and the Mine Wars of 1912–21 gave rise to considerable bloodshed. In the latter case Federal troops and aircraft were dispatched to the conflict, in which 8000 miners were engaged in armed confrontation with the forces of the mine operators. The great Ruhr strikes usually provoked an armed response from the German authorities, leading to fatalities at the hands of police and troops, as well as armed pit guards; and there was a civil war and the formation of a 'Red Army' in the Ruhr coalfield in 1920. A 'syndicalist' spirit can be detected in the 1911–13 wave of strikes in the coalfields of South Wales, where the *Miners' Next Step* was issued, in the southern massif of France before 1914, in the Ruhr between 1918 and 1923, and amongst the radical young of New Zealand's pits in the decade before the First World War. In the *El Teniente* Copper Mine in Chile there were also conflicts between formal union organizations and the popular culture of the workers, who leant towards violent, direct action. In these areas trade union advocates of caution often found it difficult to restrain an impatient rank and file; and indeed many pit stoppages have begun from an individual pit and then spread, rather than being the result of any central union command.[3]

The radicalism of miners has also embraced politics. In Britain between the wars the three most well known cases of 'Little Moscows', where industrial militancy and political ideology seemed to fuse, were in mining communities: Chapwell (near Blaydon), Mardy in the Rhondda and Lumphinnans in the Fife coalfield. In South Wales, the South Wales Miners Federation (the 'Fed') became much more than a union. Its Workmen's Institutes spread through the coalfield, ran leisure and cultural events, including film shows, laid on medical schemes and built libraries for their members. Moreover the 'Fed' had the abolition of capitalism written into its rulebook in 1917. Much later, no other group of workers contested Thatcherism more obviously than the colliers of South Yorkshire, South Wales and Scotland, not only by their industrial action but also in local politics. Ruhr miners mounted massive campaigns for the socialization of their pits in 1919 and 1920. Furthermore it was in the mining colonies of towns such as Herne and Wanne-Eickel that the German Communist Party got its largest percentage of the vote (between 60 per cent and 70 per cent in July 1932 in some districts of these towns). The German Communist Party (KPD) still mobilized significant support in these same places in 1946. In the Asturias, Spanish miners rose in revolt against governmental authorities in 1934.[4]

Other Miners

The typicality of the strike-prone and politically radical miner, however, is more than questionable. Even in the Scottish coalfields the proclivity to strike was far from uniform: whereas the average number of strikes in Lanarkshire was 12.8 a year, the figure for Fife was only 0.8 between 1903 and 1909. The work of Brian McCormick indicates the existence of 'spatial clusters' of strikes in South Yorkshire rather than their universality in the 1950s and 1960s; and the more recent research of Church, Outram and Smith has indicated the existence of a small number of collieries and places where strikes were repeatedly and densely concentrated, and others where there were no recorded strikes at all between 1893 and 1940. Between 1946 and 1973, Yorkshire, South Wales and Scotland produced the largest numbers of strikes. In this period 41 per cent of the larger disputes (78 out of 186) occurred in the Yorkshire coalfield alone. Aggregate statistics disguise this signal difference. Obviously, therefore, not all British miners were strike-prone. It is also the case that the chronology of strikes in the UK mines indicates not only periods of huge activity, such as 1911 to 1914, but also periods of quiescence, as from 1926 to the 1950s. Even in 1926 itself, few miners struck before the General Strike and few thereafter. In fact there were fewer coal strikes in this year than virtually any other. Conciliation in the coalfield was much more pronounced in Northumberland and Durham than in South Wales, more in Fife than in Lanarkshire; and, most infamously of all, the strike of 1984–5 cruelly exposed the regional divisions in the post-nationalization coal industry. Whereas the strike was largely solid in South Wales, for example, significant minorities continued to work in North Wales and North Staffordshire. In Nottinghamshire, only 2300 miners joined the strike, compared to the 24 000 who ignored the strike call and, in creating the Union of Democratic Mineworkers, revived memories of the strike-breaking

Spencer union of 1926. In Leicestershire, all but 39 men continued to win coal in 1984.

Relations between the different groups of miners in this strike became notoriously bitter and reinforced the different cultures of different coalfields.[5] During the stoppage one Thurcroft collier expressed the wish that a Union Carbide plant be built in Nottinghamshire, 'so that the next Bhopal will sterilize them ... They're not miners. The only thing that comes to mind is piles'.[6] Moreover, before 1900, the development of trade unionism in the British coalfields was very uneven. A number of regional federations came and went with the trade cycle. The miners of the North East, who possessed the most stable tradition of union membership and moderate unions, remained aloof from the Miners' Federation of Great Britain, which drew its support from the inland coalfields, and trade union density varied from one coalfield to another.[7]

It is thus clear that industrial militancy was never a necessary consequence of being a miner. If the equation of the mining occupation with industrial conflict is cast into doubt by the British material, international comparisons make it quite untenable, as Rimlinger realized a long time ago. Strikes were much less common in the continental mines than in those in the UK. In the German Saarland, for example, they were virtually unknown. Although the big Ruhr disputes of 1899, 1905 and 1912 have grabbed the attention of historians, the great majority of miners' disputes in Imperial Germany were small, often involving a single pit, and of short duration; and they usually failed in their objectives. German miners did not show the solidarity and tenacity of their British counterparts. Job changing and absenteeism were infinitely more common than participation in industrial disputes. Although miners featured prominently in the strike wave of 1919–22 in the Weimar Republic, thereafter pit protests were extremely rare. Indeed pit managers spoke of having 'reconquered' the mines in the mid- and late 1920s. What is more, rates of unionization for German miners remained below those of printers, skilled building, wood and engineering workers before the First World War. David Crew's comparison of miners' supposed militancy with the docility of foundry workers in Bochum before 1914, therefore, is somewhat misleading. In the Ruhr in 1913, just over 30 per cent of the pitforce was unionized, though union density had actually fallen since the previous year and the union organizations were themselves split along confessional and ideological lines. In 1912, 22.7 per cent of pitmen (70 000) in the area had joined the *Alte Verband*, loosely associated with the SPD, but 40 000 Ruhr miners belonged to the Christian Trade Union (Catholic) and a further 30 000 to the ZZP, a specifically Polish Union. A 'yellow' (company) union movement even managed to recruit 21 000 miners (about 5 per cent of the total pitforce in Ruhr) on the eve of the First World War. Levels of union support were desperately low in the Saar and became even lower in 1912, when the membership of the Catholic miners' union actually dropped from 17 000 to a mere 7000. The Free Union (socialist) had only 1660 members in the whole of the Saar at the same point in time. Needless to say, these organizational divisions could have serious consequences for the bargaining strength of miners. There were times when Poles alone went on strike, as in 1899 in the Northern Ruhr. At others, the Christian Unions refused to join the Free Union strikers, as in the miners' strike of 1912. In a very few cases Christian Unions struck without the support of the Free Unions.[8]

Mining districts which experienced little or no industrial conflict can, of course, be found in many other countries. In the South African Transvaal, miners' strikes were of no significance before 1900. In Natal, there were no coal stoppages at all between 1914 and 1959. In Poland, miners were absent from waves of industrial unrest in 1956, 1970 and 1976. They were little involved in the development of the *Solidarity* trade union movement and absent from the first wave of strikes in 1988, though thereafter they radicalized with great speed. In the Asturias of Northern Spain, strikes were rare and union membership even rarer before the 1920s. It was not mining but agrarian Andalusia and the textile industry of Catalonia that provided the bastions of Spanish syndicalism; French syndicalism drew its strength above all from the artisans of Paris and not from the country's coalfields, with minor exceptions in the Southern Massif.[9]

Clearly, therefore, occupational determinism (the nature of the miner's work) will not explain local, regional and national differences in the proclivity of miners to strike or join a union. It is even less successful in predicting political affiliation.

Miners and Politics

Syndicalism and communism have had an appeal amongst some groups of mineworkers, as we saw earlier, yet the political history of miners is, if anything, even more diverse than that of their industrial behaviour, and their politics have often been far from radical. At the same time as the syndicalist *Miners' Next Step* appeared in South Wales and the 'Fed' established deep roots in the valleys, Liberalism remained strong in the coalfields of Durham and Northumberland. When the TUC decided at the turn of the century to support a strategy of independent political representation for workers in the shape of the Labour Representation Committee, subsequently to evolve into the Labour Party, the miners' union refused to give its support. In fact it was the last of all the major unions to join the Congress, despite its later sponsorship of many Labour MPs. Even after 1910 many of these sponsored candidates remained on good terms with Liberal organizations in their localities. In the Asturias, the Spanish Socialist Workers' Party (PSOE) remained weak until the 1930s. In France, the miners of St Etienne tended for a time towards revolutionary anarchosyndicalism but their counterparts in the much larger coalfields of the Nord and Pas-de-Calais districts supported reformist socialism. In Peru, Chile and Bolivia, as Francisco Zapata has shown, the emergence of miners' unions remained separate from and unrelated to the development of the leftist *political* movements.[10] The sometimes violent industrial militancy of miners in West Virginia was not reflected in class-based politics. To quote Roger J. Fagge:

> the identity that emerged ... among the coalminers of West Virginia, was one based upon the struggle for civil and constitutional rights and a re-assertion of a very broad meaning of 'America' ... The battle was not one for a radical–socialist vision similar to South Wales, let alone the revolutionary transformation of society of the syndicalists.[11]

In Germany, until the privatization of the Ruhr mines in the 1860s, miners had constituted the core of a loyal labour force, which sought redress for its grievances through the presentation of petitions to the monarch. Later the politics of the mining community were divided along ethnic and confessional, as well as ideological, lines. Some miners, especially those who lived in company housing and worked for large, vertically integrated firms like Krupp, voted National Liberal before 1914, for the Nationalist Party (DNVP) in the 1920s and for the Nazis in the Depression. The same was true of workers at small, rural collieries in the Bochum area. Although other Ruhr miners became involved in insurrectionary politics in the early 1920s and some supported the KPD throughout the 1920s, as we have seen, they were not the initiators of revolution in 1918: that accolade went rather to skilled engineering workers in Berlin and Vienna, as it had done in Petrograd in the previous year and was to do in Budapest in the next.

Furthermore the SPD had been much weaker in the Ruhr than in many non-coalmining areas before 1914. In 1912, for example, the party won over 70 per cent of the popular vote in Berlin and over 60 per cent in Hamburg and Leipzig, but only 34.5 per cent in Düsseldorf and 34 per cent in the Arnsberg district (Dortmund and Bochum). What is more, the social-democratic sub-culture, which flourished in Berlin and the Saxon cities and could boast 660 000 members by 1914, was almost completely absent from the Ruhr and the ratio of party members to party voters in 1912 was much lower (12 per 100 in Bochum, 19 in Dortmund) than the national average of almost 22.8. In the Dortmund branch of the SPD in 1906, 15 per cent of its members were recruited from the mines, whereas building workers supplied 34 per cent of the membership – and this in a mining region. Metalworkers and woodworkers were also overrepresented in the social-democratic ranks. Conversely the Centre Party could rely on the votes of most Catholic miners in the Ruhr. Only amongst Catholic immigrants did the SPD make much headway in the *Revier* before 1914.

Polish miners stayed away from both German Social Democracy and the Catholic Centre Party. They voted for a Polish nationalist party, which took 7 per cent of the Bochum vote in 1912 and was even stronger in the north of the Ruhr. Masurian immigrants, who were Protestant and detested their Polish neighbours, were also missing from the ranks of Social Democracy and often voted National Liberal. In the Ruhr, Social Democrats encountered a strongly developed Catholic culture of Christian Unions and Catholic Workers' Associations. In Schalke, a mining district of Gelsenkirchen, there were some 30 Catholic clubs (for skittles, billiards, singing and dancing) as well as Catholic miners' libraries. The hold of Catholicism on miners' politics in the Saarland, where only 777 Social Democrats were to be found in 1913, was even greater, as it was in the Aachen coalfield, where the Catholic miners' union, the *Gewerkverein*, was strong. As one contemporary observer wrote of the Saar: 'here everything is sacred or kingly'. The first part of the quotation obviously refers to the strength of Catholicism; the second, however, is equally significant. It has in mind the monarchism of the region's Protestant miners, which stemmed from the existence of a vibrant Evangelical but anti-socialist culture, which has until recently been largely ignored by historians.

A similar culture existed amongst some of the Protestant miners in the Ruhr too, though here the SPD had a greater attraction, especially after the big disputes of 1889 and 1905, than it did in the Saarland. The first Evangelical workers' club in the Ruhr was set up in 1882 and subsequently a regional organization was created for the Rhineland and Westphalia. It is reported that many of the members of these clubs refused to join the miners' strike of 1889 and that they were hostile to the SPD. The Protestant workers' association in Bochum opposed the 1912 stoppage as well; and its members voted National Liberal in the main. (Interestingly, Alan Campbell notes that strident Protestantism acted as a deterrent to communist influence in some of the Scottish pits after the First World War.) Thus the politics of Ruhr miners were fractured. Division, not unity, was the hallmark of the region's largest occupational group. In one district of Bochum (Altenbochum) in 1911, for example, the social-democrat workers' club had 280 male and 100 female members, yet the Catholic miners' club had 100 members, two Polish clubs a total of 310 members and two Protestant clubs 178. Thus there was no unitary culture of work and leisure in Germany's largest coalfield.[12]

Miners' politics were not only diverse, they could also change very rapidly. In the Asturias, the PSOE was weak in the 1920s, yet a revolutionary insurrection took place in 1934. In the Ruhr, previously docile miners mounted the barricades in 1918–19, were militant until 1923, yet almost disappeared from view after 1924. In Britain, in 1911, the Independent Labour Party (ILP) candidate for the Barnsley constituency (Tom Mann) was stoned by miners at Wombwell, yet the area elected a Labour MP subsequently. In 1898, the South Wales Miners' Federation was critical of the 'ungodly' ILP, yet Keir Hardie was elected in the Merthyr constituency only two years later. Thus, the political, as well as the economic, conjuncture, which is central, for example, to Adrian Shubert's account of the radicalization of Asturian miners in the 1930s, can be as significant as occupational/structural factors. This was especially true in periods of great social upheaval and particularly when the forces of social control were temporarily undermined, as in Central Europe at the end of the First World War. We know, for example, that many of the coalminers of the Ruhr, who joined left–communist and syndicalist organizations in the 1919–23 period and became engaged in insurrectionary and violent behaviour, had been members of the 'yellow' (that is, company) unions before the First World War. Indeed the most elemental upheavals took place at the plants of Thyssen in Gelsenkirchen and Hamborn, where miners had been notably quiescent before 1914 and where the employer had been notably authoritarian. It is significant, however, that the period of radicalization remained brief and was followed with equal rapidity by a process of deradicalization. Many of the – often younger – miners involved in syndicalist and left–communist actions in the Ruhr in the early 1920s simply disappeared from the ranks of protest after 1924.[13]

Diversity and Fragmentation

That miners have behaved differently at different times and in different places is not surprising, despite many similarities in the nature of work under-ground, which, of course, underwent significant changes over time and was never identical across all coalfields. It is clear that the legislation of different states and their treatment of labour organizations can hinder or aid strike action and the process of unionization: the relative frequency of strikes and the especially high levels of union density in Britain were facilitated by a relatively liberal state, which was even prepared to encourage collective bargaining, whereas the repression encountered in Central and Eastern Europe before the First World War and in Latin America much later made union membership difficult and participation in strike action dangerous. Political cultures (the models of political action and ideology that are available to miners) were not only nationally but regionally variable, as Alan Campbell's work on Lanarkshire indicates and as the continued significance of ethnic and confessional divisions within the labour force demonstrates in the German case. What happens outside the mines is thus as crucial to the articulation and nature of miners' protest as is the experience of mining itself. Furthermore the behaviour of mine owners has had a fundamental impact on both the ability of workers to engage in struggle and the nature of their struggles, as we will see later.

However, it becomes increasingly clear not only that miners' attitudes and behaviour have differed in time and place and that exogenous influences affect levels of industrial militancy and political radicalism, but also that the occupational community of miners itself has often been more fractured than many earlier commentators imagined. This is true in various ways.

Firstly the economic conjuncture plays a role in determining solidarity or its absence. Despite images of communities hanging together through economic hardship, divisions in miners' communities, at least in Germany, have been more pronounced in times of recession, whereas solidarity has thrived best in times of full employment. My own work on the Ruhr in the 1920s and early 1930s highlights the differences between the two periods. It identifies conflicts in the Depression between the employed and the unemployed; amongst the unemployed themselves in a sometimes undignified scramble for jobs; and amongst those at work, fearful of dismissal and competing with their colleagues. Conflicts between workers occurred both within and between individual pits. It is no accident that productivity often rises in depressions. Job insecurity can also serve to increase conflict between younger and older workers when questions of redundancy are raised, whilst the unemployed can suffer various forms of isolation and atomization. Long-term unemployment in particular tends to be disruptive of family ties and gender relations. Old miners in the Ruhr remember the early 1930s as a time when miners began to steal from one another.[14]

However, even in the good times, solidarity cannot be taken for granted. Differential pay systems for hewers and hauliers can create conflicts of interest, even though there has usually been no formal system of apprenticeship in mining and thus no clear skills hierarchy in the sense that pertains in many other industries. Even if German colliers working together in a single team (*Kameradschaft*) had unitary interests, there was no reason why solidarity should necessarily develop with other teams working elsewhere underground; and in any case the growth of long wall mining increasingly reduced the solidarity of teams and made possible the individualization of pay. In the Morro Velho gold mine in Nova Lima, in the state of Minas Gerais in Brazil, the company sought to maximize output by pitting teams of *carreiros* (hauliers) against each other. Geological conditions, which could be vastly different from one pit to another, and the very different number of 'free shifts' worked (that is, the extent of short time working in recession), also implied very different experiences for colliers at different mines. Younger workers in a hewing team in Germany sometimes complained if an older and weaker colleague was included in a work team, as this threatened to reduce their earnings. In particular, the young hauliers formed the backbone of strike action in several coalfields, whilst their elders and trade union officials were more cautious. The tension between younger and older miners could, however, be less a question of conflict between the generations per se than a conflict between those in mining for their working life and those who were new to the pits and had no intention of staying there. The latter group could prove extremely difficult to organize, as in Scottish Coatbridge and the mines of West Virginia, yet could also, as Karen Hartewig shows in the case of Ruhr miners in 1919–20, provide the ranks of elemental insurrection. In the USA, on the other hand, the fact that most black labour was migrant until relatively late in the day served as a deterrent to collective action and organization, as it did too in South Africa.[15]

Gender relationships in the coalfields could aid or hinder communal solidarity, which depended upon the crucial role of women outside the pit and workgroup, as well as on the male experience of work. Women not only supplemented family income by taking in lodgers, which, argues Franz Brüggemeier, helped to socialize young and single miners in the Ruhr into communal values, but also played a major role as providers of food, washers of clothes, nurses, educators of children and seamstresses. In strikes they could sustain picket lines, as in the Collie Burn strike of 1911 in Western Australia, and in some cases, as in that of the 'Amazons' in the Pennsylvania coalfields in the 1890s, they could shame men from returning to work and strike-breaking, as they did too in some cases at Thurcroft in the British strike of 1984–5. The role of the Women's Action Group in that strike is well known. Employers' attempts to sanitize coalfields, get rid of prostitutes and create settled nuclear families, as at the *El Teniente* mine, could also backfire: by creating female economic dependence, the coalminers' wives found a commonality of interest with their husbands and a basis was created for class solidarity across gender divisions, or at least so argues Thomas Klubock.[16] However, the women of the coalfields could also have ambivalent attitudes towards industrial action and urge their husbands not to strike. (Examples of this are also to be found at Thurcroft in 1984.) The transmission of religious values to children by their mothers further served to perpetuate confessional divisions and restrain the advance of socialist parties in continental Europe.

Whatever the relationship between men and women at times of industrial action, however, the 'sexism' of miners' communities has been legendary. Union control at Broken Hill forbade the employment of married women (and, for that matter, non-locals). The 'Slav' mining communities of Pennsylvania insisted upon wives and daughters remaining in traditional roles (wives, mothers and guardians of the home). The practice of 'knocking-down' wage packets before they were handed to the wife and the invention of all kinds of reasons to explain discrepancies ('I have to pay the company for transport from the surface to the coalface') was not only common in Pennsylvania but also frequent in the UK. The classic British study, *Coal is our Life*, reveals the overlap between masculinity and class or, perhaps more properly, occupation. Work, play and politics were clearly gendered in the Yorkshire pit. The 'ethnicized identities' of Black South African miners were also, according to Tom Moldrum, 'very masculine' in their construction. In parts of the south island of Japan, of course, where females continued to work underground until 1933 in law and into the 1960s in fact, gender relations were rather different.[17]

Pitworkers could also be divided along lines of confession (religious denomination), as in the case of German miners before 1945, and even more obviously in terms of ethnicity. We have seen that the quite large numbers of Polish mineworkers in the German Reich formed their own industrial and political organizations. As a contemporary commentator in Bochum observed, 'they stay more and more together and avoid everything German'. Despite real attempts to win Polish support on the part of the *Alte Verband*, most German unionists not only resented Polish workers as potential wage-cutters and a threat to safety, but also shared a whole bag of cultural prejudices. Otto Hue, the leader of the social-democratic miners' union, compared Polish immigration into the Ruhr with the mass migrations of the Dark Ages. Foreigners became a target of abuse in economic crises and the 'Herne Polish Revolts' of 1899, which arose out of strike action, were blamed by the German trade union leadership on the 'ignorance' of the Poles. Hue spoke of Polish migrants as a 'flood-tide of refuse' flowing from the 'culturally lowest standing districts of Germany'. The Catholic miners' union commented on their 'low level of civilization' and wished to exclude non-Germans from responsible positions in the pits. Although the argument was ostensibly about mine safety, the discourse makes it clear that something other than purely instrumental concerns was at play. In the Ruhr upheavals of 1919–20, miners' anger often turned against shopkeepers accused of hoarding and price-hiking. Its expression on occasion included elements of anti-Semitism. During the Second World War the treatment of foreign slave labour in the pits of the Third Reich by German workers was brutal in the extreme – much more so, in fact, than in the steelworks and engineering plants, as the recent work of Ulrich Herbert demonstrates.[18]

Of course racial divisions and hostility were not the preserve of German colliers. In Lanarkshire, before 1914, attitudes to Irish miners resembled those of their German counterparts towards Poles. In both cases ethnic divisions were reinforced by religious sectarianism. At Thurcroft in South Yorkshire the union consistently refused to work with immigrant Poles. The employment of Poles and Lithuanians in the Scottish coalfields was also resisted in the mid-1940s and the arguments were precisely those about threats to safety and to wage levels (though – at least at the official level – without the disparaging rhetoric of 'civilization' that we saw in Germany). French miners used a variety of racist terms to characterize strike-breakers ('kaffirs', 'Prussians') and at La Mure (Isère) in 1901, armed miners forced 800 Italians to flee. In Southern France, it was Spanish miners who were disparaged by their French co-workers. In Australia, the same miners who united in collective struggle to form an egalitarian commonwealth, a 'level society', stormed a Chinese camp at Lambing Flat (New South Wales) in 1861 and joined with the mine owners to drive Aboriginals out of Tennant Creek in the Northern Territory in the 1930s. Indeed the goldfields were the bastions of Australian racism as well as sexism. At Kalgoorlie in Western Australia, there was hostility between miners of British descent and Italian immigrant mineworkers. In South Africa, apartheid structured the labour force and its conflicting interests along lines of race until relatively recently. The Rand strike of 1907 was a last-ditch attempt on the part of British skilled miners to prevent de-skilling through the employment of Afrikaans labour. The Rand White Revolt of 1922, this time composed largely of Afrikaaners, may have been a testimony to the industrial militancy of South African miners, but its immortal slogan tells all: 'Workers of the world unite and fight for a white South Africa!'[19]

The issue of labour, class and race has, of course, been at the centre of a vast amount of research on American labour and there can be no doubt that waves of immigration sometimes made the creation of a unitary class or occupational community difficult. Firstly, white miners in Arizona in the 1860s waged war on the native Indians (Yavapai and Apache). In Montana, they caused Red Cloud's War, in Colorado the Sand Creek Massacre and in the Black Hills the Sioux War. Secondly, hostility to Blacks continued to characterize many American miners, even when they had overcome their initial antipathy to 'Slavs' and other groups of white migrants. English, Welsh and German miners found it difficult to accept the Irish, though arguably this had more to do with religion than race. The native-born Protestants of the Pennsylvanian anthracite fields disliked Irish immigrants, who were castigated as 'micks', 'paddies' and, most significantly, 'papists'. In the 1850s, a wave of anti-Catholic sentiment swept the region: convents, schools and churches were sometimes torched. Initially, at least, the Knights of Labour were more successful with the Pennsylvanian Irish, whilst the Amalgamated Association of Miners had deeper roots amongst the English, Welsh and German. Whereas Polish and Irish Catholics got on reasonably well with one another, the Irish were hostile to the abolition of slavery in the Civil War, fearing that emancipated Blacks might take their jobs or undercut their wages. American Protestant miners remained firm in their suspicion of Catholics.[20]

Yet, just as gender was no inevitable source of conflict in the coalfields, and just as younger and older miners could and did unite in major struggles, such as the 1926 General Strike in Britain, so different racial groups could and did cooperate, as in the case of the US Knights of Labour and the Amalgamated Association after 1898, and in the big Pennsylvanian strikes of 1900 and 1902.

Those the white American miners called 'Slavs' (a term which was used indiscriminately to describe Hungarians and Lithuanians as well as true Slavs) were often divided along ethnic lines in the early days of settlement, but they came together to participate in miners' struggles and to create powerful communal loyalties. The myth that the immigrants and, for that matter, American Blacks were incapable of industrial militancy has been well and truly laid to rest. In the Lattimer Massacre, mentioned earlier, the 19 killed and 32 wounded included 26 Poles, 20 Slovaks and five Lithuanians. Black miners became the leaders of conflicts in the Flat Top/Pocahontas field, played a major role in the UMWA and participated in armed marches in West Virginia. In Germany, Poles might organize in a separate union, but they often participated in strikes by the side of German miners, as in 1905 and 1912. To be a Pole in the Ruhr was to be a worker: there was no conflict between class and ethnicity in this case. At Bisbee in Arizona, Anglos and Mexicans went on strike together. In the copper mines of Burra Burra in South Australia, relations between Germans and Cornishmen were cordial, as they were mostly between English, Welsh and Germans in Pennsylvania from the 1860s. As Neville Kirk has pointed out, race and class are both in a constant state of remaking and much depends upon both conjunctural factors and political developments. Kirk's account of ethnic conflict and cooperation in the USA, for example, sees conflictual relations dominating until the mid-1870s. The strike waves of 1873–7, 1884–6 and 1890–94 saw 'bonds of unity' overshadowing 'fragmentary forces'. Racial divisions returned with a vengeance between the 1890s and 1920s, only again to be replaced by more inclusive strategies and a labour offensive in the 1930s. Similar points could be made about relations between Catholic and socialist labour organizations in Imperial Germany, marked by cooperation in 1905 but conflict in 1912. In the Scottish coalfields the initial opposition to Poles tended to melt away when it transpired that the latter were reliable workers and prepared to join the existing unions.[21]

So ethnic, confessional, gender and age factors can fragment the capacity for solidarity, but they do not always do so. However the creation of the kind of communal solidarities which Campbell identifies in parts of Lanarkshire and which formed the basis of the 'Fed' in South Wales depended not only upon the existence of a relatively homogeneous labour force but also on the relatively slow expansion of the mining community, at least in comparison with developments in some of the American coalfields and the Ruhr, where, in the first case, White Americans, Black Americans and Europeans were thrown together and, in the second, Germans, Poles and distance migrants. The spectacularly rapid creation of the new labour force and its scale in both the American and German cases, where miners often had to travel significant distances to the pit and where mining colonies were often segregated along racial lines, rendered impossible the proximity of pit and home characteristic of South Wales and the 'Little Moscows'. Perhaps even more significantly, pit work was for many American Blacks but a temporary expedient. The existence of migratory labour, so important in the South African mines too, militated against a strong occupational identity. Such an identity first began to grow amongst those South African Blacks who stayed on in the compounds and did not return to the homelands.[22]

Occupational identity and solidarity was endangered by a high incidence of job changing and geographical mobility, and both of these factors characterized the Ruhr before 1914, as they did many of the American coalfields. In 1907, migrants constituted no less than 84 per cent of the mining labour force in Bochum. In the north of the Revier (Herne, Wanne-Eickel, Castrop-Rauxel, Recklinghausen), Poles may have formed a majority of miners. Many of the migrants stayed in mining but a short period of time, while many who did stay changed pit and employer with remarkable frequency. In 1913, 78 per cent of all miners in the Ruhr had been hired by and 69 per cent had departed from their pits within the previous 12 months.[23]

Structures of Control and Protest

The development of an independent communal culture of resistance was frustrated, not only by ethnic division and high labour turnover in both West Virginia and the Ruhr, but also by the power of the coal companies and the collusion of government. In West Virginia, miners often came into new company towns rather than longer established settlements. The state had the highest proportion of company towns in the whole of the USA and some 80 per cent of the state's miners lived in them. The employer was in such situations often the landlord, merchant, purveyor of food and provider of entertainment. The power this conferred upon the company was used to sack union members on discovery of their membership. Anyone who dared to strike could reckon with eviction as well as the sack. Mine guards were even known to murder union sympathizers, while judges authorized any number of injunctions to curb strike action and union membership. The companies could also often rely upon compliant governors to declare martial law.

Although the situation was somewhat more complicated in the German coalfields, it was also true that there was considerable collusion between the state authorities and the mine companies, especially during the 1912 strike in the Ruhr, when more miners were arrested than in the previous 10 years put together. The Upper Silesian miners' strikes of the 1870s saw large-scale military intervention, as did the great Ruhr strike of 1889, when there were fatalities in Bochum, Bottrop, Dortmund and Gelsenkirchen. Machine guns were turned on striking copper miners at Mansfeld in 1909 and in the central lignite fields two years later. After the closure of most unions during the period of the anti-socialist laws (1878–90), the German labour movement was still subject to a variety of forms of harassment and prosecution. In 1912, the Prussian police authorities continued to ban some workers' leisure associations and the surveillance of local union branches increased. In the state-owned pits of Silesia and the Saar, membership of the Free Unions and the SPD was proscribed. The use of legislation to control the Black labour force in South Africa was, of course, even more blatant.[24]

If the space for action and organization was circumscribed by the behaviour of the political authorities in many places, as it was in most of Latin America, for example, it was even further restricted by the policies of employers. Labour relations in mining were characterized by high levels of mutual hostility between employers and employees in many, if not most, countries. The origins of this conflict can be traced to the fact that mining companies were much more sensitive to wage costs, given the labour-intensive nature of extraction, which meant that, unlike the case of chemical or electrical industries, wages constituted an unusually high percentage of total costs (still 50 per cent in the Ruhr mines in the 1920s). At the same time, the problem of recruiting and retaining labour was an acute one for a variety of reasons: the isolated location of mines in Japan, New Zealand, Australia, Peru and Bolivia; the very rapid expansion of pits in previously rural and relatively unpopulated districts, as in the Ruhr and West Virginia; and the dirty, unhealthy and dangerous nature of the work, which led to a very high turnover of the labour force. Under these circumstances employers and sometimes governments (migratory labour laws in South Africa, privileges to miners in communist Poland) resorted to a variety of strategies to attract miners to and keep them in the coalfields. In Natal, Indian labour was tied to the mining companies first by the indentured labour scheme and then, when that broke down, by the token system. In the British-occupied zone of Germany at the end of the Second World War attempts were initially made to conscript Germans to the mines. These coercive strategies, however, were often limited in their efficacy and so states and employers resorted to other methods of recruitment.

The classic case of alternative methods was employer 'paternalism' in the mines, which was much more extensive and much more authoritarian in its continental than its British variant. Such 'paternalism' involved the provision of benefits in the shape of company housing, company pension schemes and leisure programmes. At first this 'paternalism' was aimed at attracting and keeping a core of reliable workers. Increasingly, however, and especially with the growing threat of industrial action and unionization, it developed its more sinister and authoritarian side. Those living in company housing faced rigorous control and the possibility of eviction if they came into conflict with their employers. Direct deductions from wages for pension funds and other company benefits might be forfeited in similar circumstances. The scale of these provisions in the German coalfields was massive. In 1914, 22 per cent of all Ruhr miners lived in company housing, for example, and the figure was much higher in the northern part of the coalfield. Dormitory and company housing was often segregated according to ethnicity, whilst the housing regulations of the companies aimed to create a disciplined, sober and obedient labour force. Infringements of such regulations (and of those governing work) were met with a plethora of fines, which could swallow a substantial part of the miners' wage packet, to judge from the Silesian evidence. Credit at company stores constituted a further mechanism of control, as it did in the case of the Chilean copper mine and the Morro Velho goldmine in Nova Lima (Brazil), where, incidentally, the local priest also organized pro-company workers against the independent union.

Authoritarian employers in continental Europe and the USA had recourse to a huge armoury of weapons in the battle to control labour. Bonuses were paid to spies and informers. Extensive blacklists of agitators were drawn up by the Ruhr mine owners, who set up their own labour exchanges to control the recruitment of miners. Mass dismissals of strikers were common and mining companies in France, Germany, Austria and Belgium usually refused to recognize trade unions or negotiate with workers' representatives. Whereas 900 000 British miners were covered by collective wage agreements in 1910, in the whole of the German Reich three years later a paltry 83 miners were so covered, none of whom, in fact, were coal miners. It is important to realize that it was the British and not the German situation which was atypical. Collective bargaining was rare before 1914. Its existence in the UK was a consequence of low levels of capital concentration, the smallness of firms, the absence of cartels and the weakness of employer organizations in the British coal industry in comparison with the situation of the German companies. The authoritarianism of employers in continental Europe and elsewhere was a function of their power. The power stemmed from the monopoly of local labour markets, high degrees of capital concentration and cartelization, and strong employer organization, as well as the help of the state. In the class war of the coalfields in Germany and France, the employers were the clear winners before 1914. Strikes failed, lock-outs were successful. This was one of the reasons why miners often looked to the state and nationalization as a solution to their problems. It was only when the state became more sympathetic to unions and employers were forced to recognize them that mining unionism was really able to develop in Germany's Weimar Republic; or when the labour market turned strongly in favour of miners, as at the end of both world wars in Europe and in South Africa in the 1970s and 1980s; or when the state actually encouraged collective bargaining, as was to some extent the case in Britain and its Dominions after the turn of the century.[25]

The authoritarianism of mining employers had important consequences for the attitudes and actions of miners. Although miners' strike demands were usually concerned with wages and working conditions (including – predictably – safety issues), there was often an additional, anti-authoritarian element present: demands for the sacking of supervisors in Bolivia, attacks on compound managers in Natal and on company officials in Japan. Arbitrary practice in payment was regularly as serious a cause of complaint as the actual level of wages itself. The experience of autocratic management, which was often enough to prevent collective action for a time, gave rise to elemental and often violent explosions at specific points in time, often associated with political upheaval, though these were often of short duration. As Karin Hartewig has shown, the leaning towards syndicalist styles of action was most pronounced in those Ruhr pits which had suffered most from autocratic management. What the socialization campaign of 1919–20 meant to miners in the Ruhr was primarily the control of their own pit, rather than the total reorganization of society, and it was a campaign of miners alone. In New Zealand, as Len Richardson has shown, the miners' concept of socialism in the decade before the First World War was tied in with localism. This raises an important question. Much as 'Little Moscows' may have regulated their own small localities, they were not going to change the larger world. Moreover the socialization campaigns in Berlin and the Saxon cities between 1919 and 1923, which embraced various occupations and arose from a powerful socialist sub-culture, were altogether more deeply rooted and grander than the pit-oriented movements of Ruhr miners, which aimed at control of the individual pit.[26]

This obviously begs the question of sectionalism, of the extent to which miners' solidarity was that of occupation rather than class. I think we have seen that it was often the first rather than the second. Sometimes it was neither. Miners in some districts did not cooperate with other miners, let alone with other sections of the working class. In the British coal strike of 1984, the fact that miners picketed the Ravenscraig steelworks, which were threatened with closure at the time, scarcely indicates a class, as distinct from occupational, solidarity. The same could be said of the refusal of Zimbabwe's miners' union, the most important of the country's Black unions, to join a general confederation of labour (memories of the British miners here). Like all other groups of workers, the industrial militancy and political identity of miners has been the product not only of their workplace and occupation but also of culture, power and politics outside the mine, as well as of historical conjuncture. It has ebbed and flowed. It may now have disappeared, at least in the United Kingdom.

Notes

1   Roger J. Fagge, 'A comparison of the miners of South Wales and West Virginia, 1900–1922', in Klaus Tenfelde (ed.), *Sozialgeschichte des Bergbaus im 19. und 20. Jahrhundert* (Munich, 1992), pp.116f; Donald L Miller and Richard E. Sharples, *The Kingdom of Coal* (Philadelphia, 1985), pp.217f; Donald Reid, *The Miners of Decazeville* (Cambridge MA, 1985), pp.7, 58ff, 74f; Peter Gibbon and David Steyne, *Thurcroft: A Village in the Miners' Strike* (Nottingham, 1986), p.16.

2   Clark Kerr and Abraham Siegel, 'The inter-industry propensity to strike', in Arthur Kornhauser *et al.* (eds), *Industrial Conflict* (New York, 1954), pp.189–212; Victor L. Allen, *The Militancy of British Miners* (Shipley, 1981); Paul Rigg, 'Miners and militancy', *Industrial Relations Journal*, **18** (1987), 189–200; Kenneth G.J.C. Knowles, *Strikes: A Study in Industrial Conflict* (Oxford, 1952); Roy Church, Quentin Outram and David N. Smith, 'Theoretical orientations to miners' strikes' in Tenfelde (ed.), *Sozialgeschichte*, pp.565–81. These last arguments are expanded in R.A. Church and Q. Outram, *Strikes and Solidarity: Coalfield Conflict in Britain* (Cambridge, 1998). For France, see Reid, *Decazeville*, p.3 and Samuel Cohn, *When Strikes Make Sense and Why: Lessons from French Third Republic Coal Miners* (New York, 1993), p.93; also in general Michelle Perrot, *Workers on Strike: France 1871–1890* (Leamington Spa, 1987); Edward Shorter and Charles Tilly, *Strikes in France* (London, 1974); and Roland Trempé, *Les mineurs de Carmaux*, 2 vols (Paris, 1971). On Germany, see Albin Gladen, 'Die Streiks der Bergarbeiter im Ruhrgebiet in den Jahren 1889, 1905 und 1912', in Jürgen Reulecke (ed.), *Arbeiterbewegung an Rhein und Ruhr* (Wuppertal, 1974), pp.111–48; Hans Mommsen, 'Soziale und politische Konflikte an der Ruhr 1905 bis 1925', in Hans Mommsen (ed.), *Arbeiterbewegung und industrieller Wandel* (Wuppertal, 1980), p.64; Dick Geary, 'Rhein, Ruhr und Revolution', in *Mitteilungsblatt des Instituts zur Geschichte der Arbeiterbewegung*, **7** (1984), 30–36; Franz-Josef Brüggemeier, *Leben vor Ort* (Munich, 1983), pp.180–233; Klaus Tenfelde, *Sozialgeschichte der Bergarbeiterschaft an der Ruhr im 19. Jahrhundert*, (2nd edn, Bonn, 1981), pp.397–601; Stephen Hickey, *Workers in Imperial Germany: The Miners of the Ruhr* (Oxford, 1985), pp.169–225; David F. Crew, *Town in the Ruhr* (New York, 1979), pp.159–220; Lothar Machtan, *Streiks im frühen deutschen Kaiserreich* (Frankfurt am Main, 1983). For Japan, see Kazuo Nimura, 'The Ashio riot of 1907', in Tenfelde (ed.), *Sozialgeschichte des Bergbaus*, p.791. For Latin America, see Francisco Zapata, 'The constitution of Bolivian, Chilean and Peruvian mining unions in the early Twenties', in Tenfelde (ed.), *Sozialgeschichte des Bergbaus*, pp.946–73.

3   Nimuro, 'Ashio riot', pp.789–808; Alan B. Campbell, 'Communism and trade union miliancy in the Scottish coalfields', in Tenfelde (ed.), *Sozialgeschichte des Bergbaus*, p.90; Miller and Sharples, *Kingdom*, pp.229–32; Fagge, 'Comparison', p.107; Mommsen, 'Soziale Konflikte', p.64; Klaus Tenfelde, 'Linksradikale Strömungen in der Ruhrbergarbeiterschaft', in Hans Mommsen and Ulrich Borsdorf (eds), *Glück auf, Kammeraden!* (Cologne, 1979), pp.199–223; Geary, 'Rhein', 30–36. On German syndicalism amongst Ruhr miners, see Klaus Tenfelde, 'Linksradikale', pp.202f.; Manfred Bock, *Syndikalismus und Linkskommunismus* (Meisenheim, 1969); Irmgard Steinisch, 'Linksradikalismus und Rätebewegung' in Reinhard Rürup (ed.), *Arbeiter- und Soldatenräte im rheinisch-*

westfälischen Industriegebiet' (Wuppertal, 1975), p.206; Erhard Lucas, *Zwei Formen Arbeiterradikalismus* (Frankfurt am Main, 1976). For civil war in the Ruhr, see Erhard Lucas, *Märzrevolution*, 3 vols (Frankfurt am Main, 1970–79); Hans-Ulrich Ludewig, *Arbeiterbewegung und Aufstand* (Husum, 1978); Jürgen Tampke, *The Ruhr and Revolution* (Canberra, 1978); Karin Hartewig, *Das unberechenbare Jahrzehnt* (Munich, 1993). On syndicalist unrest in Britain before 1914, see James E. Cronin, *Industrial Conflict in Modern Britain* (London, 1979), pp.93–125; Peter N. Stearns, *Lives of Labour* (London, 1975), pp.322ff; Henry Pelling, *History of British Trade Unionism* (London, 1966), pp.134–8; Roy Church, 'Edwardian labour unrest and coalfield militancy, 1890–1914', *Historical Journal*, **30** (1987), 841–57; Dai Smith, 'Tonypandy, 1910', *Past and Present*, **87** (1980), 158–84; D.M. Gilbert, *Class, Community and Colective Action: Social Change in Two British Coalfields* (Oxford, 1992); E.H. Hunt, *British Labour History 1815–1914* (London, 1981), pp.329f. For the Southern Massif, see Trempé, *Mineurs*. For New Zealand, see Len Richardson, 'The birth of mining unionism in a settler society: New Zealand, 1880–1913', in Tenfelde (ed.), *Sozialgeschichte des Bergbaus*, pp.743–7. For Chile, see Thomas Miller Klubock, *Contested Communities: Class, Gender and Politics in Chile's El Teniente Copper Mine* (Durham, NC, 1998).

4 The classic study is Stuart Macintyre, *Litle Moscows* (London, 1980). On South Wales, see Stefan Berger, 'Working-class culture and the labour movement in South Wales and the Ruhr Coalfields, 1850–2000', *Llafur*, **8**(2) (2001), 5–40; Werner Berg, 'Zwei Typen industriegesellschaftlicher Modernisierung: Die Bergarbeiter im Ruhrbegiet und Südwales im 19. und frühen 20. Jahrhundert', in Gustav Schmidt (ed.), *Bergbau in Grossbritannien und im Ruhrgebiet* (Bochum, 1985), pp.199–219; Hywel Francis and Dai Smith, *The Fed* (Cardiff, 1998); Robin Page Arnot, *The South Wales Miners* (London, 1967); Chris Williams, *Capitalism, Community and Conflict: The South Wales Coalfield, 1898–1947* (Cardiff, 1998); Fagge, 'Comparison', pp.112–14. For Ruhr miners, see the references in note 3. On the strength of the KPD in mining districts, see Ludger Fittkau (ed.), *Das 20. Jahrhundert der Gaudigs* (Essen, 1997); Eric D. Weitz, *Creating German Communism* (Princeton, 1997), pp.221–3; Dick Geary, 'Unemployment and working-class solidarity in Germany, 1929–33', in Richard J. Evans and Dick Geary (eds), *The German Unemployed* (London, 1987), pp.261–80. On KPD strength in the Ruhr in 1946–7, see Christoph Klessmann and Peter Friedemann, *Streiks und Massendemonstrationen im Ruhrgebiet* (Frankfurt am Main, 1977), pp.41–53; G. Mannschatz and J. Seider, *Der Kampf der KPD im Ruhrgebiet* (Berlin, 1962), pp.53, 208; G Schädel, Die KPD in Nordrheinwestfalen von 1945–1956 (phil.diss, Ruhr University, 1974), 47; *Jahrbuch Nordrheinwestfalen 1949* (Düsseldorf, 1950), pp.323ff. On the Asturias, see Adrian Shubert, 'A divided community: the social development of the Asturian coalfields to 1934', in Tenfelde (ed.), *Sozialgeschichte des Bergbaus*, pp.284–93; and Adrian Shubert, *The Road to Revolution in Spain* (Urbana, 1987).

5 On different proclivities to strike in the Scottish coalfields, see Alan B. Campbell, 'Communism and trade union militancy in the Scottish coalfields', in Tenfelde (ed.), *Sozialgeschichte des Bergbaus*, pp.85–104. See also Alan B. Campbell, *The Lanarkshire Miners* (Edinburgh, 1977). Patterns of diversity in British coalfields constitute a central argument of Church and Outram, *Strikes*, and of Church,

Outram and Smith, 'Theoretical orientations'. See also Brian J. McCormick, 'Strikes in the Yorkshire coalfield 1947–63', *Economic Studies*, **4** (1969), 171–97, and his *Industrial Relations in the Coal Industry* (London, 1979); Andrew Charlesworth *et al.*, *An Atlas of Industrial Protest in Britain* (London, 1996) provides a great deal of data on the regional distribution of strike action, including that of 1984–5. On the 1984–5 dispute see also Gibbon and Steyne, *Thurcroft*; Ralph Samuel *et al.*, *The Enemy Within* (London, 1986); Jonathan and Ruth Winterton, *Coal, Crisis and Conflict* (Manchester, 1989).

6   Gibbon and Steyne, *Thurcroft*, p.49.

7   See note 5's British entries; also N. Emery, *The Coalminers of Durham* (Stroud, 1992); Neil Evans, 'Patterns of protest and regional labour implantation in South Wales and the north east of England', *Tijdschrift voor Sociale Geschiedenis*, **18**(2/3) (1992, 212–30; Keith Burgess, *The Origins of British Industrial Relations* (London, 1975), pp.172–215.

8   Gaston V. Rimlinger, 'International differences in the strike propensity of coal miners', *Industrial and Labour Relations Review*, **12** (1959), 389–405; Klaus Tenfelde, 'Comparative research on the history of mining workers', in Werner Berg (ed.), *Bergbau in Grossbritannien und im Ruhrgebiet* (Bochum, 1985), pp.18–35; Klaus-Michael Mallmann, 'Realms of experience and interpretation: class building, fragmentation and the German miners' movement, 1871–1914', in Tenfelde (ed.), *Sozialgeschichte des Bergbaus*, pp.486–543; Gerhard A. Ritter and Klaus Tenfelde, *Arbeiter im Deutschen Kaiserreich* (Bonn, 1992), pp.416, 598ff; Hickey, *Workers*, pp.169–225, 236–55; Geary, 'Solidarity', p.273; Crew, *Town*, pp.159–220. On Polish miners and their unions, see John J. Kulczycki, 'A trade union for Polish miners in the Ruhr' in Tenfelde (ed.), *Sozialgeschichte des Bergbaus*, pp.593–608 and, by the same author, *The Foreign Worker and the German Labour Movement* (Oxford, 1994). Also Christoph Klessmann, *Polnische Bergarbeiter im Ruhrgebiet* (Göttingen, 1978). For a more sanguine view of German–Polish relations within the working class, see Richard Charles Murphy, *Guestworkers in the German Reich* (Boulder, 1983). For an overview, see also Stefan Berger, 'British and German socialists between class and national solidarity', in Stefan Berger and Angel Smith (eds), *Nationalism, Labour and Ethnicity 1870–1939* (Manchester, 1999). On Catholic workers, see Michael Schneider, *Die Christlichen Gewerkschaften* (Bonn, 1982); Eric Dorn Brose, *Christian Labor and the Politics of Frustration in Imperial Germany* (Washington, 1985); and William L. Patch, Jr, *Christian Trade Unions in the Weimar Republic* (New Haven, 1985). On 'yellow' miners and their unions, see Klaus Mattheier, *Die Gelben* (Düsseldorf, 1973). Also Dick Geary, 'The Prussian labour movement, 1871–1914', in Philip G. Dwyer (ed.), *Modern Prussian History 1830–1947* (Harlow, 2001), pp.135–43.

9   Philip G. Eidelberg, 'An historical overview of twentieth-century trade union organizations in the Transvaal mining industry', in Tenfelde (ed.), *Sozialgeschichte des Bergbaus*, pp.749–67; Ruth Edgecombe, 'Labour organization, accommodation and resistance on the Natal coal mines' in Tenfelde (ed.), *Sozialgeschichte des Bergbaus*, pp.768–88; Jerry Holzer, 'Polish miners' strikes and the Polish government since World War II', in Tenfelde (ed.), *Sozialgeschichte des Bergbaus*, pp.670–81. Shubert, *Road*, depicts the transition from reformism to insurrection in the Asturias. On French anarcho-syndicalism and

the diversity of French labour, see Roger Magraw, 'Socialism, syndicalism and French labour' in Dick Geary (ed.), *Labour and Socialist Movements in Europe before 1914* (Oxford, 1989), pp.48–100. Also P.N. Stearns, *Revolutionary Syndicalism* (New Brunswick, 1971); Georges Lefranc, *Histoire du mouvement syndical français* (Paris, 1937); Paul Louis, *Le Syndicat en France* (Paris, 1963). On Spanish anarchism, see M. Bookchin, *The Spanish Anarchists* (New York, 1977); T. Kaplan, *The Anarchists of Andalucia* (Princeton, 1977); R.W. Kern, *Red Years/Black Years* (Philadelphia, 1978); G.H. Meaker, *The Revolutionary Left in Spain* (Stanford, 1974); Martin Blinkhorn, 'Spain', in Stephen Salter and John Stevenson (eds), *The Working Class and Politics in Europe and America 1929–45* (London, 1990); Paul Heywood, 'The labour movement in Spain before 1914', in Geary (ed.), *Labour*, pp.231–60.

10 On the varieties of politics amongst British miners, see A. Campbell, N. Fishman and D. Howell (eds), *Miners, Unions and Politics* (Aldershot, 1996); also Howard, *Coalminers*; Evans, 'Patterns'; Charlesworth, *Atlas*. On the Asturias, see Shubert, *Road*; for France, Magraw, 'Socialism'. For Chile, Peru and Bolivia see Zapata, 'Constitution' and Jorge Lazarte Rojas, 'Crisis in the mining sector and mine workers' organizations: The cases of Bolivia, Chile and Peru', in Tenfelde, *Sozialgeschichte des Bergbaus*, pp.974–86.

11 Fagge, 'Comparison', pp.117f.

12 The barriers that Catholicism, Evangelical culture, ethnic divisions and employer paternalism erected against the spread of socialist politics in the Ruhr are traced in Geary, 'Prussian labour', pp.129–43. See also Hickey, *Miners*, pp.199–261. Campbell, 'Communism', 85–104.

13 Shubert, *Road*, has this transition as its central theme, as it is also in Geary, 'Rhein'. On Barnsley and Merthyr, see David Kynaston, *King Labour* (London, 1976), pp.148f. On Poland, see Holzer, *Polish Miners*.

14 Geary, 'Solidarity', 261–80.

15 Hickey, *Miners*, p.164; Marshall Eakin, *British Enterprise in Brazil: The St John d'El Rey Mining Company and the Morro Velho Gold Mine* (Durham, NC, 1989) and Douglas C. Libby, *Transformação e trabalho em uma economis escravista* (São Paulo, 1988). On the young and new miners in the Ruhr, see Mommsen 'Soziale', p.64; Hartewig, *Unbrechenbare*, pp.246–83; Campbell, 'Communism', 89ff; Fagge, 'Comparison', 111; Patrick Harries, *Work, Culture and Identity: Migrant Labourers in Mozambique and South Africa* (Johannesberg, 1994); Edgecombe, 'Labour', 784.

16 Brüggemeier, *Leben*, pp.52–62; Bill Latter, *Blacklegs: The Scottish Colliery Strike of 1911 in Western Australia* (Nedlands, WA, 1995), p.36; Miller and Sharples, *Kingdom*, pp.226ff; Gibbon and Steyne, *Thurcroft*, p.172. See also Jean Stead, *Never the Same Again: Women and the Miners' Strike 1984–85* (London, 1987); Joan Witham, *Hearts and Minds* (Nottingham, 1985). On the *El Teniente* mine see Klubbock, *Contested Communities*.

17 Gibbon and Steyne, *Thurcroft*, p.187; William A. Howard, 'The rise and decline of the Broken Hill industrial relations system', in Tenfelde, *Sozialgeschichte des Bergbaus*, pp.737–48; Miller and Sharples, *Kingdom*, pp.125, 192–5; Norman Dennis *et al.*, *Coal is Our Life* (London, 1956); Tim Moldran, 'Treading the diverse paths of modernity: labour, ethnicity and nationalism in South Africa', in Berger and Smith (eds), *Nationalism*, p.224; Sone Sachiko, 'Coal mining women

in Japan: cultural identity, welfare and economic conditions on the Chikuho coalfield', PhD, University of Western Australia (2002).

18   See the references to Polish workers in Germany in note 8. See especially Kulczycki, 'A trade union', 611–14, which includes the quotations. On anti-semitism and Ruhr miners in 1919–20, see Hartewig, *Unberechenbare*, pp.228–44. On the treatment of foreign workers in the Third Reich, see Ulrich Herbert, *Hitler's Foreign Workers* (Cambridge, 1997).

19   Campbell, 'Communism', 99ff; Gibbon and Steyne, *Thurcroft*, p.24; Kenneth Lunn, 'The employment of Polish and volunteer workers in the Scottish coalfield, 1945–50' in Tenfelde (ed.), *Sozialgeschichte des Bergbaus*, pp.582–92; Magraw, 'Socialism', 62–4; Roger Magraw, 'Appropriating the symbols of the *patrie*', in Berger and Smith, *Nationalism*, pp.93–120; Reid, *Miners*, p.120; David Carment, 'Mining, race and politics of the North Australian frontier', in Tenfelde (ed.), *Sozialgeschichte des Bergbaus*, pp.242–53; Terry Irving, 'Labour, state and nation building in Australia', in Berger and Smith (eds), *Nationalism*, pp.193–214; Brian Kennedy, *A Tale of Two Mining Cities: Johannesberg and Broken Hill* (Melbourne, 1984), pp.113–18; Moldran, 'Treading', 224f.

20   Robert L. Spude, 'Native Americans and gold rushes' in Tenfelde (ed.), *Sozialgeschichte des Bergbaus*, pp.213–22; Miller and Sharples, *Kingdom*, pp.149f.

21   Miller and Sharples, *Kingdom*, p.216; Fagge, 'Comparison', 116ff; Kulczycki, 'Trade union', pp.609–21; Neville Kirk, 'The working class in the United States', in Berger and Smith, *Nationalism*, pp.164–92; Lunn, 'Employment', 589f. On the multiple identities of European workers, see Dick Geary, 'Working-class identities in Europe', *Australian Journal of Politics and History*, **45**(1) (1999), 20–34.

22   Campbell, 'Communism', 89ff; Fagge, 'Comparison', 111; Harries, *Work*.

23   South Wales was clearly relatively stable in comparison to both the Ruhr and West Virginia, as the articles by Fagge and Berg, cited above, make clear. For migration figures, see Geary, 'Prussian labour', 139.

24   Fagge, 'Comparison', 110f. On the role of repression in structuring German labour's identity, see Geary, 'Prussian labour', 131–5. More generally, on the significance of the state for an understanding of collective action and identity in Europe, see Dick Geary, *European Labour Protest, 1848–1939* (London, 1981), pp.19–22, 58–65.

25   On employer strategies to attract, keep and control labour, see Dick Geary, 'The industrial bourgeoisie', in David Blackbourn and Richard J. Evans (eds), *The German Bourgeoisie* (London, 1990), pp.140–52; Klaus Wisotzky, 'Company welfare schemes in Ruhr coalmining', in Tenfelde (ed.), *Sozialgeschichte des Bergbaus*, pp.1066–82; and Mark Roseman, 'Settling the workforce', in Tenfelde (ed.), *Sozialgeschichte des Bergbaus*, pp.1102–23. Also Joel Michel, 'Mining communities and paternalistic ownership in Western Europe before 1914', in Tenfelde (ed.), *Sozialgeschichte des Bergbaus*, pp.58–84; Holzer, 'Polish', pp.670–81. See Edgecombe, 'Labour', for Natal and Lawrence Schofer, *The Formation of a Modern Labour Force* (Berkeley, 1975) for Silesia. On Brazil, see Eakin, *British*, and Libby, *Transformação*.

26   See references in note 3.

CHAPTER FOUR

Cameras in the Coalfields: Photographs as Evidence for Comparative Coalfield History

Janet Wells Greene

The published photographic record for bituminous coalfield communities in the United States in the first half of the twentieth century abounds with images which embody high emotional content as well as pointed and ideologically motivated captions. While researchers may stumble upon troves of photographs as part of the archival record when examining coalfield communities, few studies take these images seriously as historical evidence, and instead merely employ the images as illustrations.[1]

Photographs of coalfield communities provide a deceptively simple basis for comparison. Many investigations into living conditions for miners in the bituminous coalfields in America during the twentieth century gathered and published photographs to reinforce the view that there were only two kinds of mining communities: the 'model towns' constructed and operated by a benevolent paternalistic management; and the squalid hovels owned by companies with no regard for the welfare of their workers.[2] As a paradigm for comparative coalfield history, this approach has merit, in that it could lead to an in-depth comparison of mine owners and their management practices over time or in diverse locations. However, corporate records are not readily available for such large-scale investigations.

Another way to deepen our understanding of the ways in which coalfield communities might be understood and compared is to scrutinize the images as documents. Such an approach leads to fresh sources of information: photographers' notes; the captions for both published and unpublished images; and the archival documents of the organizations which employed photographers. By placing the photographer and his or her subject in this broader social or political context, it then becomes possible to explore the relationship between the creation of images and power.[3] The photographic record of US coalfield communities provides some striking examples of the way in which a seemingly simple image is actually a complex product, chosen and published for its ability to illustrate a point of view.[4]

The first photographers in the twentieth century to bring images of mining life to national attention in the USA were documentary photographers employed by agencies seeking social reforms in the coalfields in the first decade of the twentieth century. Naturally there was resistance on the part of coal operators, which was often sporadic and individualistic. As the coal industry began to resist competition from other fuels, especially oil, there were efforts on the part of the National Coal Association, an industrial alliance, to centralize that resistance, an impulse which was also present in regional coal associations.[5]

Access

The tradition of keeping cameras out of the coalfields in America had a long history. Lewis Hine, a crusader against child labour and a photographer for the National Child Labor Committee, earned the distrust of employers following his work in bituminous mining operations in West Virginia in 1908 and the anthracite coalfields of Pennsylvania in 1911. His photographic style contrasted vulnerable children and the harshness of their industrial surroundings.[6] Hine's emotionally compelling images and captions were made possible in part by his ability to elude and deceive officials of the industries which employed children. As the campaigns of the National Child Labor Committee became more visible to the public, employers began to refuse its investigators entry altogether.[7]

In the 1930s, cameras were unwelcome in the coalfields and owners posted signs against trespassers and union organizers. The *UMWA Journal*, official publication of the United Mine Workers of America, sent artists to some non-union locations, and published their drawings of run-down 'non-union' dwellings as part of organizing drives to illustrate the miseries of 'non-union' life.[8]

By 1946, even documentary photographers employed by the federal government had difficulty gaining access to mining operations. During a federal takeover of the mines following a nationwide coal strike in 1946, federal investigators and their cameras for the Medical Survey of the Bituminous Coal Industry technically had unrestricted access to the private property of the 260 mines selected for review. However, before the investigators could start their work, a series of meetings between union and company officials, state and county public health officers, and the administrators of the investigation were needed to pave the way for the investigating teams and their cameras.[9]

Photographers working for political organizations and wire service agencies such as Wide World Photos found that their access to mining operations varied with the subject matter. Photographers might be kept away or find their access limited in the case of a mining accident, but they were welcomed to photograph new housing.[10]

The owners of mining operations also welcomed cameras on other occasions. Corporations and regional coal associations hired photographers for public relations images or to document developments and improvements to their operations. Trade and industry publications sent photographers and reporters to cover stories on technological developments and production records in mining operations. Freelancers also travelled the coalfields, making portraits for miners and their families, workplace photographs for companies and employees, and documentary images for artistic or commercial purposes.[11]

Content

Images made by corporate photographers and coal associations often celebrated the mineral itself, the production process and the equipment and the administrative offices of mining operations. Larger mining operations sometimes commissioned photographs of miners at the end of their shift, or of residents in the homes and stores of company-owned communities.[12] Selections from these images were published as glossy presentations in project notebooks, corporate histories, annual reports or employee newsletters. Others became exhibits in legal documents, illustrations for reports on accidents, or were appended to press releases. Some were used for public relations purposes, while others served to aid internal reviews of mining operations or to document the progress of construction in a new or renovated mining operation.

In contrast, images by New Deal reformers in the 1930s were often published to depict the negative effects of unemployment and low wages on the workers in coal and other industries. Mining, as a rural industry with high unemployment levels and labour unrest, attracted the attention of photographers employed by two agencies of the US government, the Resettlement Administration (RA) and the Farm Security Administration (FSA). These agencies were charged with documenting both evidence of poverty and suffering in rural areas and the efforts of government agencies to implement reforms. Many of these images, like those of Lewis Hine, featured vulnerable members of society (in this case women and children) photographed in rural poverty.

Well-known photographers from the FSA and RA who photographed communities in the coalfields included Walker Evans, Russell Lee, Sheldon Dick, Arthur Rothstein, Marion Post Wolcott, John Vachon and Jack Delano. Ben Shahn, a photographer for both agencies, visited picket lines at mines in eastern Kentucky and West Virginia. Photographs by these federal photographers which were published in general publications featured paired images of the condition of housing 'before' government intervention and 'after' federal programmes were implemented. Coalfield communities made up a small number of these examples.[13]

Surprisingly, the United Mine Workers of America, the most significant labour organization of coal miners in the United States from the 1930s until the end of the century, was comparatively slow to gather and publish photographs of impoverished coalfield communities and residents. In its campaign to organize miners in 1933, the union attempted to project itself as a prosperous and respectable organization by publishing only studio portraits of its officers. In the mid-1930s, the union's editor began to use published photographs of events in coalfield communities as organizing tools. The *UMWA Journal* included images of parades, scenes of mining accidents and the more dramatic images, such as labour confrontations, generated by wire service companies. But until the end of the Second World War, the union's publication avoided photographs of impoverished miners or their families, preferring images of clean and well-dressed men, women and children to illustrate the value of joining a union.[14]

Coal miners suffered some of the worst living conditions in America among industrial workers. Improvements in mining equipment and operations did not always extend to the sanitation and housing needs in company towns. Even mining towns which began as model operations began to deteriorate as decades wore on. Coal dust, lack of maintenance, poor water systems and dense populations added to the sanitation problems of these settlements. As president of the United Mine Workers' Union, John L. Lewis fought against company control of mining towns as part of the union's organizing campaigns for more than 25 years. During the organizing drives of the 1930s, Lewis and his organizers equated sub-standard housing with non-union operations, and railed against the un-American way of life in 'non-union' company towns.[15]

Water supply was an important convenience that came to be seen as an 'American' right by the end of the Second World War, despite the fact that, in many rural and urban areas, housing for working-class people often lacked running water or indoor plumbing.[16] Most mining operations had signed union contracts with the United Mine Workers of America by 1942, but sub-standard conditions persisted in mining communities, a fact that many critics termed 'un-American'. Exceptions were the 'model' towns and those built for new production during the Second World War, but they were far outnumbered by the grim, utilitarian homes for miners clustered around tipples and railroad tracks. Following the war, the whole country began to clamour for better housing, and the mining regions were no exception. By then, Lewis had begun to claim that most miners lived in sub-standard housing, and pointed to factors such as the water supply in company-owned mining communities as an example.[17]

The following images illustrate the way in which photographs were enlisted to compare the housing and water supply in coalfield communities to an ideal 'American' standard during the 1930s and 1940s. These images also highlight the manner in which the emotional content and the meaning of the image shifts with the publication and captioning of an image, so that, in conflicts over the nature of the life in the coalfield communities, ammunition provided by the cameras could help all sides claim to be winners.

Granger, Iowa, 1936

Mrs Sesto Fiori smiled shyly at a photographer from Wide World Photos as she demonstrated the running water in the kitchen sink in her new home in 1936 (Figure 4.1). The unnamed wire service photographer also made an image of the exterior of Mrs Fiori's home in the Granger Homestead Projects in Iowa. The new house, which was built next door to her old one, was part of a community of houses built by the federal government for miners and farmers on a 224-acre tract of land purchased by the Resettlement Administration, a government programme to aid rural residents.[18] The editors of the *UMWA Journal* published photographs of many new homes for miners in the late 1930s, including those built by the federal government as well as new or refurbished housing built by mining companies. Like the article on Granger, these articles seemed to be part of press kits, which the *Journal* adapted for its own needs.[19]

When UMWA editors published pictures of Mrs Fiori and her two houses, they accompanied an article by a local union leader who compared the new house to sub-standard company housing, even though the 'old' house in the photograph was not a company house.[20] The article celebrates progress in the coal industry, and the images, by implication, illustrate the benefits of anticipated stability in the industry which union president John L. Lewis sought by his support of the federal Bituminous Coal Conservation Act.[21]

The image of Mrs Fiori at her sink is not one of poverty or suffering. She is clean and neatly dressed; the sink is white, as are the enamelled pans below the running water. This is an image which complements the optimism of the accompanying article. Mrs Fiori is not represented as helpless, but as a woman at work in her home and who is grateful for the improvements in her life. As the caption notes, it is 'the first kitchen sink she has had with running water'.

Page Fourteen   UNITED MINE WORKERS JOURNAL

Government Builds Homes for Iowa Miners At the New Granger Community

Here is the modern new home of Sesto Fiori and family at New Granger Homestead and their old home which is being torn down.

By FRANK WILSON,

President of District 13, Albia, Iowa

Having some duties to perform in our northern field on New Year's Day, I found it convenient and very interesting to visit the new Granger Homestead Community. I visited with Mr. and Mrs. Sesto Fiori in their new six-room home, which is situated on a plot of more than four acres of ground.

New Year's Day being a holiday for the miners of District No. 13, I found Mr. Fiori and son, Geno, at home, and for the same holiday reason, all the children were also home. Mrs. Fiori was in the midst of preparing the mid-day meal, and demonstrated that she was thoroughly enjoying the use of this modern kitchen, with its sink, built-in cabinet and other conveniences. Brother Fiori took pleasure in showing me the home, going from room to room and pointing with pride to the cosy sitting room with its furniture very appropriately and effectively arranged, the spacious dining room, the modern kitchen, roomy bedrooms, each with a separate clothes closet, the spotless bathroom, with everything in order. We also visited the basement, examined the furnace and water system.

I thoroughly enjoyed my visit and the information which was so readily given. Brother Fiori and I exchange reminiscences concerning coal camp houses, in which we both have lived, recalling that at about this time of the year it was not unusual to awaken in the morning and find everything in the house, including food for breakfast, completely frozen—all by way of contrast to this comfortable modern home.

Both Mr. Fiori and his son, Geno, are members of Local Union No. 840, District No. 13, United Mine Workers of America, and work at the Dallas mine. Sesto Fiori, the father, came to this country from Italy in 1912, lived for some time in Detroit, Mich., later entered the Enterprise mine near Des Moines, and at the same time joined the United Mine Workers of America. About 1917 he moved to the Dallas mining camp, some three miles north of his present home. The Fioris have twelve children, nine of whom live at home, and each one expressed happiness in their new surroundings.

Brother Fiori and I soon found we had common interest in the New Deal under President Roosevelt, the halting of the industrial destruction, and the attempt at revival under the NRA, and now the anticipated stability of the coal industry through the Bituminous Coal Conservation Act, with increase in wages and reduction of hours. All of which we agreed was made possible by the untiring efforts of our great International President and his able associates.

The Fiori home is one of fifty erected by the Federal Resettlement Administration, near the town of Granger, Iowa, about twenty-five miles northwest of Des Moines. The Government purchased a tract of some 224 acres of land

Mrs. Sesto Fiori at the first kitchen sink she has had with running water.

Figure 4.1 Images from a wire service provided evidence of improved housing in 1936. Reproduced with permission of the UMWA *Journal*

Camera Wars

In 1945, water and sanitation in mining camps became a primary focus of postwar contract negotiations, and photographs became weapons in the battle between the companies and the union. The UMWA accused the companies of insanitary living conditions; the companies insisted that such matters were both untrue and outside the framework of collective bargaining agreements. In preparation for the 1946 national contract negotiations, the union sent its own photographers into the field to gather evidence on water supply and sanitation. Not to be outdone, more progressive coal operators and organizations mounted a defensive campaign, and hired public relations agencies which dispatched cameras to gather evidence to counter the union's claims.[22]

Ironically, photographs provided evidence which allowed each of the combatants to claim victory. The union's images showed poverty, the company's images chronicled progress. Yet, however simple these representations seem, they actually mask some complex facts, as the field notes of a subsequent investigation by the federal government reveal. Companies with superb housing facilities also had many sub-standard ones set aside for unskilled workers or for racial and ethnic minorities. Some companies which had built 'model' towns also owned other, sub-standard communities. A few of the communities cited as the most 'modern' were built by government programmes for wartime housing, not by the coal companies. And finally, some of the worst conditions were in deindustrialized mining communities, where the loss of the company meant the loss of all community services.

Shawmut, Pennsylvania, 1945

The focus of the Wage Conference in 1946 was to be the union's demand for a Health and Welfare Fund for miners, to be financed by a royalty on coal production. In preparation for the Joint Wage Conference, each district of the United Mine Workers was asked to supply photographs and supporting information on housing, wages and sanitation conditions in mining camps for the bargaining sessions. The photographs that were sent to the UMWA headquarters in Washington were taken by photographers engaged by the leadership of each district. Some were clearly influenced by the work of New Deal documentary photographers. Mack Hughes, a photographer from Lexington, Kentucky, took pictures for UMWA District 30 which included an image of a begrimed miner pouring his bathwater from a tea kettle into a zinc washtub on his kitchen floor.[23]

An unnamed photographer took pictures in Pennsylvania's District 2 following a series of wildcat strikes to protest against insanitary conditions in the camps owned by the Shawmut Mining Company in the summer of 1945. The strikes were also part of a protest over the loss of the doctor, who had resigned her position after the company refused to grant her demands for better sanitation in its camps. Members of Local Union 4036 refused to go to work without a doctor on duty. The company had allegedly neglected repairs to its houses for 40 years. When the *UMWA Journal* published accounts of this conflict from June to December of 1945, there were no photographs to accompany the articles about the 'sewage-ridden' towns of the Shawmut Mining Company.[24]

Perhaps one reason for the lack of published images of this situation was that the photographs made available to the union were unable to convey the desired information or elicit an emotional response from the viewer. Miner Lewis Cooper and his family, African-Americans and residents of Shawmut, posed for a photographer hired by District 2 (Figure 4.2). The images at Shawmut included a group portrait of Cooper, his wife and six of their children standing at the rear of their unpainted house in Shawmut. The union's notes for this image indicate that the house lacked indoor plumbing and running water, but the family members all appear to be clean and tidy in this image, which certainly undercut its value as a public relations piece.[25]

Kenvir, Kentucky, 1946

When the union did publish an image to represent its demands for better sanitary conditions, the editors enlisted an image of a small blond female child, sitting alone and barefoot on the ground, her face dirty, her hair dishevelled (Figure 4.3). The unidentified child appears on the cover of an issue of the *Journal* following a series of articles about insanitary conditions in mining camps, including a story about a miner's child who suffered third-degree burns when she fell into a burning slate dump near her house on company property in Kenvir, Kentucky. There are no photographs of the burning slag heaps, although one article was accompanied by a photograph of a wooden privy. But, to win support for the Health and Welfare Fund, the editors enlisted this image of a child to signify the purity of children and the danger to them posed by conditions in mining camps.[26] Who is the child on the cover? Who took the picture? We don't know, but her image, used here in 1946 as a generic example of a dirty coal-camp child, represents the first time the UMWA ever published an image of a miner's child as dirty or unkempt.

The union's decision to enlist images of its members and their families in the campaign for a health and welfare fund was a radical departure from the visual tactics of the past.[27] The union's editorial leadership decided that the emotional weight of images of poverty, dirt and disability would be a public relations asset during the Joint Wage Conference of 1946, and published a

Figure 4.2 This Pennsylvania miner's house had no running water in 1945. UMWA Collection, Special Collections Library, The Pennsylvania State University Libraries. Reproduced with permission

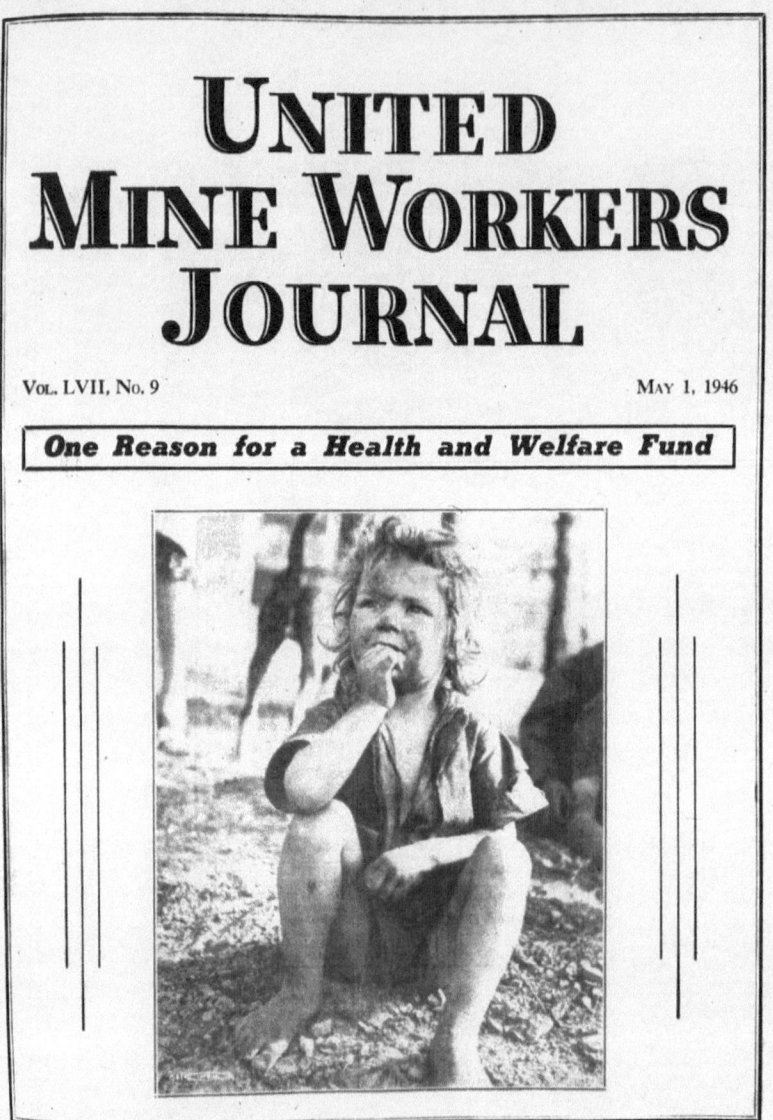

Figure 4.3 Not until 1946 did the union adopt the style of social documentary photographers. Image courtesy of the UMWA *Journal*

series of photographs of miners who were amputees in addition to the image of the dirty female child to enhance the union's cause during the public portion of the talks. Unsurprisingly, representatives of the coal companies argued that the union's claims were neither true nor appropriate subjects for collective bargaining.[28]

When a nationwide coal strike ensued, the federal government seized control of the mines in the spring of 1946 and ordered an investigation which included the gathering of photographic evidence. Consequently there were several groups of photographers in the coalfields in the summer and early fall of 1946. There were cameras employed by the union, the federal government and the coal operators, as well as cameras belonging to the press. All groups recorded images, some of which were later published to defend the political positions of their creators.

The government appointed a team of investigators for the Solid Fuels Administration for War (SFAW) of the US Coal Mines Administration-Navy (CMAN), a wartime branch of the US Department of the Interior. The investigators were naval officers specializing in civil engineering, public health, community recreation, sanitation and housing. They were charged with gathering facts and images for a medical survey of communities in the bituminous coal industry.[29]

The coal industry mounted a somewhat fragmented defence against the union's accusations and the government's investigation. There were more than 8000 independent coal producers in America, organized in regional associations. Some had sophisticated public relations departments, others had nothing. The Bituminous Coal Institute (BCI), the public relations arm of the National Coal Association, attempted to speak for the entire industry and to cover up the most egregious management practices in the industry while exhorting coal producers to modernize both their operations and their image. The BCI and the NCA saw the union's claims as very damaging to their efforts to promote coal as a modern fuel in the marketplace competition against oil. Photographers for these industry groups concentrated on the modern equipment and improved dwellings for miners which had been instituted during the war years.[30]

Press photographers followed investigators into the fields, but most of their attention was focused on the photo-opportunities created by the medical survey's director, Rear Admiral Joel T. Boone, and his entourage.[31] The mission of the investigating teams was to conduct an objective evaluation of a sample of 260 mining communities chosen at random from the 8000 bituminous mines in the USA (Figure 4.4). However, the travels of Rear Admiral Boone included tours of mining communities selected for his review as examples of the 'best' and the 'worst' in any given region. Russell Lee photographed Boone and his hosts on these tours; he also took photographs of the sites they visited.[32]

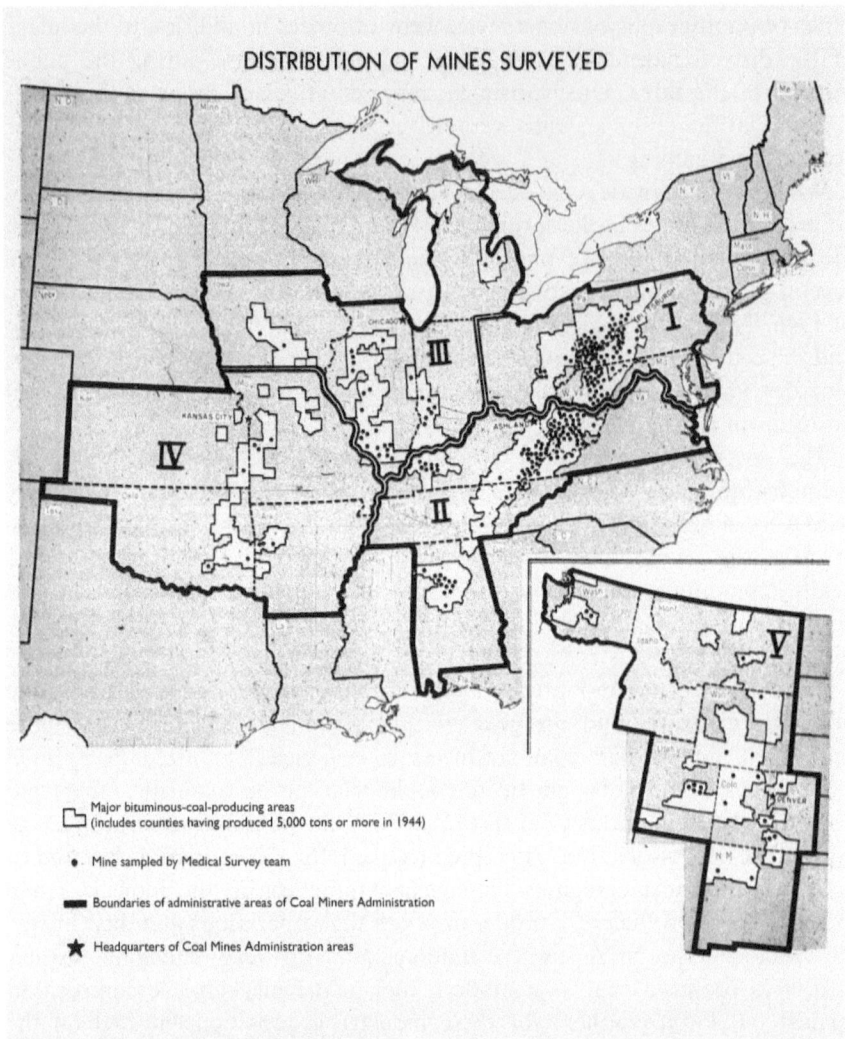

Figure 4.4 Bituminous coal regions of the United States in 1946. Map adapted from *A Medical Survey of the Bituminous Coal Industry* (U.S. Government Printing Office, 1947)

The government's investigation amassed nearly 4000 images; more than half of them were produced by Lee, who photographed many of the miners and their families who lived in the towns visited by Admiral Boone. He also took pictures of miners and their families in dwellings located adjacent to the areas visited by Boone, and travelled to southern West Virginia and eastern Kentucky on his own in late September, after his official work as Boone's public relations photographer had ended. Only a few of these locations were part of the sample of communities selected for the medical survey.[33]

As a result, there exists an unusually large body of documentary images by Lee and others for coalfield communities in the USA in 1946 which illustrate that dwellings for miners and their families, and the communities in which they lived, had many differences as well as commonalities. Their survey forms, narrative reports, photographs and captions in the archives of the medical survey document a wide variety of conditions in coalfield communities, including the ownership of mines, geographic and regional variations in the topography of towns, segregation of housing by race and occupation, and variations in management practices in relation to maintenance, public health and interaction with local officials in government and public health.[34]

Both union and company officials were concerned that federal investigators would not understand what they were seeing. Most locations had more than one type of housing. Even the most dismal communities had better homes for the mining officials such as the superintendent, company doctor, mine electrician and store manager. But only a caption could distinguish between them. Likewise, images alone could not convey the ownership of a mining community or the fact that the miners and their families in these images had brought the cameras into the coalfields by participating in a nationwide work stoppage in the coal industry.

E.C. Perkins, former coal operator and president of Appalachian Coals Incorporated, a coal sales agency, insisted that many of the dwellings photographed by the Medical Survey were not actually part of company property, but belonged to people who lived outside the camp. Perkins toured the southern Appalachian coalfields accompanied by a photographer from the trade journal, *Black Diamond*, and published an illustrated article which disputed the union's claims with the words, 'seeing is believing'. One of the illustrations for the article was a photograph from Red Jacket Coal Company.[35]

The Red Jacket Coal Company at Coal Mountain, West Virginia, had also been surveyed and photographed by a medical survey team headed by Commander E.F. Philpott, a medical doctor and an expert in public health. Their field notes and captions noted that there were two distinct levels of housing owned by the company. One group of houses, called the 'white camp', for mining officials, had indoor plumbing and fenced yards. The other group of houses, called the 'green camp', had no indoor plumbing and were less well maintained. The water supply was from a tap on the back porch (Figures 4.5 and 4.6).[36]

Another community which received attention from several groups of photographers was Fourmile, Kentucky, which had made national headlines when the Belva mine exploded in December of 1945, killing 24 miners. After the explosion, the entry to the mine was sealed, with 20 bodies still inside.[37] The area was known for its poverty. Russell Lee arrived there on 4 September 1946, and photographed dozens of families at work in the kitchens and yards of their ramshackle houses. One of the more striking images was his portrait of a barefoot woman getting water for washing from a creek near her home (Figure 4.7). Not only was this community without adequate water supply but, according to the field notes and captions written by Lee and other investigators, the homes were without electricity as well, since the mining company which had supplied it shut down following the explosion at the Belva mine in December of 1945. Commander J.R. Shronts of the Medical Survey also visited Fourmile, and photographs by both men included views of the Belva mine. The cause of the poverty represented in the images by Lee and Shronts was deindustrialization, not employment in the mines.[38]

The Medical Survey report was published in 1947, following a year of tense labour relations in the USA. The director of the Medical Survey chose not to publish the real locations or names of the places investigated, and the report's editors used journalistic rather than scientific language. It was illustrated with more than 200 of Lee's photographs, captioned as 'representative' or generic images, rather than as images of specific locations or named coal companies. The report was written and illustrated in a comparative mode, and some sections were illustrated with paired images of examples of the 'best' and the 'worst' types of conditions. For images of the 'worst', the report enlisted pictures of places which were actually not part of the survey, including some of the Lee images from Fourmile, Kentucky.

Lee's images were more interesting and emotionally compelling than the pictures taken by the Medical Survey teams,[39] yet both Lee and the teams produced captions and field notes which were rich in detail. Much of that material was not included in the report. For example, the images published from the deindustrialized communities in Fourmile, Kentucky, omitted from their captions the information that the mines were destroyed in an explosion more than a year earlier.

While the report and its images were journalistically successful, they were actually misleading in their representation of the relationship between mine owners and coalfield communities. The images published in the comparative framework of the *Medical Survey of the Bituminous Coal Industry* reinforced the one-dimensional stereotype that mining towns were either 'good' or 'bad', depending on their relation to the ideal of corporate paternalism. They also

Figure 4.5 'White Camp' houses for company officials had indoor plumbing. Photograph from the Medical Survey of the Bituminous Coal Industry. RG-245-MS. National Archives. Still Pictures Division, Washington, D.C.

Figure 4.6 'Green Camp' houses for miners had no indoor plumbing. RG-245-MS. National Archives. Still Pictures Division, Washington, D.C.

Figure 4.7 Water supply in an abandoned mining camp in 1946. RG 245-MS 2324-L. National Archives. Still Pictures Division, Washington, D.C.

allowed both the companies and the union to claim victory in the 'camera wars'. Historians have relied on the published report for information about coalfield communities, and Lee's images have become famous as illustrations of mining life. Yet the complex data and the images collected by the teams, which could support a more nuanced comparison of selected communities in relation to their geography, ownership, productive capacity, age of operations, workforce, race relations and relationship to local government and health officials in the bituminous coalfields, remain unpublished.[40]

Conclusion

Cultural historian Peter Burke has argued that in order to understand images one must 'study the purposes of their makers'. Researchers must be aware of the intentions of the photographers and their patrons or clients, question the purposes of both the image and its content, and, above all, be aware that images are 'historical agents which both record and influence'. Moreover, he notes, images which have been constructed to amuse, deceive or convince lose their context over time. Out of context, they can appear as unproblematic illustrations of 'facts' to the unwary researcher.[41] Burke's caveat is especially useful to the study of images of coalfield communities in which the emotional dimension of images makes the use of visual evidence for historical research particularly challenging for scholars. As Walter Benjamin has noted, photographs provide evidence that stirs the viewer and captions provide the direction for that emotion.[42] As the representation and publication of coalfield images demonstrate, the social meaning of images shifts with their publication, disguising the fact that the images are often artefacts of the political struggles which created them.

Notes

1   Eric Margolis, 'Mining photographs: unearthing the meanings of historical photos', *Radical History Review*, **40** (1988), 33–49.

2   Peter Roberts, *Anthracite Coal Communities: A Study of the Demography, the Social Educational and Moral Life in the Anthracite Regions* (New York, 1904); Nettie P. McGill, *The Welfare of Children in Bituminous Coal Mining Communities in West Virginia*, US Department of Labor, Children's Bureau, bulletin no. 117 (Washington, 1923); United States Coal Commission, *Report of the United States Coal Commission*, Senate Document 195, 68th Congress, 2nd session (Washington, 1925); United States Coal Mines Administration, *A Medical Survey of the Bituminous Coal Industry* (Washington, 1947); President's Commission on Coal, *The American Coal Miner: A Report on Community and Living Conditions in the Coalfields* (Washington, 1980).

3 Janet Wells Greene, 'Camera wars: images of [American] coal miners and the fragmentation of working-class identity, 1933–1947', unpublished PhD dissertation, New York University (2000); Lili Corbus Bezner, *Photography and Politics in America: From the New Deal to the Cold War* (Baltimore, 1999); Laura Wexler, *Tender Violence: Domestic Visions in an Age of US Imperialism* (London, 2000).

4 Peter Burke, *Eyewitnessing: The Uses of Images as Historical Evidence* (New York, 2000).

5 During the Second World War, the National Coal Association formed the Bituminous Coal Institute to serve as its public relations arm.

6 George Dimock, *Priceless Children: Ideal American Photographs 1890–1925* (Weatherspoon Art Museum, 2001).

7 Hugh D. Hindman, *Child Labor: An American History* (London, 2001), p.95.

8 'Here is how one anti-union coal company handles its non-union employees', *UMWA Journal* (15 September 1939), 16; 'We are 100% non-union', *UMWA Journal* (1 October 1939), 17; 'Girls roughly treated', *UMWA Journal* (1 March 1935), 18.

9 [Carbon copy of letter of introduction for Russell Lee] from Joel T. Boone, 3 August 1946. Papers of Joel T. Boone, Box 99 'Coal'. Library of Congress, Manuscript Collection; Press Release, Department of the Interior Information Service, Coal Mines Administration, 31 May 1946. The Papers of Julius Krug, Box 42, 'Medical Survey', Library of Congress, Manuscripts Division; Greene, 'A task full of dynamite: photographs and the Medical Survey of the bituminous coal industry, 1946', in 'Camera wars', 266–347.

10 Janet Greene, 'Fighting on two fronts: representations of accidents and injuries in the *UMWA Journal* during World War II', and 'Seeing is believing: postwar crisis management and public relations photographs in the American bituminous coal industry', in 'Camera wars', 117–96, 348–426.

11 Janet Greene, 'Images of poverty and progress in the *United Mine Workers' Journal*, 1933–1941', in 'Camera wars', 23–116.

12 For example, *History of the Consolidation Coal Company, 1864–1934* (New York, 1934).

13 Maren Stange, *Symbols of Ideal Life: Social Documentary Photography* (New York, 1989), pp.108–14; Alan Trachtenberg, 'From image to story: reading the file', in Carl Fleishauer and Beverley Brannon (eds), *Documenting America, 1935–1943* (Berkeley, 1988), pp.43–71; Lawrence Levine, 'The historian and the icon', in Fleishauer and Brannon (eds), *Documenting America*, pp.32–7. See the microfilmed clippings file for the use of these photographs, in Written Records of the Farm Security Administration, Historical Section – Office of War Information Overseas Picture Division, Washington Section Collection, Prints and Photographs Division, Library of Congress.

14 Greene, 'Camera wars', 23–117.

15 John L. Lewis, *The Miners' Fight for American Standards* (Indianapolis, 1925); 'Bradley's miners are isolated from the rest of the US', *UMWAJ* (1 November 1939), 10–12; Melvyn Dubofsky and Warren Van Tine, *John L. Lewis: A Biography* (New York, 1977).

16 Greene, 'Camera wars', 208–65.

17 Greene, 'Front porch politics: miners and photographers in the UMWA campaign for a health and welfare fund', 'Camera wars', 197–265; Gwendolyn

Wright, *Building the American Dream: A Social History of Housing in America* (New York, 1981), pp.220–39.

18   Frank Wilson, 'Government builds homes for Iowa Miners at the New Granger Community', *UMWA Journal* (1 March 1936), 14; Dorothy Schwieder, *Black Diamonds: Life and Work in Iowa's Coal Mining Communities, 1895–1925* (Ames, IA, 1983), pp.73, 77, 105–6.

19   'This [Bellingham Coal Company, Washington State] doesn't look like a coal mine, but it is', *UMWAJ* (1 October 1939), 14; 'Kopperston [West Virginia] is a modern town', *UMWAJ* (1 July 1940), 14.

20   Wilson, 'Government builds homes', p.14.

21   James P. Johnson, *The Politics of Soft Coal: The Bituminous Industry through the New Deal* (London: 1979), pp.141, 219–23; John G. Clark, *Energy and the Federal Government: Fossil Fuel Policies, 1900–1946* (Chicago, 1987), pp.171, 270–71.

22   Actually, the corporate response was a bit slower, and only got under way after the spring of 1946, when the federal government seized the mines and negotiated a contract on behalf of the operators. See Greene, 'Camera wars', 348–426.

23   UMWA Archives, Penn State University; Greene, 'Camera wars', 197–265.

24   'L.U. 4036 urges Congressional inquiry into housing conditions in coal camps', *UMWAJ* (15 June 1945, 5); 'Court removal of Shawmut receiver asked; miners awarded unemployment compensation', *UMWAJ* (15 July 1945), 12; 'Wacky or not, Wick has the answer,' *UMWAJ* (1 September 1945), 12; 'Miners ask us to probe 40-year Shawmut receivership, smash "reign of oppression"', *UMWAJ* (15 September 1945), 15; 'F.B.I. launches investigation of 40-year Shawmut receivership at miners' request', *UMWAJ* (1 October 1945), 14; 'Shawmut miners put "reform" petition before court after company evicts dr.', *UMWAJ* (1 December 1945), 5.

25   Lewis Cooper and Family, Shawmut, Pennsylvania, District 2 Surveys; Miner Family Photographs, UMWA Archives, Penn State University.

26   'One Reason for a Health and Welfare Fund', *UMWAJ* (1 May 1946), cover; 'Miners reveal intolerable insanitary conditions in Kentucky Coal Camp of 3000', *UMWAJ* (15 April 1946), 12. See also Frank Hodges, President, Local Union 4493, Kenvir, Kentucky, to John L. Lewis (3 April 1946); E.L. Reed to John L. Lewis (5 April 1946); C.M. Enochs, Powers Dixon and Dennis Burns to John L. Lewis (5 April 1946); telegram from Abe Vales, President, District 19, UMWA to John L. Lewis (6 April 1946); John L. Lewis to Frank Hodges, 13 April 1946); John L. Lewis to C.M. Enochs, Powers Dixon and Dennis Burns (13 April 1946), President's Office, Correspondence, District 19, 1944–1948, UMWA Archives, Pennsylvania State University.

27   'Tory Congressmen would cut their relief!', *UMWAJ* (1 February 1939), cover; 'Magic carpet', *UMWAJ* (15 August 1941), cover; 'A poor stick in bad weather', *UMWAJ* (1 February 1933), cover; Greene, 'Camera wars,' 88–94.

28   'Affidavits reveal the tragic story of inadequate compensation for miners', *UMWAJ* (15 April 1946), 7; 'Sam Harris, whose picture appears on the front cover, is a living example of inadequate compensation for miners', *UMWAJ* (15 May 1946), 14.

29   *Medical Survey of the Bituminous Coal Industry* (Washington, 1947), p.244.

30   'Memorandum from Hamilton Wright Organization, Inc., 'Bituminous coal industry goes to work for Uncle Sam' (23 May 1946), Universal Newsreel

Library, National Archives, Motion Picture, sound and Video Branch; 'Modern District 8 mining communities', *Black Diamond* (7 December 1946), 28; [Bituminous Coal Institute], 'Bituminous coal mining towns: pictorial story of their progress' (Washington, 1946); Greene, 'Camera wars', 328–436.

31   Jacob Scher, 'Times visits a coal town', *Chicago Times* (8 June 1946), 19; 'Boone survey bares life of coal miners', *Chicago Times* (8 June 1946); 'An American standard mine!', *Chicago Times* (9 June 1946), 9; papers of Joel T. Boone, 'Scrapbook', Box 100, Library of Congress, Manuscripts Division; 'Navy begins check of mine conditions: first investigations made in Muse, Near Canonsburg', *Pittsburgh Post-Gazette* (13 June 1946).

32   Greene, 'Camera wars', 269–80.

33   Greene, 'Camera wars', 460–580.

34   RG 245, Solid Fuels Administration for War, Medical Survey Group, National Archives.

35   E.C. Perkins, 'District 8 producers accept modern competition's challenge', *Black Diamond* (24 October 1946), 48.

36   RG245-MS 3083-90-P, Coal Mountain, Wyoming County, West Virginia, September 6, 1946; [US] National Archives, Still Pictures Division, Washington, DC.

37   RG 245-MS 1330-S, 1333-S, 1335-S, Still Pictures Division, [US] National Archives, Washington, DC.

38   See RG245-MS 2250L, 2252-L, 2259-L, 2249-L, 2254-L, 2324-L, 2254-L, 2219-L, 2230-L, 2295-L, Still Pictures Division, National Archives; 'Mine explosion traps 30 to 50 in Kentucky; Fire Hampers Rescue', *New York Times* (26 December 1945), 1; 'Rescue crews nearing miners trapped in blast', *New York Herald Tribune* (26 December 1945); 'Brought out alive', *Baltimore Sun* (29 December 1945); 'Two chairs empty in Branstutter home', *UMWAJ* (15 January 1945); Eddie Doughterty, 'The saga of Fourmile, Kentucky', *Chicago Sun* (29 January 1946); 'History written in blood of Fourmile mining area', *Chicago Sun* (30 January 1946); 'Mine rescue ordeal bares true heroism of Fourmile', *Chicago Sun* (31 January 1946); 'Fourmile trudges ahead, part of another world', *Chicago Sun* (2 February 1946).

39   Janet Greene, telephone interview with Jean Lee, 12 May 1994 (author's possession); 'Suggestions for taking official photographs', typescript, July 1946, National Archives RG 245, Box 2675, Folder 'Medical Data.'

40   Greene, 'Camera wars'.

41   Peter Burke, *Eyewitnessing: The Uses of Images as Historical Evidence* (New York, 2000), pp.30, 103, 145.

42   Walter Benjamin (1979), 'The work of art in the age of mechanical reproduction', in Walter Benjamin, *Illuminations: Essays and Reflections*, trans. H. Zohn (New York, 1969), p.393.

CHAPTER FIVE

A Mining Film without a Disaster is like a Western without a Shoot-out: Representations of Coal Mining Communities in Feature Films

Bert Hogenkamp

Among the working-class occupations shown in feature films, mining has stood out in popularity. Of course, films have been made that feature workers in other trades. Railway workers, for example, have been represented in films ranging from *La Roue* (Abel Gance, France, 1921) to *The Navigators* (Ken Loach, UK, 2001), while even typesetters have had their moment on the silver screen courtesy of Jean Renoir's *Le crime de Monsieur Lange* (France, 1936). But only mining has provided a setting that has appeared in a sufficient number of feature films and television drama series to generate a distinctive film genre (some would call it a sub-genre) of its own.[1] It will be suggested here that film makers have drawn from a relatively limited array of images and storylines when attempting to represent the experiences of mining communities. This consistency is all the more striking given both the temporal and the spatial diversity of those experiences. With the oldest productions dating from the beginnings of the cinema, mining films cover every phase of the medium's development right up to the present day.[2] They also encompass the rise and fall of the coal industry itself. Indeed, notwithstanding the decline of the coal trade, the mining film is as popular as ever with producers, film makers, scriptwriters and audiences, as evidenced by the success of productions such as *Germinal* (Claude Berri, France, 1993), *Brassed Off* (Mark Herman, UK, 1996) and *Billy Elliot* (Stephen Daldry, UK, 2000).

Many of these mining films have been set in the 'classic' coalfields of Europe and North America. However, others have been located in mining areas of Bolivia, Chile, China, India, Japan and Australia. Yet, despite the distinctiveness of such coalfields, when represented on cell0luloid, they can appear remarkably similar. Whether one looks at productions emanating from capitalist studios in Hollywood or nationalized studios in the Soviet Union, or at films made by small independent companies operating on shoestring budgets, the story is (often literally) the same.

Three Sources of Inspiration

In seeking to explain the similarity of mining films, it is useful to note that there were three main sources from which film makers drew their inspiration. Firstly, there was a literary tradition that started with Emile Zola. In his novel *Germinal* (1885), the French writer managed to concoct a gripping mix out of the four classic elements of earth, water, fire and air. Zola described the mining community in the north of France as if it was a village in the interior of Africa or New Guinea. He was not alone in looking upon miners as 'Others': a threatening mob in times of strike, or a band of superheroes who risked their lives when rescuing their comrades after a pit disaster. A substantial number of coal mining films have been based on successful novels, such as A.J. Cronin's *The Citadel*, Richard Llewellyn's *How Green was My Valley* and, indeed, *Germinal* itself.

Secondly, there is a long visual tradition going as far back as the sixteenth century. Georg Agricola's *De re metallica* (1566) is undoubtedly the most famous illustrated study on early mining techniques. However, his was just one book amongst many dealing with this subject at this time. Most used illustrations that appealed to the imagination of readers who were not directly familiar with mining.[3] In the nineteenth century, miners inspired artists with a social conscience. Vincent van Gogh was one such. He encountered miners during his time spent as a lay preacher in the Belgian region of Borinage. In fact, from the 1870s to the 1960s, the Borinage attracted a steady stream of Dutch, Belgian and French painters (for example, Jan and Charley Toorop, Cécile Douard and Harmen Meurs), sculptors (including Constantin Meunier), writers (Egon Erwin Kisch, André Gide), photographers (Dolf Kruger, Jean Loup Sieff) and film makers (Joris Ivens, Henri Storck, Yves Allégret and Paul Meyer). Their style and approach may have been quite different, but there is a remarkable consistency in the subjects that they depicted: miners on their way to and from work; men and women looking for coal on the slagheaps and carrying huge sacks of coal on their back; and the misery caused by strikes and, on occasion, the workers' militancy. Other coalfields have similar 'visual' histories.

Finally, the illustrated press played an important role in disseminating these images, through their use, first, of lithographs and, later, of photographs.[4] They provided their readers with a plethora of memorable images that stood as a shorthand for the mining experience. Certain pictures became especially familiar: the miner on his way to work; miners' wives gathered anxiously at the pithead awaiting news in the wake of a disaster; the mass demonstration of striking miners. These images were supplemented by countless articles written by journalists and 'social explorers' who had visited coalfield areas, either to report upon newsworthy events such as a strike or pit disaster or to pen more general investigations. A number of the authors, including names such as Pierre Hamp, Egon Erwin Kisch, J.B. Priestley and George Orwell, had their reports published in book form.[5]

All three sources played their part in creating a means of 'seeing' mining communities and their influence can be discerned in the types of images produced by film makers. While historians have been remarkably slow to recognize the importance of feature films as primary sources, over the last two decades a number have begun to appreciate their significance as cultural artefacts. Certainly they must be approached in a sophisticated manner. As the German historian Anton Kaes has put it, films are 'complex fictional constructs [that] offer ambivalent perspectives and contradictory attitudes that resist simple explanations and call for multiple readings'. Nevertheless he is convinced that they are valuable pieces of historical evidence because of their ability to 'unlock the viewers' hidden wishes and fears, liberate fantasies, and give material shape to shared moods and dispositions. Films can thus be seen as interventions in cultural and political life'.[6] Kaes's book, *From Hitler to Heimat*, can be seen as a direct sequel to a path-breaking study by his compatriot, Siegfried Kracauer, *From Caligari to Hitler*, published in 1947. In this work, Kracauer set out to prove that 'through an analysis of the German films deep psychological dispositions predominant in Germany from 1918 to 1933 can be exposed'. He gave two main reasons for this contention: firstly, that films are never the product of an individual, but of a group of people, 'suppressing individual peculiarities in favor of traits common to many people'; and secondly, that 'films address themselves, and appeal, to the anonymous multitude'.[7] Kracauer's study has been a source of inspiration to many, but others have dismissed his readings of feature films produced in the Weimar era as too mechanical. As historians Jeffrey Richards and Anthony Aldgate have pointed out, 'the relationship between film-maker, film and audience is now seen to be more sophisticated, a two-way process operating in areas of shared experience and shared perception'.[8]

Bearing such insights and reservations in mind, this chapter will explore the ways in which mining communities have been represented in feature films from a number of different countries. Unlike Hermann Barth's study of the German film, *Kameradschaft* (Georg Wilhelm Pabst, Germany, 1931), no attempt will be made to examine the 'psychological strategies' of the films.[9] Neither will I be looking into the history of their practical realization, as, for example, John Sayles and Pierre Assouline have done for *Matewan* (USA, 1987) and *Germinal* (Claude Berri, France, 1993), respectively.[10] Instead I want to concentrate on the nature of the cinematographic representations of mining that have been created by the film makers.

The Quest for Realism

What most mining films have in common is that, even if they have (largely) been filmed in a studio, they consistently aim at 'realism'. Film makers try to convince their audiences that they are looking at, for example, the 'real' valleys of South Wales, or the 'real' Ruhr area. This has often been one of the yardsticks which film critics have used when judging a mining film. Graham Greene, for example, as film critic for *The Spectator*, criticized *The Stars Look Down* (Carol Reed, UK, 1939) for its narrative structure ('too much story drowning the theme'), but praised its realist setting ('constructed authentically though it is of grit and slagheap, back-to-back cottages, and little scrubby railway stations').[11]

Nevertheless, whereas film makers have invariably striven towards 'realism' with regard to their films' geographical settings, they have often been less interested in giving an accurate representation of the work carried on underground. It is perhaps understandable that, in the slapstick comedy, *Pick and Shovel* (also known as *The Miner*, George Jeske, USA, 1923), Stan Laurel's idea of working underground amounted to little more than swinging his pick axe into the floor – or more usually the backside of his fellow workers – rather than into the coal seam (a seam so thick that it would have been the envy of any miner who saw the film). His 'unproductive' behaviour was part of the humorous slapstick routines. But there was no reason for German actor, Emil Jannings, to do exactly the same in *Algol* (Hans Werckmeister, Germany, 1920), as this film purported to offer a realist representation of the work underground. This is an extreme example, but it must be said that rare indeed are the mining films or television dramas in which the representations of underground work correspond to the reality. *Grube Morgenrot* (Wolfgang Schleif and Erich Freund, East Germany, 1948) was one such exception. Filmed in the Soviet Zone of Germany, this production set out to show how, in the days of the Weimar Republic, the management of the 'Aurora' pit gambled on increases in productivity at the expense of safety in order to keep the company solvent. This led to the dreaded pit disaster, which caused many deaths and subsequent unemployment for the surviving miners. The film's message was that only under socialism could the priorities be put right.

While *Grube Morgenrot* depicted the techniques of production in detail and made it clear to audiences how these affected safety underground, one can think of various reasons why this was not done more often. Most film makers lacked the knowledge of what coal miners did at the coal face. Moreover safety precautions made it hard for them to film underground. And to reconstruct a mineshaft in the studio (as Ernö Metzer did for *Kameradschaft* so successfully, but at great expense) was not always financially feasible. There is a parallel here with the visual arts: nineteenth-century 'realist' artists such as Constantin Meunier and Cécile Douard painted miners above ground on their way to or from work, because they were not allowed underground (indeed, even if they had been, there would have been no light for them to do their work).[12] However, significantly, audiences have seemed unconcerned by the unrealistic representations of the work process. Shots of the ascending or descending cage, seen in virtually every miners' film, are apparently all that is required to prove the miners' productivity. Indeed, there is no stereotype of the 'lazy miner' (Stan Laurel excepted, of course), a cinematographic stereotype that exists with regard to other working-class occupations. Any miner who enters the cage is, by definition, a productive miner.

Disaster

While generally unconcerned with the details of miners' working practices, the location of that work – the coalface – has invariably been used by the film maker for a different purpose: that of presenting danger, disaster and escape. A mining film without a disaster is like a Western without a shoot-out. A typical dramatic scheme contains a number of elements. In the first place, the various characters are introduced both in the pit and at home, so that the spectators can empathize with them. Then there is the accident and the havoc that it causes both above and below ground. As a result, everyone's attention is drawn to the mine. After various failed attempts, some form of contact (often by telephone) is established with a group of trapped miners, whose spirits are raised accordingly. The rescue attempt is hampered and spirits drop. Just when rescue seems imminent, an unexpected setback makes the rescue uncertain, once again depressing the morale of the men. At this stage, the natural leaders of both the rescue party and the trapped miners show their mental and moral strength. This brings about the successful recovery of the miners who are united with their families. Most of these elements can be found in what is possibly the best known disaster and rescue film, *Kameradschaft* (Georg Wilhelm Pabst, Germany, 1931). However, this film also contains another theme: the solidarity between miners from two countries, Germany and France, that had been each other's enemies during the First World War. The message that class solidarity is stronger than nationalism is illustrated when a German rescue team enters a French pit that has been rocked by an explosion. When a French survivor sees the Germans coming through the galleries with their gas masks on, he imagines himself to be back in the trenches and panics. The Germans rescue the French survivors, and the image of war is replaced by one of peace, as a French hand is seen shaking a German one.[13]

   Kameradschaft is a good example of how a disaster can bring about a reshuffling of the dramatic elements. Such a reshuffling can be quite simple, as, for example, in *The Brave Don't Cry* (Phil Leacock, UK, 1952) and *Életjel*

(Fourteen Lives, Zoltán Fábri, Hungary, 1954). In both these films, a disaster leads to the repentance of prodigal sons and their return to their home communities. In other films, the dramatic changes are more complex. For instance, in the French film, *Grisou* (Maurice de Canonge, France, 1938), the coal miner, Demuysère, has an adulterous wife, 'la Loute'. Both her, and her secret lover, the overseer Tony, constitute a disruptive presence in the heart of the northern French mining community. When Demuysère is injured after an outbreak of firedamp, which Tony had failed to report, the two have to leave the village. In *The Toilers* (Reginald Barker, USA, 1928), the opposite happens. After a pit disaster two estranged workers, Steve and Joe, are reconciled. But in this film the woman who caused the original rift is also responsible for bringing them together again. The orphan, Mary, found in the snow on Christmas Eve, seems to possess divine qualities, for during the rescue action she dreams that Steve is alive.

The Family

While 'la Loute' was not fit for the role of miner's wife, Mary seems predestined for it; how could it be otherwise, given her name? Only a handful of miners' films do not start from the presumption that the family is the pre-eminent source of the miner's well being. A crucial part is played by the mother figure. It is her job to ensure that the mining community reproduces itself physically and ideologically. Thus not only is she responsible for giving birth to children and nuturing them, she also has to raise them to become good miners or good miners' wives. (That said, it should be noted that the converse also applies sometimes. Hence, in certain films, the mother ensures that, through her sacrifice, her offspring will escape the pit, get a 'decent' job or marry above their station, outcomes that potentially undermine the community.)

The classic miners' film about the family is of course *How Green Was My Valley* (John Ford, USA, 1941), an adaptation of Richard Llewellyn's best-selling novel. The film's theme is the destruction of the original, preindustrial family ties. It is narrated by the youngest son, Huw, who, despite the whole family's hopes and expectations for him to escape the pit, becomes a miner too. Gwilym Morgan, the father, is a righteous, stern man who sticks to the principles taught to him in a preindustrial society. But it is his wife, Bethan, who is the real source of authority in the house. In a typical scene she waits outside the front door for her husband and sons to drop every penny of their wages into her apron. This symbolic act is followed by another: the men undress and wash away the coal dust from their bodies in a tub. A number of films contain such bathing scenes. For example, in *Blue Scar* (Jill Craigie, UK, 1948), actress Rachael Thomas, who played the archtypical Welsh

'Mam' in so many stage productions and films, washes the back of her husband after his return from the pit.

From the family to other areas of social and cultural reproduction is but a simple step. Miners' pastimes, whether in the form of leek growing, the keeping of homing pigeons or the playing of unusual games such as lacrosse, have always made a great impression on the outside world. A good example is *Brassed Off*, which is based on the very notion that a typical mining community in the north of England will have its own brass band. In *Grisou*, too, a brass band was shown, with Demuysère playing the trumpet. Generally speaking, film makers had great difficulties in accommodating miners' leisure in the dramatic structure of their films. Mostly it was used to add local colour and as a means of enhancing the authenticity of the film. There are a few examples, though, where leisure itself provided the drama. In *Proud Valley* (Pen Tennyson, UK, 1939), a film set in the Welsh valleys, singing is made to stand for mining itself. It is because the wandering tramp, David Goliath (played by Paul Robeson), has such an impressive voice that he gets a job as a coal miner. The sequence where he adds his voice through the open window of the hall where the local choir is rehearsing refers to another recurring dramaturgic theme in miners' films, that of getting into coal mining. This has been a classic theme ever since Emile Zola described the arrival of Etienne Lantier at the Voreux pit in his *Germinal*. If one is not born a miner, one does not simply take the job, as would be the case with any other occupation; instead, one has to undergo a ritual of integration into the mining community. In David Goliath's case, this means becoming a member of the choir. On the day the choir is going to visit the Eisteddfod, its leader gets killed in an accident that results in the closure of the pit. Ultimately, David Goliath sacrifices himself while carrying out operations to reopen the pit. The film ends with the crowd singing a hymn to commemorate him.

Strikes

Strikes form another important dramaturgic element of the miners' films. Being no longer underground as a result of a strike means that the miners become visible to the outside world. At the same time they lose their identity as productive workers (the cage is no longer going up and down). And in films that are hostile to strikes, stoppages mean that the miners lose their personality and become part of the dreaded mob. A good example is the Warner Brothers production, *Black Fury* (Michael Curtiz, USA, 1935). Joe Radek, a carefree miner of Polish origin (played by Paul Muni), falls prey to a Pinkerton agent, after his girlfriend, Anna, has left him. The agent has been sent to stir up trouble in the coalfield and Joe is his ideal tool. One night, a drunken Joe stumbles into a trade union meeting and persuades the men to strike. Not fully aware of what he is doing, Joe thinks he is somehow fighting to get Anna back. The film shows a complete lack of understanding about how strikes were run; there are no pickets and no organization for self-help. Instead, the strikers become a mob that attacks scab labour. It is only by redeeming himself and becoming an individual that Joe can personally bring the strike to an end, enabling his fellow-workers to do likewise and resume their job. On the other hand, films that are sympathetic to strikes often show how the conflict leads to the miners becoming more class-conscious. Take, for example, the Polish film, *Perla w Koronie* (Pearl in the Crown, Kazimierz Kutz, Poland, 1971), in which the leading character, Jan, and his colleagues stage a stay-down strike with great dignity in order to prevent the closure of their pit. Jan, like Joe, his American–Polish counterpart, is young, impetuous and apolitical, but his actions are guided by the notion of solidarity with his colleagues. Unsurprisingly the notion of the mob is completely absent in this film.

But, in a few films, strikes offer an opportunity to highlight the role of women in the mining community. The best example is undoubtedly *Salt of the Earth*, made in 1953 by the blacklisted Hollywood director, Herbert Biberman.[14] Although the film is set in a zinc-mining rather than coal-mining town in the American state of New Mexico, there is no reason why it should not be included in the body of films discussed here. In a series of stark images, the film shows how the miners outwit the officials and representatives of the mining company. A leading role is taken by the women, who form a picket line when an injunction bans their husbands from doing so. When arrested and put behind bars in the local gaol, they drive the sheriff and his men to distraction. Finally, they prevent the eviction of the leading couple of Esperanza and Ramon Quintero from their company-owned home after they had fallen into arrears with the rent. In the end, the company is forced to settle with the men. In a memorable role, Mexican actress Rosaura Revueltas convincingly plays the vulnerable (because she is pregnant) yet determined Esperanza. She is the opposite of the two mother figures in the strike sequence of *The Stars Look Down*, a film based on a novel by A.J. Cronin. In that, one openly criticizes the roles played by her husband and son in a desperate strike to prevent the opening of an area of the pit where there is danger of flooding. Another wife inadvertently sparks off a food riot, after being refused a cut of meat by the local butcher. The riot leads to the imprisonment of the leader of the strike.

Unions

The strike in *The Stars Look Down* is organized by the men without the consent of their union. In a few telling shots the union official is shown siding with the boss and his son. In contrast, in *Salt of the Earth*, the strike is shown as a successful joint operation by the workers and their union. The union even encourages the women (who are of course non-members) to get actively involved in the strike. Because union activities are always connected with dramatic events such as mine disasters, strikes and pit closures, the miners' films rarely show the day-to-day work that the unions perform. This is also the case in *The Molly Maguires* (Martin Ritt, USA, 1969), a Hollywood production that was completely devoted to a forerunner of the unions. The Molly Maguires, a secret organization formed by Irish-born anthracite miners, was active in the Pennsylvanian coalfield in the 1860s and 1870s. Rather than concentrate on the function of the society, the film tells the story of trust and betrayal between two men, the leader of the Molly Maguires and a company detective who has been sent to uncover the identities of the society's leadership. Along the way, themes such as child labour and the ubiquitous pit disaster are introduced. However, the main purpose of the film is to present the workers as being incapable of discovering the true nature of their employment. As a consequence of their ignorance, they are reduced to executing only childish acts of revenge, obstructing production and targeting management and members of the police.

Although unlikely at first sight, *The Molly Maguires* bears some resemblance to the miners' films made by Leonid Lukov in the Stalin era. In this incarnation, the miners are now members of the Party or the Komsomol (the union being subordinate to these organizations) and call themselves Stachanov workers rather than the Molly Maguires. Their opponents, the Trotskyists, are the saboteurs and wreckers. In *Ital'janka* (The Italian Woman, Leonid Lukov, Soviet Union, 1931), for example, the Komsomol secretary, Alëša, wants to disprove the persistent rumour that there is firedamp in the local pit. To do so, he decides to set an example by descending first. He is killed by a kulak before he gets the chance. The kulak tells everyone that Alëša is a dodger and has gone into hiding, but in the end he is exposed and production at the pit is resumed. Like *The Molly Maguires*, the central theme of this and other Lukov films is still trust (this time in Stalin and in the Party) and betrayal.

The modern trade union official became a figure to be pitied in a number of British mining films. In *Above Us the Earth* (Karl Francis, UK, 1977) he is shown as a victim of the relentless mechanism of bureaucracy, whereas he should really have been defending the interests of the rank-and-file membership when a pit is closed in South Wales. That the trade union official in *The Price of Coal* (Ken Loach, UK, 1976) fares little better is no surprise, given Loach's conviction (demonstrated in so many of his films) that, when it really matters, ordinary workers are always betrayed by their union representatives.

Getting Out

For many miners, their escape route from a life of toil underground was provided by debilitating diseases such as silicosis. Film makers have used such cruel, work-related illnesses as a means of conveying the exploitative, and all-consuming, nature of mining. In *Above Us the Earth*, Karl Francis powerfully shows how silicosis ruins the life of the main protagonist, Windsor Rees. And the Dutch film, *De Anna* (Erik van Zuylen, 1983), sees the playing out of a struggle between a retired face worker who is suffering from silicosis and the former owner of a closed mine who has plans to reopen. Against such a context, it is little wonder that many have seen machines as a potentially liberating force. In *Der Herr der Welt* (The Master of the World, Harry Piel, Germany, 1934), for instance, a group of miners are allowed to retire to the countryside and live as smallholders courtesy of a benevolent engineer who invents robots to do the dangerous work underground. Few, in reality, could hope to be so fortunate.

Perhaps a more realistic ambition, and one that recurs time and again in mining films, centred on the possibility of climbing out of one's class. In *The Stars Look Down*, young David Fenwick is destined to become a miners' MP (in A.J. Cronin's novel he is actually elected to Parliament). He wins a scholarship and goes to Tynecastle (a thinly disguised Newcastle) to study. When his father dies, he becomes a miners' leader and elaborates his belief that nationalization of the mines is the only way to improve the lot of the rank and file. In the French film, *Le Point du Jour* (Louis Daquin, France, 1949), Marie argues violently against her fiancé who wants her to give up her job on the surface of the local pit after their marriage. At the same time, she tries in vain to convince her mother that it is in the best interest of her younger brother to learn 'a trade', instead of following his father's footsteps down the mine. In *Blue Scar*, the miner's daughter, Olwen, moves to London to pursue a career as a singer. When her sweetheart, Tom, comes to visit her on the occasion of her first appearance on radio, he discovers that, though it may suit Olwen, big city life has little to offer to him. His future lies with the (recently nationalized) mining industry. Therefore Tom returns to his South Wales mining village, where he eventually settles down and marries.

Not until the years after the Second World War were films produced in which the closure of pits is offered as a means of getting out of mining. *Déjà s'envole la fleur maigre* (Paul Meyer, Belgium, 1960) was one of the first to envisage a future without mining. The film starts with the arrival of an Italian family at a local railway station in the Borinage. Mother and children have come from the home country to join the breadwinner. But the economic prospects are gloomy, and a growing number of pits in the region are being closed. Interestingly, director Paul Meyer uses the film to reverse many of the familiar images. When the miners leave the cage, their talk is so dominated by the current pit closures that it is impossible for the viewer to see them as 'productive' workers. Futhermore, the slag heaps are no longer the metaphorical fields in which the gleaners of the industrial age gathering tiny pieces of coal (as does Paul Robeson in *Proud Valley*, for example). Instead they are the playgrounds for children, who use baking tins for daring races downhill. Most significantly, there is no pit disaster in the film. In a kind of interior monologue Domenico, an elderly Italian worker who is about to leave the region, gives meaning to the landscape: 'Borinage, paysage, charbonnage, chômage' ('Borinage, landscape, coal mine, unemployment').[15]

Conclusion

The extent to which so many of the same dramaturgic elements can be found in films made under such different conditions and in so many different coalfields is striking indeed. They offer variations on the same themes, but in this case it is the coalfields that form the variations, while the themes show remarkable consistency. Still the listing of dramaturgic elements leaves us with a number of questions. First of all, would these representations have been different if the miners themselves had helped to shape them, instead of the outsiders that the film makers inevitably were? Active involvement of miners in the film-making process was a rare occurrence. Apart from acting as consultants (for example, former Welsh miner, Jack Jones, helped out on the script of *Proud Valley*), I have found only one example of direct involvement by miners themselves. The 50-minute-long silent film, *Black Diamonds* (Charles Hanmer, UK, 1932), was produced by a group of Yorkshire miners. It tells the story of an old miner who invites a film producer to visit a pit to experience for himself the realities of life as a coal miner. While it does not touch on some of the dramaturgic themes discussed above (there is, for example, no mention of strikes, of unions or of getting into and out of the industry), *Black Diamonds* nevertheless features a pit disaster. And particular attention is paid to the leisure activities of the miners. Of course, it is difficult to draw any firm conclusions on the basis of a single film.

A further question concerns the changes in the representations in the post-industrial era compared to those in the heydays of the mining industry. Looking at films like *Déjà s'envole la fleur maigre* and *De Anna*, one would be tempted to agree that there have been fundamental changes. On the other hand, films like *Germinal* and *Brassed Off* show how strong the old images are and how successfully they can be recycled. Then there is the question of difference and similarity in the way the various coalfields have been represented. As I have pointed out, film makers are always trying to convince their audiences that they are looking at a 'real' coalfield, be it in South Wales, the north of France or in Silesia. But most of the dramaturgic elements that were used are remarkably similar, with the pit disaster as the most likely common feature. Still exception must be made for a small group of German mining films. In the so-called *Kumpel* series, a group of soft-porn films from the 1970s, miners from the Ruhr area were presented as sex maniacs that were as good (indeed, even better) in bed as they were at the coal face.[16] I do not know of any equivalent that celebrates the masculinity of the miner in such a blatant way by equating his capacity to produce coal with his sexual prowess. Could it be that these films offer an open invitation for a Freudian rereading of the complete body of mining films?

Notes

1 In my book, *Bergarbeiter im Spielfilm* (Oberhausen: Karl Maria Laufen Verlag, 1982), I have listed 144 titles of feature films (television drama series not included) that are set in a mining community. They were produced in 24 different countries. Included are films dealing with the industrial extraction of not only coal but other minerals too, as long as this is being done underground. Therefore a substantial number of mainly American films – usually Westerns – dealing with gold mining as an individual pursuit have not been included. Since 1982, a fair number of films on the subject have been made. At the same time, research on the early history of the cinema has unearthed titles that I was not aware of when I compiled the filmography. Some television drama series (including *How Green Was My Valley*, featuring Stanley Baker, and the German series *Rote Erde*) were seen by large audiences and much loved and discussed. If, therefore, productions from these three categories were to be included, the total would easily amount to more than 200 titles.

2 One of the earliest examples of a mining film is the 1905 Pathé production, *Au Pays Noir*.

3 See, for example, Gerhard Heilfurth, *Der Bergbau und seine Kultur: Eine Welt zwischen Dunkel und Licht* (Zurich, 1981).

4 See Heilfurth, *Der Bergbau und seine Kultur*, passim.

5 Pierre Hamp, *Gueules noires* (Paris, 1936); Egon Erwin Kisch, *Eintritt verboten* (Zurich, 1934); J.B. Priestley, *English Journey* (London, 1934); George Orwell, *The Road to Wigan Pier* (London, 1937).

6 Anton Kaes, *From Hitler to Heimat. The Return of History as Film* (London, 1989), p.x.

7 Siegfried Kracauer, *From Caligari to Hitler: A Psychological History of the German Film* (Princeton, 1971 edn), p.5.

8 Jeffrey Richards and Anthony Aldgate, *Best of British: Cinema and Society 1930–1970* (Oxford, 1983), p.10.

9 Hermann Barth, *Psychagogische Strategien des filmischen Diskurses in G.W. Pabsts Kameradschaft (Deutschland, 1931)* (Munich, 1990).

10 John Sayles, *Thinking in Pictures: The Making of the Movie Matewan* (Boston, 1987); Pierre Assouline, *Germinal: L'aventure d'un film* (Paris, 1993).

11 John Russell Taylor (ed.), *The Pleasure Dome: Graham Greene, The Collected Film Criticism 1935–1940* (Oxford, 1980), p.265.

12   Compare with the catalogues for the exhibitions *Arbeit und Alltag: Soziale Wirklichkeit in der belgischen Kunst 1830–1914* (Berlin, 1979); *Art et Société en Belgique 1848–1914* (Charleroi, 1980).

13   *Kameradschaft = La Tragédie de la Mine / Drehbuch von Ladislaus Vajda, Karl Otten, Peter Martin Lampel nach einer Idee von Karl Otten zu G.W. Pabsts Film von 1931* (Munich, 1997).

14   Compare with *Salt of the Earth: Screenplay by Michael Wilson, Commentary by Deborah Silverton Rosenfelt* (New York, 1978).

15   *Cinéma Wallonie Bruxelles: du documentaire social au film de fiction* (Grâce-Hollogne, 1989).

16   I have been able to trace four titles, but there may have been more: *Lass jucken, Kumpel* (Franz Marischka, 1972), *Der Kumpel lässt das Jucken nicht* (Franz Marischka, 1974), *Lass jucken, Kumpel: Dritter Teil* (Franz Marischka, 1974) and *Lass laufen, Kumpel* (Franz Marischka, 1981).

CHAPTER SIX

Modernity or 'Slaves of the Lamp'? Independence and Control in Two State Coal Mining Communities in Victoria, Australia

Meredith Fletcher

The state of Victoria in south-eastern Australia has some of the most extensive brown coal deposits in the world, but very few black coal reserves. At the beginning of the twentieth century, Victoria's fuel and energy source was black coal imported from New South Wales, the state to the north of Victoria. Black coal fired the factory furnaces, it raised the steam to generate electricity and it powered the trains. Supply was always uncertain, interrupted by strikes in both the New South Wales coalfields and the transport industry. In an attempt to make Victoria energy-independent, the state government initiated two coal mining enterprises. It established a coal mine at Wonthaggi in 1909 to mine some of the state's limited black coal reserves and it began developing the Yallourn project in 1921. This was a massive undertaking where an open cut mine was established and the brown coal was used to generate a statewide electricity supply and also produce briquettes as an alternative fuel supply to black coal.[1]

Although very different enterprises in matters of scale, technology, expectations and investment, the two mining developments grew from the same imperative to make Victoria energy-independent, and they were both state government enterprises. This has led historians to discuss and compare the two ventures as examples of state socialism in Victoria.[2] However, the aim of this chapter is to contribute to a broader understanding of the way identities are forged in coal mining communities and to consider issues of control and independence in state enterprises. It will be suggested that Wonthaggi had an impact upon the development of the much more extensive state enterprise at Yallourn, established a little over 10 years later. Moreover, conclusions will be drawn about the influence that modernity had upon the state's attitude towards and representations of its coal mining communities as it sought to create a modern, efficient and harmonious industrial enterprise.

To understand the identities that developed at Yallourn and Wonthaggi, my discussion is mostly limited to an analysis of the towns, rather than their industrial development and works. The chapter is concerned with issues such as town planning and landscaping, houses and ownership, local government, cooperatives and liaisons and how they contribute to systems of control or forging an ethos of independence. An essential component is also to explore the relationship between coal mining identities and the control of public and private space.[3]

Identity and Independence at Wonthaggi

The Victorian government's first venture into state coal mining followed a protracted strike in New South Wales. In 1909, hasty decisions were made to mine the black coal deposits in the Powlett River area near the coast overlooking Bass Strait. The coal was in an isolated part of Victoria with no road or rail facilities, and the site was covered with thick coastal scrub. Naming the new enterprise Wonthaggi, the government outlined its modest aims for the state coal mine. The coal would be supplied to the Victorian railways to power the state's steam locomotives, with administration of the mine transferring to the Railway Commissioners. All planning decisions were made quickly in an effort to start mining the coal.

A tent city was set up among the swamps to house the miners while the government outlined plans to build a model township at the state coal mine. Initially the government planned to keep control over the town by offering leaseholds for house sites and business premises, but this policy soon foundered in the face of the logistics of making the mine operational in a short period.[4] Miners and business operators in the town also lobbied strongly against leaseholds. Reflecting the Australian tradition of high home ownership aspirations, they wanted to own their homes and shops.[5] The leasehold policy was abandoned and the 100 'model' three- and four-roomed cottages that the Mines Department had built for employees to rent were sold, as were the business allotments. Wonthaggi was made a municipality with a town council. Although the government owned the coal mine that was Wonthaggi's *raison d'être,* it relinquished control of the town and the institutions that developed there.

The working conditions at the state coal mine were among the most dangerous in Australian mines, as the coal deposits were difficult to access, with fractured and dispersed seams.[6] Coming to work in these dangerous conditions were former gold miners from the declining goldfields in Central Victoria and recently arrived British migrants. Both miners and migrants brought British models of political and social organizations to the new field. Cooperatives formed by the Victorian Coal Miners' Association became major institutions in the town and included a hospital, dental surgery, dispensary, cooperative store and theatre. An alliance was forged between the union and business operators in the town who were united by opposition to the mine management. Businessmen were convinced that the state-controlled mine was putting a brake on the town's development and they felt no community of interest with the town's main employer. Both miners and business operators criticized the government policy of restricting Wonthaggi coal to supplying the Victorian railways and not marketing it further afield. Serving together on the borough council, miners and business claimed a mutual commitment to the town. At a meeting of Wonthaggi citizens at the beginning of the four-month strike in 1934, leading Wonthaggi businessman and borough councillor A. Frangeroid declared: 'If McLeish [manager of the state coal mine] does not moderate his tactics to the borough and its residents he will be very sorry. I, for one, will not take his intimidation lying down.'[7] His comments illustrate the antagonism between the business community and the mine management and the reasons for an alliance between the miner residents and the business operators. Shopkeepers extended credit during strikes and made their shops available as food distribution points.[8]

Political influence through local government, security through freehold properties, strong unionized institutions and a union/business alliance contributed to the independence and self-sufficiency that became an important part of Wonthaggi's identity and that sustained mining families through multiple strikes. The miners' creative use of their environment also contributed to Wonthaggi's independent ethos. Eight kilometres from the town is a spectacular stretch of cliffs and coves now known as the Bunurong coast. From the time the mines opened, Wonthaggi miners built simple huts on the beaches, often with materials 'obtained' from the state coal mine. The beauty and freedom of a stretching coastline and pounding surf were appreciated by men who worked in darkness underground. They fished in the coves and off the rock platforms and brought home abundant catches of whiting and bream, crayfish (lobster) and abalone; miners' helmets with carbide lamps were great assets for night fishing. But during strike periods, activities at the huts took on a greater significance. They made an important contribution to an alternative economy that sustained many Wonthaggi families. Striking miners moved to the huts, where they trapped rabbits and collected wood for distribution in Wonthaggi. They fished in earnest and supplied families with the fish they caught. The sought-after crays were sold to shopkeepers to buy supplies or exchanged with local farmers for vegetables.[9] Recreational in the good times, the huts provided a resource during strikes that contributed to the miners' independence at Wonthaggi.

Identity and Control at Yallourn

As well as its limited black coal reserves, Victoria had expansive deposits of brown coal. In the late nineteenth century, several small-scale attempts had been made to mine this resource but the high moisture content of the coal made it an inferior product. The Victorian government then turned to British and German experts for advice on utilizing brown coal as an alternative to importing black coal from over the border, and was informed that brown coal could provide Victoria with a statewide electricity supply.[10] During the First World War, when Victoria's energy supplies were stretched to the limit, black coal supplies from New South Wales often came to an abrupt halt because of a combination of mining and transport strikes. Looking ahead to the postwar years, the state government endorsed a plan in 1917 to develop an open cut mine near the banks of the Latrobe River, build a power station and generate electricity for the state. Although massive in scale, this scheme had similar origins to the state coal mine at Wonthaggi. It would be a government initiative, owned and operated by the state, and it would contribute to making Victoria energy independent. The new project was called Yallourn, a mixture of two Aboriginal words meaning 'brown fire'.

The planning for the new undertaking reveals the impact that the modestly sized Wonthaggi state coal mine had on Yallourn. At Wonthaggi, the government had relinquished control of the town. The identity at that new coal mining community, a mix of independence and militancy, was forged by its residents. Mine management had alienated its natural allies, the business community. Town and workplace had developed as separate, antagonistic entities. This would not be the case at Yallourn, where the newly formed State Electricity Commission (SEC) would, through its ownership of the town, control public and private space.

Influencing the SEC commissioners in their plans for Yallourn in the early 1920s was a mixture of idealism and pragmatism, overlaid by the *zeitgeist* of modernity. As the influential magazine, *The Bulletin*, reported in an article on Yallourn, there were 'no slaves of the lamp descending into the bowels of the earth, to lie for hours picking at the diamond hard roof of the tunnel above them'.[11] That was Wonthaggi. At Yallourn there was technology, modernity and the rule of engineers. In his book on modernity, *All That Is Solid Melts Into Air*, Marshall Berman notes the existence of a 'machine aesthetic' after the First World War where the engineer reigned supreme.[12] As David Noble has discussed in *America By Design*, in the 1920s engineers had moved into managerial positions where they turned their attention from the engineering of things to the engineering of people.[13] Using their engineering training, they adopted techniques to create a disciplined, loyal workforce for the corporate order. Modernist architect Le Corbusier endorsed the machine aesthetic, dominated as it was by harmony, unity and functionalism. He wrote approvingly of the production line of the Ford factory where he observed that 'everything is in collaboration, unity of views, unity of purpose, a perfect convergence of the totality of gestures and ideas'.[14] And Le Corbusier elevated the engineer. He looked to the engineers to 'fabricate the tools of their time'.[15]

The town of Yallourn, then, was a tool fabricated by engineers. In the new state coal mining venture, the town would be a component of the smoothly running site of production that was materializing along the banks of the Latrobe River. It would also operate as a machine. Through town planning and the model housing that the SEC was designing, ideal living conditions would be instituted and the town would serve as a site of production. It would produce a contented and compliant workforce.

Under the influence of the 'machine aesthetic', the town of Yallourn became a 'machine for living'. From the SEC's early planning documents, it was clear that the town and works would function as one.[16] The SEC would own the town, including the houses and businesses. No union/business alliance could develop at Yallourn. Unlike Wonthaggi with its town council, there would be no local government in Yallourn. Instead, the town would be governed by a senior SEC engineer with the title of 'general superintendent', a name redolent of the workplace. Under the rule of engineers, the hierarchy of the workplace would be duplicated in the living place. 'Staff' were seen as the natural town leaders, 'wages' as the populace. Through its control of public and private space, the SEC could determine who lived there. The SEC's governance of the town and ownership of the land also gave it the power to decide which institutions and community facilities could be established at Yallourn. With such a degree of control over the town, the SEC would forge Yallourn's identity.

With experiences at Wonthaggi serving as a constant reminder, the SEC comprehensively designed the town and, through town planning and landscaping, began creating Yallourn's identity and image. Yallourn was not to resemble a mining town. It would be a model town and a garden city, designed on garden city principles borrowed from Britain. The model houses in the town would have five types of room – three bedrooms, a living room, kitchen, bathroom, laundry – water supply and electricity, and they would be built on leafy streets surrounded by parks and pleasant vistas. The cohesive landscaping and architecture would represent the unity and harmony for which the SEC was striving.

The success with which the town was dissociated from images of a mining town was evident in a glowing description of Yallourn that appeared in an article in *Australian Home Beautiful* in 1929, when the town was eight years old. 'Yallourn is a surprising place,' wrote columnist Easter Soilleux. 'A brown coal mine conjures up visions of a collection of shabby depressing houses strewn along a straggling main street with a dusty general store providing the necessities of life.' But not so at Yallourn, she reported. Here, workers lived in 'delightful model homes' that were set in 'the most attractive of model streets ... with the consequence that living in congenial surroundings the workers are contented and happy'. Yallourn's gardens and lawns, she noted, were 'the freshest of the fresh and the greenest of the green'. She interviewed an SEC official who informed her that the 'weeding and mowing of lawns and the training of sweet peas and climbing roses help to keep industrial relations happy and sweeten and strengthen community life'.[17]

The SEC had successfully planted and cultivated a garden town that did not resemble images of a mining town. In this way, it was integrated with a workplace where use of words such as 'mine' and 'mining' were assiduously avoided. In the early days of the mine development, before extensive mechanization with imported German dredgers, the mine was always referred to as the open cut. The men working there were not called 'miners' but were classified as 'labourers on coal', to avoid the militancy associated with Wonthaggi and New South Wales coal miners, and also to pay them less.[18] And coal was never 'mined' at the Yallourn open cut. Coal 'winning' took place there, again avoiding any reference to mining. So at Yallourn, labourers on coal were engaged in winning coal from the open cut. Then they went back to their model homes in a garden city at the end of a shift. Meanwhile, at Wonthaggi, miners were employed to mine coal from the state's black coal mine.[19]

And as the SEC vigorously designed and implemented a garden city image, it also decided on the institutions in the town that would further develop an image that distanced Yallourn from mining towns like Wonthaggi. The Salvation Army serves as an example of this. In the early 1920s, the SEC had been keen to allocate prominent corner blocks to the churches and was soon in touch with the major denominations in Victoria, such as the Anglicans, Presbyterians, Methodists and Catholics, to discuss nominal rentals and building specifications for their churches. But when Salvation Army officers approached the SEC for land to build a hall in the model town (and told officials that they intended carrying out a mission similar to their work at Wonthaggi) they received a setback. The SEC officials decided there was no place for the Salvation Army at Yallourn. Certainly, they conceded, the Army did valuable work in a mining town like Wonthaggi, but the status of the average worker at Yallourn would be 'much higher than Wonthaggi'. The Army's 'special social work' would not be needed at Yallourn. An organization like the Salvation Army, associated with battlers and alleviating poverty, did not sit comfortably with the image the SEC was imposing on the new town.[20] The Salvation Army could build a hall in Wonthaggi, but not at Yallourn. The Salvation Army band could play stirring hymns in the streets of Wonthaggi, but not at Yallourn. The SEC hierarchy decided where and how Yallourn residents could worship, influenced by the image and identity they were creating in the town.

Yallourn and Wonthaggi

To understand how the SEC established control at Yallourn shows not only the influence the other state coal mining venture had on the realization of the SEC's new town, but also the important contribution of institutions to the ethos, self-sufficiency and independence of coal mining towns. At Yallourn, for example, not only was there no local government, there was no resident-controlled hall where people could assemble, there were no cooperatives, there was no independent newspaper where opinions could be expressed, there was no opportunity to start a club or society without SEC sanction. A comparison between institutions and organizations at Yallourn and Wonthaggi illustrates the implications that loss of control have for a mining community.

For over 30 years, Yallourn had no public hall. A facility such as a public hall could only be built to the SEC's timetable and with its sanction. No matter how much voluntary work and fund raising the residents were prepared to do to build a hall, they were powerless because the SEC owned the land. Wonthaggi had its Union Theatre, a vibrant cooperative where pictures were shown, dances and balls were held, concerts and performances were staged and, importantly, political meetings and union meetings took place. When it came to finding venues for social activities, Yallourn people had to be resourceful. Take the movies, for example. Two of the town's churches, St John's and St Therese's, doubled as substitute picture theatres. During the week, chairs faced the screen at the back of the church. On Sundays they were turned around to face the altar. At St Therese's, the confessional doubled as a projector box.

The Union Theatre at Wonthaggi illustrates the benefits that accrued to mining-town residents who could control their own spaces. During strikes, the Union Theatre became the nerve centre: important meetings were held there, strike pay was allocated there and the theatre also became a food depot and a food distribution point.[21] At Yallourn, there was no venue that any of the unions controlled, so the SEC decided which political meetings could take place. When the Yallourn branch of the Australian Labor Party organized three speakers from Melbourne to address Yallourn residents on 'Russia' one Friday night in 1932, organizers soon found that their normal meeting venue was denied to them. The general superintendent had also informed the police of the topic for the meeting. Without a venue, an audience of 250 people assembled in the Yallourn town square, as did the speakers, and an augmented contingent of police. When the police stopped the speakers from addressing the crowd, the organizers decided to reconvene the meeting outside SEC territory. A line of cars and walkers filed out of the square and set out in the night to the crossroads on the Princes Highway. As a Melbourne newspaper reported: 'In the darkness, relieved only by dimmed headlights of the cars and under the tall gum trees, the three speakers delivered their addresses from a tree stump.'[22]

Cooperatives were important institutions at Wonthaggi. They provided medical, dentistry, hospital and pharmacy services, as well as the theatre and a store. Cooperatives greatly contributed to the health and well-being of the mining community. They were also vital institutions during strike periods, contributing to the independence and strength of Wonthaggi miners during disputes with the mine management. Wonthaggi's cooperative store lent money for strike pay and also provided credit for its members.[23] But the cooperatives also contributed to the self-sustaining culture and self-determination at Wonthaggi which was suppressed at Yallourn. There, the Yallourn general store, owned and operated by the SEC, was one of the most unpopular facilities in the town. Unlike other company towns, though, it was not set up as a profit-making exercise. In fact the store's profits were ploughed back into the town to fund community projects. But it rankled with townspeople that they were not consulted on ways in which the money was being spent for the benefit of the community. Deputation after deputation of Yallourn citizens approached the general superintendent about converting the store to a cooperative with resident involvement, but to no avail. The store was an important tool for the SEC during strikes when credit was refused, but strict maintenance of control over the store indicated that it had other advantages. The SEC's paternalistic control denied any decision making or responsibility to most residents. The store balance sheets were kept a close secret and the general superintendent and managers shied away from resident involvement, convinced the business would be 'jeopardised by amateur management'.[24] Meanwhile at Wonthaggi, 'amateur management' success-fully and efficiently operated many of the town's most important and complex institutions.

The role of the local newspapers differed greatly in the two towns, with implications for the contents of the newspapers, the spreading of information and the expression of opinions. At Yallourn, the local paper, the *Live Wire*, was owned and operated by the SEC, and financed with profits from the Yallourn general store. A 'policy statement' from the general superintendent made it clear that the newspaper was also a tool of the SEC in its attempts to control public and private space. No criticism of the SEC was to be included in the newspaper, editorials had to be checked by the general superintendent and letters and reports were to be sent to him for approval before publication. 'In all matters excepting the form and substance of Store advertisements,' wrote general superintendent Bridge in 1933, 'the will of the General Superintendent shall prevail.'[25] Through the newspaper, the SEC also exercised a punitive role, as the experiences of the Yallourn Civic Association showed. After it had angered the general superintendent by organizing an unauthorized public meeting, the residents' group was immediately denied space in the *Live Wire* and could not advertise its meetings or publish reports of its activities.

Comparison of the content of the two town newspapers, the *Live Wire* at Yallourn and the *Powlett Express* at Wonthaggi, during strike periods, shows their different aims and roles as coal mining town newspapers. The two weeks before Christmas in 1926 were a time of tension and anxiety in Yallourn as the works closed down during a strike. The strike was widely reported in Melbourne newspapers because of the possibility of thousands of workers being stood down just before Christmas as a result of interruption to electricity supplies. In the Yallourn newspaper, there was no mention of the dispute or deserted coalfields. Instead the *Live Wire* reported on the progress of the fundraising Queen Carnival and a meeting of the Methodist Guild, and listed the cricket and tennis scores. It featured a major article on tree planting and landscaping of Broadway, the main entry into Yallourn, and included advertisements for the pictures that would be screened at St John's on Saturday night. There was also a large section on Christmas gift suggestions from the Yallourn general store.[26]

The progress of the four-month strike at Wonthaggi from March to July in 1934 dominated the *Powlett Express*. Even the number of free haircuts that were provided for Wonthaggi men and women was reported in the *Express*. The newspaper's detailed coverage of the strike included full accounts of the many meetings, speeches, expressions of opinion and events that took place. The full transcript of a speech given by Agnes Chambers of the Wonthaggi Women's Auxiliary at a large Melbourne meeting, where she went to elicit support for the strike, is an example of the newspaper's detailed coverage. Not a word of Agnes Chambers's speech was left out as she spoke of living and working conditions at Wonthaggi and how they affected women and families. She appealed to the women of Australia to help in the fight against 'tyranny' at Wonthaggi:

> Do the women of Australia understand what the life of a coal miner is? Do they know one man in four meets with an accident in these mines? We women living at home while our husbands are toiling in the bowels of the earth each year see the accident list growing greater and greater each day. When they leave for work we are not sure whether they will return uninjured. These accidents our husbands tell us are largely due to the system of 'speeding up'. Because our men are determined to call a halt to this sacrifice of life and limb the Government threatens to take from us our homes. .... The children will suffer privation through no fault of their own, and their suffering will be caused by a Government that is determined to break the spirit of men who are fighting against low wages, bad working conditions and the victimisation of their comrades.[27]

At Wonthaggi, the newspaper could include criticism of the management and work practices of the mine. It could report on tyranny and injustice. It could publish opinions and attitudes. At Yallourn, the newspaper carried directives from the management and spread SEC policy. Through the pages of the newspaper, residents were exhorted to 'display a greater community spirit' so that the town would 'realize the aims of its founders as the best Garden Town in Victoria'. A Christmas message from the chairman of the SEC expressed hopes that 'this festive season will usher in that air of harmonious co-operation which will have so great a bearing upon the consummation of the vital national purpose for which the works of the Commission were undertaken'.[28]

Civil Society and State Control at Yallourn

Bushfires are a constant danger in south-eastern Australia. Fanned by blistering northerly winds, they can often cause great loss of life and property. When a bushfire swept around the outskirts of Yallourn in 1944 and set the open cut alight, threatening Victoria's wartime electricity supply, the royal commission investigating the fires also ignited a fiery debate on the implications of state control and civil society. In his report, the royal commissioner exposed 'suffocating paternalism' and the lack of 'freedom, fresh air and independence' in the model town. He questioned whether 'social responsibility' could flourish in a town like Yallourn, where people could be called on to defend one of the nation's most valuable assets.[29]

The release of the report fanned public debate and discussions at a level that had not been tolerated at Yallourn. Overnight, an uncensored *Live Wire* hosted a vigorous letters column discussing conditions at Yallourn. Public meetings were called without SEC sanction to discuss issues such as local government, home ownership and the right to establish cooperatives. In those heady days, Wonthaggi was held up as a model. ('There's no excuse for anyone in Wonthaggi to be without a good set of teeth,' an elderly Wonthaggi miner assured his Yallourn audience as he outlined the benefit of cooperatives in his town.[30]) Before the bushfires, the discontent that had been simmering in the town centred on the inadequacies of the Yallourn general store with its lack of resident involvement and the blandness of the *Live Wire*. After the fires and the publication of the royal commissioner's report, Yallourn was awash with discussions on democratic rights, freedom of speech, community action and self-determination. In a response to the royal commission, the government proposed a referendum where Yallourn residents could vote for local government.

But when Yallourn residents turned up at the Housewives' Hut to cast their votes, they rejected local government and voted instead for a town advisory council that virtually left control of the town in the hands of the SEC. In the wake of the royal commission, they had gained better shopping facilities, control over the newspaper had been loosened and there was more opportunity for public debate. This seemed to satisfy the majority of the residents. Their rejection of local government and the self-determination it engendered shows an acceptance and endorsement of the identity that the SEC had imposed and nurtured at Yallourn. Above all, it was also an expression of loyalty to the SEC. It was an acceptance that Yallourn was a special town with a special mission and a special way of life. Yallourn residents were 'loyal servants', as the Yallourn Civic Association expressed it in 1944. The sentiment persisted. 'I consider the SEC have been Good Masters,' wrote one elderly resident in the 1970s as she was leaving the town after a 50-year residence. 'We will feel their loss as we would a "Good Father".'[31] Looking back over a lifelong connection with the SEC, living and working at Yallourn, another elderly resident asserted: 'I am still proud of the fact that I was a cog in the wheel of such an undertaking as the State Electricity Commission.'[32] SEC and residents agreed: the town and works were integrated. The town was part of the machine.

Implications of Independence and Control

The early abandoning of plans for a model township and relinquishing of control of the town that developed at the state coal mine at Wonthaggi served as a cautionary precedent for the second state coal mining enterprise in Victoria. The two identities that developed at Yallourn and Wonthaggi, one fostered by the employers and one fostered by the residents, affected the independence, self-determination, lifestyles and industrial power of the mining communities. They also contributed to the fates of the two towns.

Today, Wonthaggi in South Gippsland is a prosperous regional centre that provides facilities and administrative services for its agricultural hinterland and the nearby rapidly developing coastal resort towns. At first constituted as a small borough council, Wonthaggi is now the nucleus of a large rural shire, Bass Coast Shire. There is no coal mining at Wonthaggi. The state coal mine has been closed for nearly 40 years. In the postwar years, the railways turned from steam to diesel to power their locomotives and mining in the dangerous, fractured coal seams was scaled down. But the independent Wonthaggi was not subservient to the coal. After the government's early relinquishing of control of the town, the town had diversified. Unlike Yallourn, Wonthaggi did not remain an introverted industrial enclave separated from its hinterland, and was able to survive the mine closure. Today, sections of the state coal mine are open to the public and the former mine has become the preserve of consultants, preparing strategies to market Wonthaggi's coal mining heritage. The independent spirit of the mining town has been celebrated in a feature film, '*Strikebound!*'. The mine whistle, which has been relocated to a central park in the town, sounds every day at noon, often to the discomfort of unwary picnickers.

Along the Bunurong coast, several kilometres from Wonthaggi, there is no evidence of the miners' huts that played such an important role in sustaining the mining families during long strike periods. The area is now a marine and coastal park and reflects the important tourist industry that partly sustains the town and changing environmental attitudes. Only the names of the settlements – Flat Rocks, Eagle's Nest, Harmer's Haven, Shack Bay – remain.

One hundred kilometres away, a winding drive over the Strzelecki Ranges, lies Victoria's Latrobe Valley, a network of towns, open cut mines and power stations that provides Victoria's electricity supply. If you stand at the Yallourn lookout, you can see about 3000 hectares of spreading open cut mine and the massive cooling towers of the Yallourn W power station that was built in the 1970s. Coal mining and power generation continue apace, as they do at the power stations and open cuts at Morwell and Loy Yang that can be seen in the distance, post-Second World War developments that were built in an effort to satisfy Victoria's escalating energy demands.

But what you *cannot* see from the lookout is the town of Yallourn. You cannot see the patchwork quilt of bright rooftops, the autumn colours of the garden city's trees or the green of the town's beautiful parks and sporting reserves. In the 1960s, the town became surplus to SEC requirements. By then, most SEC workers lived in the three towns that were absorbed into the newly industrialized Latrobe Valley. The SEC no longer needed a town that 'produced' contented workers.

In the 1920s, Le Corbusier had applauded the 'engineer's morality' in getting rid of tools that had reached their use-by date, in throwing the 'out of date tool on the scrapheap'.[33] This was disturbingly prophetic for Yallourn. The SEC engineers had created a model garden city that was increasingly out-of-step with developments in the postwar years. The town as machine had as much value for the SEC as an outmoded power station. It was superseded technology. And not only was it expensive to maintain, easily accessible coal lay underneath the houses, shops, town buildings and gardens. The SEC that owned the town and had created its identity now determined its future and decreed that Yallourn should be dug up for coal. The tool had reached its use-by date and was thrown on the scrapheap. Model houses and gardens were redefined as 'overburden' and the dredgers moved in. Yallourn was submerged into the blackness of its open cut.

Wonthaggi survives with its history of independence and self-determination, its coal mining heritage and a diversified economy. Yallourn was subservient, not to the community that lived there, but to the coal that lay underneath.

Notes

1   For more information on the origins of these state industries, see E.W. Russell, 'The state coal mine and state socialism in Victoria', unpublished PhD thesis, Monash University (1980); C. Fahey, *Wonthaggi State Coal Mine: a History of the State Coal Mine and its Miners* (Wonthaggi, 1987); Andrew Spaull, 'The origins and rise of the brown coal industry 1835–1939', unpublished Master of Commerce thesis, Melbourne University (1966); Meredith Fletcher, *Digging People Up For Coal: a History of Yallourn* (Melbourne, 2002).

2   Russell, 'State coal mine'; Andrew Reeves, 'Industrial man: miners and politics in Wonthaggi, 1909–1968', unpublished MA thesis, Latrobe University (1977); Frederick Eggleston, *State Socialism in Victoria* (London, 1932).

3   This article is influenced by Henri Lefebvre, *The Production of Space* (Oxford, 1991).

4   Philip Harper, *The Wonthaggi Coalfields: a Story of the Men and the Mines* (Melbourne, 1987), pp.9–14; Bill Hayes, 'Communism and left-wing alternatives in Wonthaggi', unpublished honours thesis, Monash University (1996), 18–21.

5   For a series of essays that explore the notion of the Australian home and high ownership aspirations, see Patrick Troy (ed.), *A History of European Housing in Australia* (Melbourne, 2000).

6   Bill Hayes, 'The Wonthaggi Miners' Women's Auxiliary', *Gippsland Heritage Journal*, **19** (1996), 2–9.

7   *Argus*, 13 March 1934.

8   Hayes, 'The Wonthaggi Miners' Women's Auxiliary'.

9   Bill Hayes, 'Miners' Huts on the Bunurong Coast', *Gippsland Heritage Journal*, **17**(3) (1994), 3–11.

10 Cecil Edwards, *Brown Fire: a Jubilee History of the State Electricity Commission of Victoria* (Melbourne, 1969), pp.15–16.

11 *The Bulletin*, 23 February 1928.

12 Marshall Berman, *All That Is Solid Melts Into Air: the Experience of Modernity* (New York, 1989), p.26.

13 David Noble, *America By Design: Technology and the Rise of Corporate Capitalism* (New York, 1979), pp.261–78.

14 Charles Jencks, *Le Corbusier and the Tragic View of Architecture* (Harmondsworth, 1987), pp.55, 123.

15 Le Corbusier, *Towards a New Architecture* (London, 1983), p.8.

16 John Monash, 'Town of Yallourn: notes on a policy for the construction and governance of the proposed town' (24 January 1921), Monash Papers, SEC Archives.

17 *Australian Home Beautiful*, 10 May 1929.

18 J. Vines, *A History of the Morwell Open Cut* (Morwell, 1996), p.9.

19 Fletcher, *Digging People Up For Coal*, pp.57–8.

20 Ibid., pp.93–4.

21 Hayes, 'Communism and left-wing alternatives', 32–44.

22 *Sun News Pictorial*, 17 September 1932.

23 Hayes, 'Communism and left-wing alternatives', 39.

24 Fletcher, *Digging People Up For Coal*, p.104.

25 J. Bridge, 'Yallourn Live Wire', March 1933, SEC Archives.

26 *Live Wire*, 16 December 1926; 25 February 1926.

27 *Powlett Express*, 6 July 1934.

28 *Live Wire*, 16 December 1926; 25 February 1926.

29 L.E.B. Stretton, *Report of the Royal Commission to Inquire into the Place and Origin and the Causes of the Fire which Commenced at Yallourn on the 14th Day of February 1944* (Melbourne, 1944), pp.3, 5.

30 Fletcher, *Digging People Up For Coal*, p.117.

31 Ibid., p.179.

32 Ibid., p.214.

33 Le Corbusier, *Towards a New Architecture*, p.17.

CHAPTER SEVEN

A Comparison between the Richmond Coal Basin and Pennsylvania's Anthracite Fields: Slave Labour, Free Labour and the Political Economy*

Sean Patrick Adams

At the onset of the nineteenth century, Richmond, Virginia emerged as the likely centre of America's coal trade. Coal mines in the small, but rich, field (known as the Richmond Basin), a few miles up the James River, had been in operation for decades. 'The owners of the coal mines of Virginia,' Pennsylvania's Tench Coxe noted in 1794, 'enjoy the monopoly of all the supplies for the manufacturers of the more northern states, who live in the sea ports; a demand which is increasing rapidly.' Seaboard cities would become dependent upon Richmond Basin coal, the nationally renowned engineer Benjamin Latrobe argued in 1808, and 'even now the smiths within 10 miles of our sea ports, require in order to carry on advantageous business, a supply of Virginia coal'. Alexander Hamilton and Albert Gallatin cited the Richmond Basin as the single most important source for domestic and industrial mineral fuel. To the informed observer of the coal trade, it appeared that the rich bituminous mines of the Old Dominion's Richmond Basin would serve the new nation's hearths and furnaces for years to come.[1]

At the same time that the Richmond Basin's reputation soared, the prospects for the rich anthracite fields of nearby Pennsylvania sank into near oblivion. In 1803, Philadelphians spread nearly 30 tons of Lehigh anthracite coal on their sidewalks in place of gravel. After watching it smother a fire in a trial run, potential customers decided that they had no better use for it than as paving material. A decade later, George Shoemaker took nine wagon loads of anthracite to Philadelphia and gave seven of them away. 'The result was against the coal,' one observer recalled, 'those who tried them, pronounced them stone and not coal, good for nothing, and Shoemaker an imposter.' Although virtually all of the anthracite coal deposits in the United States lie within the boundaries of Pennsylvania, early consumers deemed 'stone coal' a difficult, if not impossible, mineral fuel to use in hearths and furnaces. Even Philadelphia's waterworks, whose two steam engines for pumping the city's water supply burned both wood and coal, shunned anthracite and burned Virginia bituminous in their furnaces.[2]

A few decades later, however, 'stone coal' ruled American fuel markets. In 1836, the *North American Review* reported that the growth of the anthracite trade had 'converted the wildest waste into the theatre of active life, given a fresh stimulus to individual enterprise, created an inexhaustible source of wealth in which it lay, and opened a new commerce and a new bond of fraternity to the whole Union'. The massive growth of the industry occurred in a short period of time; in the course of a single decade, production levels in the anthracite regions climbed from about 60 000 tons raised in 1825 to over half a million tons in 1835. By 1842, anthracite colliers raised more than one million tons of coal in a single year and demand for the once spurned mineral was increasing rapidly. The coal industry of eastern Virginia, in contrast, saw a relative decline during this period. By 1835, Richmond colliers sold over 96 000 tons of bituminous coal, a significant increase from the nearly 35 000 tons raised a decade earlier. But in the same year that the production of anthracite reached one million tons, shipments from the Richmond Basin actually decreased to 68 750 tons.[3]

Why did the Richmond Basin, so close to the seaboard's trade routes and its bituminous coal already in common use, lose its advantageous position in the American coal trade? How did anthracite coal overcome its initial problems to emerge as a vital force in Pennsylvania's industrial economy? This chapter examines the competition between the Richmond bituminous region and the Pennsylvania anthracite fields in the early American coal trade during the first three decades of the nineteenth century in order to shed some light on these questions. Richmond Basin colliers found the expansion of their trade constrained by their location in a political economy dedicated to the production of agricultural goods using slave labour. The nation's early anthracite trade, on the other hand, flourished in Pennsylvania's free labour economy as an alliance of entrepreneurs, boosters and state officials forged an effective campaign to promote anthracite. Without a scientific community or promotional effort to rival Pennsylvania's, the bituminous coal trade outside of Richmond lost its advantageous position in seaboard markets. Stripped of its 'first mover' advantage, Richmond coal continued to decline in importance as the nineteenth century progressed and anthracite coal became the domestic and industrial fuel of choice in most large American cities. These divergent paths were not the natural result of geological or market forces, but can largely be attributed to political circumstances and, more specifically, the institutional context in which coal was mined in each state.

The First Mover: the Richmond Basin Coal Trade

When Richmond colliers sought to expand their trade in the early nineteenth century, they were quite willing to use slave labour to increase production. The hiring or renting of black labour was not always successful, however, as agricultural work always took precedence over mining. Richmond Basin colliers scrambled to fill out their labour force with hired slaves, free blacks and unskilled white labour. As a result, labour costs remained relatively high throughout the late eighteenth and early nineteenth century. For example, Harry Heth of Chesterfield County, one of the region's most prominent colliers, employed a number of his own slaves in his Black Heath mines and hired additional hands from surrounding plantations. The inherent hazards of coal mining, such as fires, shaft collapses and floods, made many owners reluctant to hire out their slaves for such dangerous work. As a result, Heth and other Virginia miners paid very high prices for hired slave labour, if they could get hands at all. One planter in 1812 refused to replace a slave that had fled Heth's coal mines and returned home. 'I have been trying to hire hands for you ever since I saw you & have not as yet procured a single one,' one friend wrote to Heth in 1819, '& I am afraid I shant be able to get one single one in the neighborhood.' The high cost and scarcity of slave labour remained a particular expense of raising coal in the Richmond Basin throughout the early nineteenth century.[4]

Labour scarcity cut into potential profits for Richmond Basin miners by raising costs; engineering problems could do the same. The challenges of digging shafts, pumping water out of flooded tunnels and shafts, and providing proper ventilation plagued colliers of all nations during this time. Early mines in the Richmond Basin were open-air diggings that measured up to 30 feet deep and often filled with water before they could be fully worked. Samuel Paine wrote to his partner in 1801, 'we are sinking a pitt at the extreme point of the commencement of the old works and if we escape the water of the old works, I see nothing to prevent our doing as well as we have a right to expect for another year'. Rather than waste valuable time and money on pumping water out of these works, most miners abandoned them. Neglected pits and tunnels on one's land became a regular hazard for Virginia miners.[5]

As colliers sunk deeper shafts into less accessible seams, they required some form of pumping system to clear the mines of water. Harry Heth attempted to contract with local engineers to construct steam engines for pumping water from his mines, but these efforts proved ineffectual. Heth eventually sought help from outside the state in order to build a pumping system at his Black Heath pits. In 1813, he contacted Oliver Evans, the noted steam engine manufacturer of Philadelphia, about sending someone to design and implement a steam-powered system for pumping water from his mines. Evans agreed to build him an engine, but refused to send any of his employees to work in Virginia's slave labour economy. 'The workmen here have embraced a prejudicial idea of the customs of your country,' Evans wrote, 'they think that if a master mechanic goes into your employ, and will refuse to work a task himself but keep himself clean and talk big and help himself freely to brandy, wine &c, that you will treat him as a gentleman.' Aside from the problems of free and bonded workers mixing together, Evans balked at the idea that slave labour could even operate his steam engine. 'I fear [you] have wrong ideas if you think slaves can keep a steam engine in order,' he wrote. 'A man must be free before his mind will expand so much.'[6]

Without a local, or even national, technological community to rely upon, Virginia colliers sought help from overseas. In 1814, Heth wrote to a friend in London that 'This d___d war has all but ruined me,' but apparently the war of 1812 could not keep him from inquiring about a British engineer for his mines. Eventually he employed two Scottish miners and contracted the British firm of Boulton & Watt to construct his steam engine pumping system that finally worked. Although a manager at Black Heath predicted that the new steam engine 'will afford a vast quantity of the most elegant coal', the charges for shipping machinery from distant manufactories and the high cost of training skilled workers to maintain steam engines made British technology expensive to implement in Richmond Basin mines. Heth himself said that, although he had invested $7000 in steam engine technology by 1816, he was not confident in its ability to cut costs adequately. Without an indigenous community of mechanics to draw upon, Virginia colliers found the application of effective technology to their mines to be rife with delays, unanticipated expenses and other setbacks.[7]

In addition to raising labour costs and hindering the adoption of technology, Virginia's slave labour economy frustrated the attempts of colliers to implement an effective transportation network in the Richmond Basin. The James River ran through the coalfield and provided water access to Richmond's tidewater port. But coal transport along the improved route of the James River Company involved a number of costly and damaging transfers throughout the nineteenth century, which reduced the quality of the coal. Without an integrated system of staiths and keels that had proved so successful in the British coal trade, numerous transfers of coal from one wheelbarrow to another pulverized it into a less useful commodity. Urban retailers sometimes received shipments from Richmond in the form of a fine powder. Chesterfield County colliers complained to the legislature that the existing system delayed and damaged their shipments as they worked through the various locks and transfers: 'the Quality of the coal is so naturally injured,' miners argued, 'that it can never gain a sufficient character in the northern markets to offer an inducement to us to use [the canal]'. These problems cut into the cost savings of shipping coal via the James in the years following the War of 1812.[8]

The James River Company, not surprisingly, became a main target for Richmond Basin colliers who constantly complained to Virginia's legislature that the directors favoured agricultural interests at their expense. In 1818, a group of colliers from Chesterfield labelled the James River Company an 'odious monopoly, by affording the Planter, the Farmer, and Merchant of the upper country, a choice of markets'. At the same time that coal trade suffered, traffic in tobacco, wheat, corn and other products from nearby plantations increased along the James route. The Chesterfield colliers submitted testimony that confirmed the physical problems of carrying coal on the James River canal. According to one carrier, coal barges should hold 180 to 240 bushels each, but the low level of water in the James only permitted carrying 60 or 70 bushels at a time. Veteran collier Orris Paine testified that boats carried less than 70 per cent of their capacity on average.[9]

The James River never evolved into an efficient means of transport for the coal trade, even after state officials took a controlling interest in the firm and widened the navigation route in the 1820s and 1830s. Toll revenues indicate that, although the company was willing to carry some coal, it still derived the majority of profits from the products of the plantations upriver. In 1827, for example, the company collected $47 279.87 in tolls on their descending navigation. Tobacco produced 60 per cent of the revenue and agricultural commodities such as wheat, flour and corn provided 22 per cent. Coal was a significant source of revenue, amounting to 13 per cent of toll revenues, but it hardly dominated the James River Company's interests. Two years later, despite various improvements along the route, the JRC still favoured agricultural products, as tobacco produced 42 per cent of the tolls, other agricultural goods 38 per cent, and coal 16 per cent. Until the 1820s, moreover, the tolls on the James River Canal were not based on the distance travelled and therefore acted as a subsidy for the farmers and plantation owners further up the James River. The company eventually stopped the practice of paying one flat toll rate for carriage, but by then the flow of traffic had been well established – a flow that largely neglected the needs of the local coal trade.[10]

The Richmond Basin coal industry of the early nineteenth century developed into a contradiction of national opportunities and local failings. Although urban consumers and political economists entertained lofty expectations for Richmond coal, the outlook from a regional perspective was not as optimistic. In the years following the War of 1812, Richmond coal looked to be one of America's most promising industries, but constraints on transportation and mining technology instead made Richmond bituminous vulnerable to replacement at a time when the area's trade might have expanded. Marginalized in a region dominated by plantation agriculture, coal-mining interests summoned neither the economic nor the political weight necessary for rapid expansion. In both transportation networks and labour markets, colliers faced indifference or outright opposition to their interests. Simply put, the inability of the Richmond Basin to capitalize on its fortuitous position seems directly tied to the area's commitment to a slave-based plantation economy.

From 'Stone Coal' to Anthracite

While Richmond colliers struggled to expand production, 'stone coal' was going through a remarkable transformation. The first step in anthracite's rise was a scientific reevaluation of its heating properties. In a series of highly publicized experiments in the years following the War of 1812, Marcus Bull and Benjamin Silliman demonstrated that anthracite burned hotter and longer than wood or bituminous coal under experimental conditions. But the practical use of anthracite remained a major barrier in the 1820s. 'The difficulty of consuming small quantities of anthracite coal, in open grates,' Marcus Bull wrote, 'must operate to prevent its general introduction into use, unless this difficulty can be removed.' Anthracite required a high temperature to ignite and once lit it needed to be separated from its ashes in order to keep the surface of the coal exposed to the air. Many fireplaces, stoves and forges in the early 1820s, most of which burned wood or coal on an open grate, simply could not meet these requirements. Experiments thus demonstrated the potential superiority of this fuel in the laboratory, but the practical problem of how to get people to use it remained the primary challenge for boosters of anthracite.[11]

A Philadelphia-based campaign to change consumer preferences towards anthracite coal along the eastern seaboard flourished in the years following the War of 1812. Jacob Cist, a local anthracite dealer, started the ball rolling by providing demonstrations of anthracite's bright flame and even bribing local blacksmiths to use it in their hearths. Philadelphia's Franklin Institute for the Promotion of the Mechanical Arts sponsored a series of industrial exhibitions from 1824 to 1838 and awarded gold, silver and bronze medals for technological innovations. The highest accolade, the gold medal, was explicitly reserved for improvements in the utilization of Pennsylvania's coal and iron resources. The Institute's *Franklin Journal and American Mechanic's Magazine* also published articles that highlighted innovations in grates, stoves and fireplaces and often included explicit instructions on how to burn anthracite in the home, along with a more technical description of the apparatus. By the early 1830s, local newspapers published articles and testimonials explaining how to ignite anthracite in fireplaces and stoves.[12]

Promoting anthracite constituted one challenge, but getting it to market profitably was another matter altogether. Philadelphia had two major connections to the anthracite coal region and the political circumstances of their creation are important to the story of anthracite's rise. The Lehigh Coal and Navigation Company (LCNC) improved the Lehigh River to link the eastern portion of Pennsylvania's Middle Anthracite Field to the Delaware River and ultimately to Philadelphia. Endowed with both mining and transportation privileges, this firm eventually dominated the Lehigh anthracite region through the ownership of its own mining tracts. The second private anthracite canal, the Schuylkill Navigation Company (SNC), used a 114-mile long system of improvements on the Schuylkill River to link the mines of the Southern Anthracite Field to Philadelphia. Coal traffic lagged during the early operation of the SNC, and commodities such as lumber, flour and stone constituted the main business of the canal during the 1820s. Moreover, a provision on the firm's charter restricted the SNC from owning its own coal-bearing tracts. In 1820, the LCNC's first shipments of anthracite coal down the Lehigh amounted to a modest 365 tons and in 1825, the coal traffic on the SNC amounted to a mere 6500 tons. This trickle of 'stone coal' to Philadelphia hardly revolutionized the Pennsylvania anthracite trade at first, but by 1830 the combined shipments of the SNC and the LCNC exceeded 130 000 tons annually.[13]

The rapid increase in anthracite traffic can be attributed to a healthy competition between the anthracite colliers using the SNC and the LCNC, which emerged as a result of different mining systems in the Schuylkill and Lehigh regions. Despite constant fears among Schuylkill area miners that the LCNC would manipulate coal prices by alternately starving and flooding the market, traffic estimates suggest that competition forced both firms to maximize traffic. Since the SNC depended upon tolls, not sales, for its revenue, it needed to attract a high volume of traffic in order to earn profits. Its directors kept tolls relatively low, therefore, in order to increase coal traffic along the SNC. With more and more Schuylkill region coal arriving in Philadelphia during the late 1820s and early 1830s, the LCNC needed to increase production or risk losing the initial advantage of Lehigh coal among Philadelphia consumers. Rather than tightly control production, the LCNC and other mining firms in the Lehigh region leased out their coal lands to individual miners. Both lessees and the LCNC profited from high production levels, and a system in which the LCNC regulated production in order to maintain high prices never materialized. The result was a steady increase in the supply of anthracite coal reaching Philadelphia from both regions, attended by a steady decrease in price. Without the hegemonic control exerted by slave-holding interests as occurred with the James River Company, canals in the anthracite fields competed fiercely to expand their shipments of anthracite to Philadelphia.[14]

As the capacity of the anthracite canals increased, colliers in Pennsylvania used free labour to increase production from the most accessible anthracite seams. In the Schuylkill region, miners usually raised coal under a short-term lease from the land's owner, which encouraged short-term practices that focused upon winning as much coal as possible at minimum expense. The LCNC hired its own labour to raise coal, but also sub-contracted a portion of their holdings out to small-scale miners. Unlike miners in the Richmond Basin, who used slave labour, Pennsylvania colliers often raised the coal themselves or hired free white labourers. Some of these hands were recent immigrants from England (particularly Cornwall) and Wales and had practical experience in the art of mining, whilst others were simply unskilled labourers attracted by the boom economy of the anthracite regions. Whatever method anthracite miners used to raise their coal, free labourers in Pennsylvania raised more of it year after year. As long as miners worked seams that lay close to the surface, expanding production was possible by simply hiring more hands to raise the coal. Without the added expense and trouble of either purchasing slaves or hiring them away from their owners faced by Richmond Basin colliers, anthracite miners could expand their labour force both rapidly and cheaply.[15]

By the early 1830s, lower prices and reliable shipments allowed anthracite coal to emerge as an important commodity and certainly aided its growth in urban markets. Philadelphia became the first major city to burn anthracite extensively and emerged as the centre of a spreading anthracite trade. Perhaps the crowning achievement in the campaign to promote anthracite was its successful introduction in New York City, particularly in the period after 1829, when the Delaware and Hudson Canal began shipping anthracite from the Northern Anthracite Field to New York City. By 1832, over 50 000 tons of anthracite coal were sold in New York City at an average retail price of $10.65 per ton, which accounted for 38 per cent of the total fuel sales in New York – a significant increase in only two years. The most impressive aspect of this upsurge in anthracite consumption in New York City is that it began at a time when 'stone coal' was still significantly more expensive than Virginia bituminous.[16]

Pennsylvania's political institutions also nurtured anthracite's growth through various programmes of protection and promotion. In essence, 'stone coal' became the 'Commonwealth's fuel'. In 1830, some legislators suggested that Pennsylvania's debt could be lessened with a tax on anthracite and bituminous coal of 25 cents a ton. By the time the legislature convened in the winter of 1830, an anti-coal tax movement was well in place and petitions flooded into the legislature. Opponents of the coal tax deployed a strong, and ultimately convincing, rhetoric that decried its 'injurious effect upon the manufactures and domestic policy of the commonwealth', and which explicitly linked the future of Pennsylvania with the future of anthracite.

This argument proved effective, as Pennsylvania's legislature rejected any significant coal tax in February 1831.[17]

Pennsylvania's state government also promoted its anthracite industry by coupling the state's growing iron trade to 'stone coal'. In 1836, the legislature passed an act 'To encourage the manufacture of Iron with Coke or Mineral Coal, and for other purposes.' This act allowed any iron-making firm that met certain requirements to receive a corporate charter so long as they burned coal, rather than charcoal, as fuel. Enacting a general incorporation law was a potentially controversial way to sponsor anthracite iron, since the right to incorporate manufacturing enterprises in Pennsylvania was jealously guarded by the legislature during the 1830s. The usual practice of creating corporations involved the passage of a charter by the state legislature, a political process that many public officials found distasteful and inherently corrupt. In fact, Governor Ritner announced in 1836 that corporations often reflect a 'depraved appetite' for speculation, 'foster and perpetuate the thirst for gain without labor', and paralyse individual entrepreneurs. The bill received minor opposition in the legislature from anti-charter forces and established charcoal iron interests, but supporters cast its intention as a bounty on the fusion of coal and iron and therefore avoided accusations that the act would create monopolistic corporations.[18]

Anthracite iron smelting came to Pennsylvania in 1840, when David Thomas brought Welsh hot blast technology into practice at the Lehigh Crane Iron Company, a firm that had been chartered in 1839 under the General Incorporation Act. The Allentown firm's innovation created a stir in ironmaking circles, and iron furnaces for smelting ore with anthracite began to appear across eastern and central Pennsylvania. In 1841, only a year after the Lehigh Crane Iron Company's success, Walter Johnson found no fewer than 11 anthracite iron furnaces in operation in the state: at least three had been created under the 1836 Act. That same year, an American correspondent of London bankers cited savings on ironmaking of up to 25 per cent after the conversion to anthracite and noted that 'wherever the coal can be procured the proprietors are changing to the new plan; and it is generally believed that the quality of the iron is much improved where the entire process is effected with anthracite coal'.[19]

What did the Richmond Basin's coal industry do while anthracite won over the hearths and furnaces of the eastern seaboard? In a nutshell, it could not respond at all. Virginia had no counterpart to Philadelphia's Franklin Institute, and its scientific community showed little or no interest in coal. When anthracite boosters reported that Richmond Basin coal would spontaneously combust, there was no reply from scientists in the Old Dominion. In 1828, the *Register of Pennsylvania* reported a number of cases of coal igniting itself in storage rooms, and each time it was careful to note that 'Virginia coal' was the culprit. Ten years later, the *American Journal of*

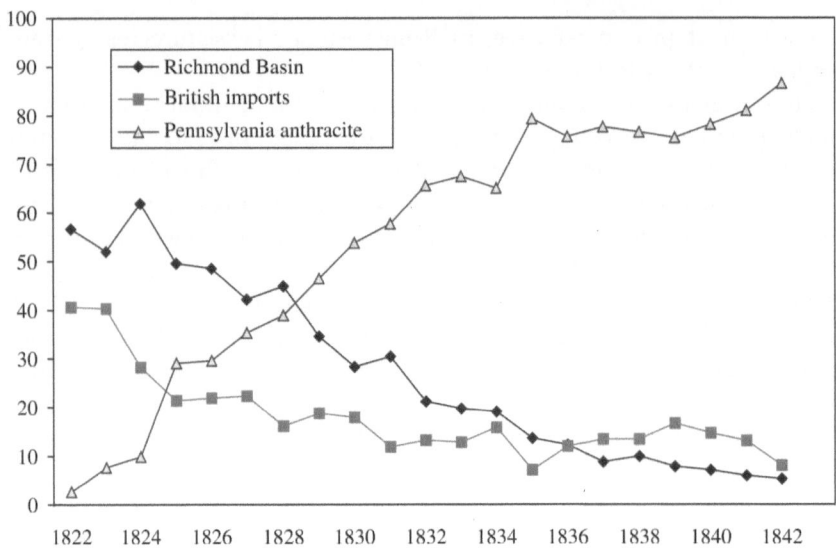

Sources: *Hunt's Merchant's Magazine and Commercial Review*, **8** (June 1843), 548; Alfred Chandler, 'Anthracite coal and the beginnings of the "Industrial Revolution" in the United States', *Business History Review*, 46 (1972): 141–81.

Figure 7.1   Percentage of seaboard coal consumption, by origin, 1822–42

Science reported a similar incident with the headline, 'Another Case of the Spontaneous Combustion of Virginia Coal'. Rather than provide concrete solutions for their state's colliers, Virginia politicians also did little to help. And, as demonstrated in Figure 7.1, the failure to respond to Pennsylvania anthracite's rise caused the Richmond Basin's market share to drop precipitously in a few short decades.[20]

Conclusion

In the grand scope of American industrialization, the story of Richmond bituminous and Pennsylvania anthracite might appear insignificant, even trivial. The replacement of one bulk commodity for another gives even the most faint-hearted economic historian little pause, but consider the impact of coal upon each state's development. The campaign to promote anthracite coal symbolized Pennsylvania's strong support of industrial growth, whereas Virginia's indifference towards her coal industry established an entirely different pattern. In examining the diverse paths of northern and southern states in regard to industrial development, previous historians have attributed sectional divergence to entrepreneurial culture, labour systems and even differences in soil and climate. The competition between anthracite and bituminous coal in the Early Republic suggests that institutional contexts, particularly state-level political economy, played a large role in the divergent paths of northern and southern states in the antebellum era.[21]

Of course, the comparison of 'northern' or 'southern' patterns of development inevitably leads to a discussion of the compatibility of slavery to industrial pursuits such as the coal trade. Many previous studies of southern 'distinctiveness' rightly suggest that slavery played an enormous role in shaping developmental paths, but fail to make the important distinction between the use of *slaves* in industrial endeavours and the impact of *slavery* upon a state's framework of political and economic institutions. Put quite simply, operating in a slaveholding society raised costs for colliers, and most policy makers in Virginia found the promotion of industrial growth to be, at best, distasteful and, at worst, threatening to the interests of slave-based agriculture. In the Richmond Basin, slavery facilitated a bifurcation of entrepreneurial initiative and the wider net of scientific, economic and political support required for industrial expansion. Slavery, while not the immediate culprit in the Richmond Basin's demise, thus undermined the ability of colliers in eastern Virginia to meet the nation's rising demand for mineral fuel.

In the Pennsylvania anthracite industry, elements of entrepreneurship and institutional support dovetailed. The use of free labour in anthracite mining certainly gave colliers in Pennsylvania a certain advantage over those in the Richmond Basin in head to head competition, but perhaps more important is the institutional support for 'stone coal' during the 1820s and 1830s. Beginning with the Franklin Institute and ending with the enlistment of state government in the anthracite campaign, Pennsylvania institutions actively promoted the use of anthracite for both economic and patriotic reasons, a promotional campaign that the political economy of slavery would likely have undermined in Virginia.

The campaign to promote anthracite coal in the 1820s and 1830s thus laid the groundwork for the strong support of industrial growth that would characterize Pennsylvania for the remainder of the nineteenth century. Private entrepreneurs, scientific institutions and state government all contributed in various ways to win anthracite's place in urban markets. Colliers of the Richmond Basin, in contrast, encountered major barriers when they most needed to expand production. They operated within a political and economic framework that limited their ability to secure adequate labour, incorporate new technology and expand transportation facilities for coal shipments. These critical precedents created initially different institutional contexts, not only for the future of the statewide coal trades in Pennsylvania and Virginia, but also for the economic growth of these two important American states throughout the antebellum period.

Notes

*   This chapter uses material from my book *Old Dominion, Industrial Common-wealth: Coal, Politics, and Economy in Antebellum America* (Baltimore, 2004).

1   Tench Coxe, *A View of the United States* (Philadelphia, 1794), pp.70–71; Letter of Benjamin Latrobe, in the appendix to 'Report of the Secretary of the Treasury, on the Subject of Public Roads and Canals; made in Pursuance of a Resolution of the Senate, of March 2, 1807', *House Reports, 17th Congress, 1st Session, 1821–1822* (Washington, 1822), p.64; Alexander Hamilton, *Official Report on Publick Credit* (Washington, 1790), pp.249–50; Albert Gallatin, 'Report on Roads and Canals', in *American State Papers, Miscellaneous*, vol. 1, *Documents, Legislative and Executive of the Congress of the United States* (Washington, 1834), p.760.

2   *United States Rail Road and Mining Register* (Philadelphia), 6 September 1856; *The Register of Pennsylvania* (Philadelphia), 1 August 1829; H. Benjamin Powell, *Philadelphia's First Fuel Crisis: Jacob Cist and the Developing Market for Pennsylvania Anthracite* (University Park, PA, 1978), p.25.

3   Howard Eavenson, *The First Century and a Quarter of American Coal Industry* (Pittsburgh, 1942), pp.426–36.

4   Richard Brooke to Harry Heth, 7 June 1805, Harry Randolph to Heth, 1 June 1812; H.B. Christian to Heth, 1 March 1819; Heth Papers, University of Virginia, Charlottesville, Virginia. For extended discussions on the market for hired slaves in the Virginia coal and iron industry, see Ronald Lewis, *Coal, Iron and Slaves: Industrial Slavery in Maryland and Virginia, 1715–1865* (Westport, CT, 1979), pp.81–103; and Charles B. Dew, *Bond of Iron: Master and Slave at Buffalo Forge* (New York, 1994), pp.67–70.

5   Samuel Paine to Richard Morris, 27 October 1801, Morris Family Papers; B. Randolph to Harry Heth, Manchester, 24 August 1810, Heth Papers, UVA.

6   Harry Heth to David Meade Randolph, 22 June 1814, Heth Papers, UVA; George Easterly to Harry Heth, 3 June 1810; Agreement Between Daniel French and Harry Heth, 30 November 1811; Harry Heth to Thomas Taylor, 31 January 1813, Oliver Evans to Harry Heth, 26 June 1813; 14 July 1813, Heth Papers, UVA; Oliver Evans to Harry Heth, 20 June 1815, Heth Family Papers, Virginia Historical Society.

7   'Account of the coal mines in the vicinity of Richmond, Virginia, communicated to the Editor in a letter from Mr. John Grammer, Jun.', *American Journal of Science*, 1 (1819), 127; Henry Heth to Harry Heth, 24 February 1818; Harry Heth to Thomas Railey and Brother, 11 August 1816, Heth Papers, UVA.

8   J.P. Plesants to Harry Heth, 8 July 1811; Thomas B. Main to Harry Heth, 2 December 1815, Heth Papers, UVA; Chesterfield County Petition, 15 December 1824, Legislative Petitions File, Library of Virginia.

9   Chesterfield County Petition, 23 December 1818, Legislative Petitions File, LVA; For more on the early history of navigation along the James River, see Wayland Fuller Dunaway, *History of the James River and Kanawha Company* (New York, 1922) and Carter Goodrich, 'The Virginia System of Mixed Enterprise: a Study of State Planning of Internal Improvements', *Political Science Quarterly*, 64 (1949), 355–87.

10   Statistics are from 'Second Auditor's Report, List of Articles Brought Down James River to Richmond', appended to the *Virginia House Journal, 1827–28*

(Richmond, VA, 1828) and 'Summary of Statements of Receipts and Expenditures on Account of the revenue from the Improvements under the Direction of the James River Company', appended to the *Virginia House Journal, 1829–30* (Richmond, VA, 1830); Dunaway, *The James River and Kanawha Company*, p.326.

11 Benjamin Silliman, 'Anthracite Coal of Pennsylvania, &c. Remarks upon its Properties and Economical Uses', *The American Journal of Science*, **10** (February 1826), 331–51; Marcus Bull, 'Experiments to determine the comparative qualities of Heat, Evolved in the Combustion of the Principal Varieties of Wood and Coal, used in the United States, for Fuel; and, also, to Determine the Comparative Quantities of Heat lost by Ordinary Apparatus, made use of for their Combustion', *The Franklin Journal and American Mechanics' Magazine*, **1** (May 1826), 257–93; the quotation is from Bull, 'Experiments to determine', 285.

12 *The Franklin Journal and American Mechanics' Magazine*, **9** (June 1830), 136; **12** (July 1831), 5–6; **13** (January 1832), 165–6; **15** (March 1833), 173–4; (April 1833), 237; **16** (July 1833), 89; (December 1833), 405–7; 421–3; Frederick Moore Binder, *Coal Age Empire: Pennsylvania Coal and its Utilization to 1860* (Harrisburg, PA, 1974), pp.19–21.

13 For more on the early history of the LCNC and SNC, see Theodore B. Klein, *The Canals of Pennsylvania and the System of Internal Improvements* (Harrisburg, PA, 1901), pp.4–7, 11–12; Edward Gibbons, 'The building of the Schuylkill Navigation System, 1815–1828', *Pennsylvania History*, **57** (January 1990), 13–43; and John Hoffman, 'Anthracite in the Lehigh Valley of Pennsylvania, 1820–1845', *United States National Museum Bulletin*, **252** (1968), 91–141.

14 Hoffman, 'Anthracite in the Lehigh Valley', 119–21. For a more detailed account of the anthracite canals, see Chester Lloyd Jones, *The Economic History of the Anthracite-Tidewater Canals* (Philadelphia, 1908).

15 H. Benjamin Powell, 'The Pennsylvania anthracite industry, 1769–1976', *Pennsylvania History*, **47** (1980), 11; Clifton Yearley, *Enterprise and Anthracite: Economics and Democracy in Schuylkill County, 1820–1875* (Baltimore, 1961), pp.51–5; Hoffman, 'Anthracite in the Lehigh Valley', 103–6.

16 *Register of Pennsylvania* (Philadelphia), 2 April 1831; *New York Daily Advertiser*, reprinted in the *Register of Pennsylvania* (Philadelphia), 4 May 1833. A good comparison of prices from 1827 to 1831 is found in Arthur Cole, *Wholesale Commodity Prices in the United States, 1700–1861*, vol. 2, *Statistical Supplement* (Cambridge, MA, 1938), pp.224–40.

17 *Pennsylvania Archives*, 4th ser., vol. 5, pp.879–86. See also Philip Shriver Klein, *Pennsylvania Politics, 1817–1832: A Game Without Rules* (Philadelphia, 1940), pp.336–7; *Miners' Journal* (Pottsville, PA), 21 March 1830; *Pennsylvania House Journal, 1830–31*, **1** (Harrisburg, PA, 1831), 93; *Report of the Committee of Ways and Means, Relative to the Finances of the Commonwealth. Read in the House of Representatives, January 19, 1831* (Harrisburg, PA, 1831), p.15; *Miners' Journal* (Pottsville, PA), 26 February 1831; *Pennsylvania House Journal, 1830–1831*, **1**, 3–5.

18 *Laws of Pennsylvania, 1835–36* (Harrisburg: Theo Fenn, 1836), pp.799–803; *Pennsylvania Archives*, 4th ser., vol. 6, 287–8. The roll call vote is in the *Pennsylvania House Journal, 1835–36, Volume I* (Harrisburg, PA, 1836), pp.1414–15; 1423–4.

19 *United States Commercial and Statistical Register* (Philadelphia), November 1839, p.368; April 1840, pp.230–31; July 1840, p.32; November 1841, p.203; Walter R. Johnson, *Notes on the Use of Anthracite in the Manufacture of Iron. With Some Remarks on its Evaporating Power* (Boston, 1841), pp.32–65; *United States Commercial and Statistical Register* (Philadelphia), March 1841, p.207. For a partial list of the corporations created under the 1836 Act, see Records of the Corporations Bureau, Pennsylvania State Archives, Harrisburg, Pennsylvania.

20 *Register of Pennsylvania* (Philadelphia), 15 November 1828; *The American Journal of Science*, **33** (January 1838), 200; William Pope to Harry Heth, 4 April 1812, Heth Papers, UVA.

21 Most works that attempt to explain divergence actually focus upon the southern economy rather than presenting the problem in a comparative analysis. See, for example, Thomas Doerflinger, *A Vigorous Spirit of Enterprise: Merchants and Economic Development in Revolutionary Philadelphia* (New York, 1986), pp.335–64; Eugene Genovese, *The Political Economy of Slavery: Studies in the Economy & Society of the Slave South* (New York, 1965); Gavin Wright, *The Political Economy of the Cotton South* (New York, 1978); Fred Bateman and Thomas Weiss, *A Deplorable Scarcity: The Failure of Industrialization in the Slave Economy* (Chapel Hill, NC, 1981); and Frederick Siegel, *The Roots of Southern Distinctiveness: Tobacco and Society in Danville, Virginia, 1780–1865* (Chapel Hill, NC, 1987). Significant exceptions to this trend that employ a comparative analysis include John Bezis-Selfa, 'Forging a new order: slavery, free labor, and sectional differentiation in the mid-Atlantic charcoal iron industry, 1715–1840', PhD thesis, University of Pennsylvania, 1995; and John Majewski, *A House Dividing: Economic Development in Pennsylvania and Virginia Before the Civil War* (New York, 2000).

CHAPTER EIGHT

Nigerian Coal Miners, Protest and Gender, 1914–49: the Iva Valley Mining Community

Carolyn A. Brown

Historians of both European and colonial African labour have, all too often, failed to appreciate the connections between colonial workers and their metropolitan counterparts. The history of the Nigerian coal miners, a small but strategically placed segment of the colonial working class, suggests that we need to relax the conceptual barriers between colony and metropole if we are to comprehend fully the ways in which the miners struggled with capital in the workplace, in their homes and in their communities. This chapter, a study of the Enugu miners, shows how colonial labour policy, informed by the British state's bruising encounters with its miners at home, tried to prevent the formation in Africa of the type of coal mining villages that had proven so conducive to the growth of a militant class consciousness in the British coalfields. It also highlights how the Enugu miners were able to resist the attempts at social engineering that took place as colonial policy makers tried to reconstruct the African family.

The Enugu Government Colliery, 1914–39: the Genesis of the Colonial Workplace

The Enugu Government Colliery was located in south eastern Nigeria, and was the British state's first experiment with a nationalized coal mining industry. When it opened, during the First World War, its coal replaced that of Wales, which was the primary energy source for the steamers plying the West African coast. The colliery consisted of three separate mines. The first, Udi Mine, opened in 1915 and the second, at Iva Valley, in 1917; Obwetti Mine was opened shortly thereafter. All three functioned until 1936, when the old Udi Mine was closed. For the period of this study, the Iva and Obwetti mines were the principal workplaces. The mines were the first capitalist enterprise in this area of Nigeria and transformed rural social and gender relations so significantly that mine work became a central pivot of masculine identity in the labour source villages.

Most workers in the Enugu Colliery were Igbo, one of Nigeria's three largest ethnic groups. They lived in autonomous villages within a highly decentralized sociopolitical structure without chiefs or a cohesive pre-colonial state. Before the opening of the colliery most labour was performed by families in self-sufficient farms which produced foodstuffs both for this region and for communities in the south that focused on palm oil, a principal export for German and British merchants. Two-thirds of the colliery's workers were 'locals' who lived on the eroded Udi escarpment within 15 miles of Enugu where severe land shortages prevented reliance on subsistence farming.[1] When the colliery opened, mine work replaced slavery.[2] It also ended the pre-colonial migration of young men who moved to work on farms elsewhere.[3] Among the last third was a second group of 'locals', who lived on the fertile plateau to the east of Enugu. For them, mine work competed with commercial farming, made especially lucrative with the development of Enugu. They saw mining as a job of last resort: 'A man might as well be buried alive.'[4] The few employed in the colliery would only work in the less dangerous surface jobs while the former group worked underground. In this way work became an important factor shaping a colonially imposed regional identity.

Men used income from mine jobs to fulfil rural masculine norms. As young unmarried men, they came to the mines to earn income to pay bride price. Later, as married men, they worked to earn the fees to join prestigious rural associations called 'title societies' through which they could exert political influence within the village.[5] Rural masculinity was not just physical but also moral, with many contradictory norms:[6] aggression, perseverance, stoicism, restraint but determination, achievement through hard-fought effort, bravery and courage, and generosity. A man had a social obligation to 'uplift' his family, village and village group. This ranged from helping the less fortunate with resources from one's household to sponsoring some development project in one's village. Either action brought notoriety as a 'big man' and benefactor and established important ties of obligation of benefit in village politics. Thus, while individual success and achievement were compelling male norms, a responsible man was expected to contribute to the common good.[7]

Miners also used their incomes to buy prestige goods such as bicycles and certain types of Western clothing and to dispense patronage and largesse to their neighbours and family. They financed social rituals that were important for village cohesion and personal status. To these miners, being a 'responsible' worker did not necessarily mean being an industrially disciplined man. They were discharging largely rural obligations. One miner who worked in the early years cited these obligations as an incentive for entering the mines:

> I joined the mine when I felt like. In those days what normally happened was that whenever the villages wanted to celebrate certain feasts, people will rush to the

mine to get money; that is, to work for sometime and get money for their feast. When you must then you retire home to celebrate the feast with the money you got from mining work. When the money finished you go back again . . . [8]

The industry also fostered the development of an urban identity in the colonial city of Enugu, which was established as an administrative and commercial centre. There, under the influence of missionaries and 'foreign' African men from the Westernized elite, inhabitants came to see themselves as more 'modern' than their rural counterparts. They fused eclectic gender ideologies of the British middle and working classes with the complex gender roles of the African clerks and artisans who had converted to Christianity.[9] These ideologies were subject to profound contradictions. From its establishment in 1914 through to the nationalist period, the city was known as 'Red Enugu' because of the popularity of radical nationalist agitation among its working class and the militant protest traditions of the coal miners. The city evolved from a collection of mining camps in 1914 to become a centre of regional government in 1938.

The Contested Terrain of Labour Camps: the Violence of Early Class Formation

When the mines opened during the First World War there were three camp sites: 'Coal Camp' and two 'unofficial' camps, Ugwu Alfred (Alfred's Hill) and Ugwu Aaron (Aaron's Hill), which were administered by 'native' labour contractors.[10] Africans lived in 'bush' camps in the valley near the Udi escarpment, and Europeans in temporary housing high on a ridge overlooking the future site of Enugu. In the opening years the camps were run by powerful men, designated as colonial 'chiefs', whose abuses were protected by a weak colonial state. The men who became 'chiefs' were the result of a social formation created by centuries of involvement in the slave trade. There were significant social hierarchies in the countryside. The imprint was visible in the early twentieth century both in the existence of large slave communities and in the presence of wealthy merchant capitalists who had sold and employed slaves. After initial resistance, they welcomed the opportunity to collaborate with British colonialists. The transition from slave owner/merchant to labour recruiter was smooth and often overlapped as they conscripted slaves to be 'workers' through forced labour. Nearly all of those chiefs who supplied labour to the mines had close connections with the internal slave trade. The suppression of the international market in slaves in the mid-nineteenth century was somewhat compensated for by an increased internal demand for slaves to produce those exports designated as 'legitimate' crops: for example, palm oil, kernels and, later, groundnuts in the north. The area adjacent to the mines was one of the few areas in Africa where a rigidly institutionalized slavery system had evolved. Once people were designated as slaves they were virtually unable to escape. They and the freeborn poor were a vulnerable group victimized by corrupt chiefs who conscripted them for labour in the early phase of industrialization. Thus, for many subalterns, mine work transformed them from one subordinate social category to another.[11]

When the Iva Mine opened in 1917, the first camp, Iva Valley No. 1, had 400 bush houses and was about three and a half miles from Enugu township and one and a half miles from the mine entrance. The second camp, Iva No. 2, located on the Iva River, began as a prison for some 400 men who worked stacking coal at the mine entrance. The prison closed in 1925 and the colliery assumed control over the camp. The prison was converted to labourers' quarters and was then called Prison Camp. The Iva camps differed from those under control of chief recruiters in that their people fiercely resisted any attempts to put them under appointed urban chiefs. In fact, by the 1920s, both the camps and the township had become centres of anti-chief resistance and a refuge for young men and women who were persecuted by the chiefs in the countryside. Despite the attempts by collaborative chiefs to convince authorities that they should be given jurisdiction over the urban and labour camp areas, the inhabitants were effective in preventing this extension of an 'invented' tradition.

The first coal settlements reflected the heterogeneity of the workforce, the militarism of the early years (south-eastern Nigeria was only subdued in 1912–14), and the nascent state's reliance on the cooperation of corrupt local chiefs who became crucial allies both in the recruitment of labour and in its control and housing. The violence of the conquest combined with these new groups of local rulers to create a context of overarching repression that made it difficult for workers to constitute themselves into communities and to develop organizations to articulate their concerns. Initially the form of resistance was desertion; colonial officials found that it was one thing to dragoon labourers to the worksite and quite another to keep them there.

The workforce was an eclectic collection of forced and voluntary unskilled labour, prisoners, unskilled contract workers, voluntary clerks and artisans and a handful of European managers and skilled workers.[12] A collaboration of convenience emerged in which the colonial officials turned a blind eye to the methods the chiefs used to 'recruit' workers and the methods that they employed to maintain order in the camps. The chiefs dragooned their villagers to the mines, using paramilitary forces and taking advantage of British administrators' erroneous assumptions that the authoritarian chief was an appropriate leader for a rural African society. Protected by the state from the wrath of the communities they oppressed, these chiefs 'invented' elaborate traditions of power that had no basis whatsoever in pre-colonial Igboland.[13] The state, for its part, gave them the authority to run the labour camps, paid them to supply food and gave them wages to distribute to their workers. The abuses were predictable and severe, provoking, on at least one occasion, an uprising that almost crushed the new colonial state.[14]

One could hardly call the labour camps 'communities', as most men were forced into the workplace, often absconded at the first opportunity and were most often victimized by the chiefs who were given carte blanche to govern them as they willed. But by the end of the decade the powers of these recruiters had been weakened both by the attraction of wage employment and by the opportunities village men had to run away and seek out jobs on their own in the township. While the unskilled workers were subjected to the repressive policies of the chiefs in their camps, clerks and artisans were able to organize protests and to adapt 'traditional' organizational structures to support their position as workers. One indication of the syncretic nature of urban gender construction during the First World War was the establishment of all-male urban improvement unions by colliery and railway clerical workers and artisans. These men considered themselves to be a source of 'modernization and enlightenment' for rural villages. They opposed the chiefs and developed a critique of the type of neo-traditionalism that the state was building around 'chiefs'. Originally called *Nzuko*, or 'meetings', these organizations combined consultative village political traditions with self-help initiatives to socialize new immigrants, funnel money into rural improvement projects, and to assist the state in preserving urban order.[15]

The value of the *Nzuko* as a blueprint for a modern associational structure was recognized by coal workers and they transformed the *Nzukos* into workers' organizations that met to discuss 'food shortages, token cheating, bribery and corruption in the obtaining and retaining of jobs and in the allocation of work on hard or soft coal faces'.[16] These meetings, called *Nzuko ifunanya*,[17] were structured by job category with each class of men holding their own meeting under a president and secretary. They functioned as a negotiating body and held periodic meetings with the management. Nonetheless they had clear conceptions of who could be trusted to represent their position. In important respects this pattern of worker consultation with management was an indigenous model of worker representation in the capitalist workplace. The *Nzuko* fiercely defended their autonomy and reacted violently towards management spies or 'snitches'. Occasionally meetings were held in the mines, where secrecy was maintained by physically stopping Europeans from entering. 'Boss boys' and other staff were also prevented from attending meetings. The *Nzuko* used oaths to ensure solidarity and to deter informers.[18] They were just the type of workers' organization that the Colonial Office feared most – autonomous and potentially volatile groups that operated in semi-secrecy.[19]

Space and Power: the Contested Nature of the Colonial City

In the first three years of the First World War, the city of Enugu did not formally exist. There was only a military outpost at Udi and a cluster of labour camps near the mine entrance and at the site of the railway station. During this period, when conditions were so rudimentary, political authorities like Sir Frederick Lugard, the then governor-general of Nigeria, was preoccupied with establishing order.[20] The racialized system of authority implicit in colonialism was a crucial element in the ideology of power. But, for race to become an important signifier of power in a recently conquered area with so few Europeans, colonial 'whiteness' had to be given meaning and authority. These ideas influenced the design of the city. Assertions of colonial power were especially important when there was so little difference in the material conditions under which colonizer and colonized lived. Both lived in 'bush' houses, with thatched roofs and dirt floors.[21] But even before the town was formally designed with spatial boundaries symbolizing the social boundaries between 'European colonizer' and 'African subject', Africans had to interpret the political meaning of European settlement high on the Udi escarpment. In case the significance was missed, there were the 'hammock boys'. The underground manager, overmen, mechanics and electricians lived on the escarpment. Every working day, from this period until after the Second World War, the men had to carry them in hammocks to the mines.[22] So dramatic was the impact of this ritual of racial power and privilege on the men's consciousness that, 30 years later, nearly every miner cited this practice as an example of racism in the mines.[23]

Between 1914 and 1917, even before the settlement was legally recognized as a township, hundreds of men were drawn to the site by the unrest accompanying the conquest in the countryside and in anticipation of the arrival of the railway and mining operations. Additionally, there were the artisans and clerical workers who had been trained in Church Missionary Society schools in Onitsha, Owerri and, to the west, in Lagos.[24] They, and those unskilled labourers who were dragooned by chiefs, were the first settlers in Enugu. The clerks and artisans created a cosmopolitan community with links to the Lagos elite that had mobilized so vigorously against the conquest and discrimination in government employment.[25] Many were Yoruba converts who held their own weekly church services, while others were Igbo clerical workers from Owerri province, about one hundred miles south, and from Onitsha town. They made Enugu a largely Igbo city of 'strangers' who were drawn from the more developed and earlier colonized areas of Nigeria. Even in the early days, the contours of an African political culture were visible. It was distinguished by eclectic concepts of 'freedom' which for many meant autonomy from oppressive rural chiefs. Urban men created a culture that, by its very existence, challenged Lugard's imaginary Africa which had no place for city dwellers. The *Nzuko* also attracted unskilled workers who adopted their critique of rural authoritarianism. Those miners living in the camps were able to create a society free from the chiefs, a society that approximated to ideals of 'freedom'. The impact of these new ideas was reflected in the discourse of protest in the 1930s, which is discussed below.

The formal layout of the town was overseen by a senior health officer with frequent interventions by Lugard. Lugard's ideas on urban design were representative of early twentieth-century British theories about urban planning which were a reaction to the nineteenth-century industrial city.[26] They emphasized 'health, light and air', in a type of 'environmental determinism' which sought to resolve 'social pathologies' by the manipulation of physical space.[27] Segregation, largely but not exclusively racial, was an important colonial modification of these general principles.[28] As an architect of empire, Lugard clearly understood the importance of creating the Manichean worlds of the colonizer and colonized. The order of the European reservation was a symbol of the cultural superiority of British society. It is contrasted with the disorder, dirt and chaos of the African area, the 'Native town'. Africans 'needed' the British to save them from themselves. Lugard's most elaborate musings on the spatial requirements of this racialized authority were expressed in intricate and obsessive detail in a memorandum on 'Townships', where he explained the racial protocol of colonial Nigeria to his political officers.[29]

Cultural, racial and class boundaries were especially important in spaces shared by Europeans and Africans, colonizers and subjects, whites and blacks. In those spaces where Africans and Europeans were in daily contact (the city and the colonial workplace) such 'markers' had political significance. Lugard's ideas of racial segregation exemplified a second characteristic of British colonial design, the obsession with health, or 'the sanitation syndrome'.[30] He detailed a mile wide *cordon sanitaire* between the 'nearest European dwelling' and the 'native area', a separation that he wrote of in his *Political Memorandum*. In an exemplary statement of this racialization of early health theories, he used the discourse of public health and social hygiene:

> The first object of the non-residential area is to segregate Europeans, so that they shall not be exposed to the attacks of mosquitoes *which have become infected with the germs of malaria or yellow fever*, by preying on Natives, and especially on Native children, whose blood so often contains these germs. ... Finally, it removes the inconvenience felt by Europeans, whose rest is disturbed by drumming and other noises dear to the Native.[31]

But Lugard, like most colonial officials, found that it was one thing to conceptualize a colonial city to represent power relations that colonial rule *wanted* to create and quite another to make those power relations *actually exist*. As African men and women came into Enugu they also had ideas about the type of city that should exist and they did not always coincide with the vision that colonialists held. Enugu, like any colonial city, was a contested space whose reality was produced out of the struggle between African colonial subjects and the agents of the state.[32]

Colonial policy regarding workers' settlements had, at its base, political concerns. In the early decades of colonial rule, officials feared that Africans settled in camps and townships were dangerously autonomous and potentially disruptive. If they cut their connections with the rural economy in periods of economic slump, they would be unemployed, rootless and, as such, ripe for political radicalization. Colonial officials were imprisoned by these assumptions and, even though the *Nzuko* helped them to police the city, old conceptions of transgressive Africans proved resilient. It was difficult for officials to see African society as capable of self-generated change, while the notion of an 'urban African' seemed a contradictory concept. Nonetheless, officials appeared oblivious to the reality around them: a generation of Africans had been born and was growing up in the townships and camps. They were no longer peasants.

The colliery *Nzuko* were an effective means of organizing and executing workers' protest. In the 1920s, workers' protest shifted from desertion (that is, a rejection of capitalist production relations) to more conventional forms of worker agitation. Strikes, sabotage and stay-aways occurred with increasing frequency and effectiveness in the inter-war period. Workers protested against abusive supervisors and the corrupt interpreters, as well as the deplorable conditions under which they worked. When the Depression hit the industry, many workers returned to the village to await a recovery (which occurred in the mid-1930s).

The most successful period of protests came in the late 1930s. By that time the miners ('pit boys') had earned a reputation for being particularly recalcitrant and difficult to control. Labour historians often associate mining with a type of rough-hewn masculinity.[33] In Enugu, the possession of mining skills, the danger of the workplace and 'the aggressive celebration of physical strength' were key contributing factors to the masculine work culture, as was the case in European coal mining. Hewers, proudly called 'pit men', had a self-identity which stressed that they were skilled workers and were independent, self-improving men with a work culture that resisted attempts at supervision.[34] These values defined a social hierarchy within mining communities both in Enugu and in Britain. In Enugu, the hewers' self-identity was encouraged by an extraordinary degree of workplace autonomy, the absence of mechanization and the fact that they were allowed to function as labour contractors.[35] Consequently it was no coincidence that the agitation of the late 1930s began with the miners.

The *Nzuko* were an institutional expression of this culture of combativeness and solidarity that had been encouraged by the workers' masculinity.[36] The leaders were men with 'good mouth',[37] steadfast in resisting unjust authority, and diligent in attacking problems. Their authority over the workers depended on their ability to deliver what the men demanded, whether increased wages, prosecution of an abusive boss or better housing. Management saw the leaders as demagogues, men of 'energy, personality and slight extra knowledge of affairs', who manipulated the 'ordinary unsophisticated local men',[38] but to the men they were an important element in their strategies to resist authoritarian abuses in the workplace.

The economic recovery, and the gathering tensions in prewar Europe, gave the workers an opportunity to make the Colonial Office take notice of their plight. In March 1937, the hewers called a strike to recoup wage losses during the Depression. But this time they attacked the conditions in the camps by asserting that they were immoral. They argued that the camps exposed their families to unhealthy, sub-standard conditions, and that they were beneath their status as working men. At that time the colliery camps were free and offered 'foreign' workers an important alternative to high-cost rental housing in Enugu. But the camps were built in the style of range-type housing used in other parts of the empire for indentured workers. They were a row of rooms 12 feet square and 10 feet high, separated by partial partitions and topped with corrugated roofs. They were insufferably hot during the day and had an average of 5.5 inhabitants.[39] Although this style of housing had been declared unacceptable for family life by both the International Labour Organization and the British government, it was left to the workers themselves to alert colonial authorities. The houses violated Igbo marital norms because children and parents slept in one room. The men considered the quarters to be demeaning and immoral. Perched at the top of the production hierarchy, the hewers felt morally compromised by the conditions in which they were forced to place their families. They were especially outraged by the necessity of boarding a single 'pick boy' with their family or of sharing a room with another family. In a petition they reminded the government that they wanted the 'new construction of quarters with more accommodative rooms than the present ones enough to live in, that is to say, quarters of providency of sufficient rooms ... to accommodate a workman and his wife alone'. They also recognized that such overcrowding was a health hazard and complained that 'our current mode of living is rather an inconvenience and far out of sanitation so well'.[40] Officials had to admit that overcrowding meant that they had failed to establish control over the pace of urbanization. The Colonial Secretary, W.G.A. Ormsby-Gore, was outraged that a government-owned enterprise should tolerate such conditions.[41] With war on the horizon, colonial labour policy was too important to be entrusted to local officials who proved especially resistant to the new trends in labour and social policy. The Colonial Office would have to step in.

The Colliery during the Second World War

The war demanded that the state make extra efforts to contain the contradictions of colonial society. Urban overcrowding was severe as Enugu's population grew from 13 000 in 1939 to over 40 000 by 1945. Clearly the city needed to be regulated and, as officials paid more attention to urban life, a series of conditions came under scrutiny. State authorities attacked the colliery for its daily deployment of labour and its wasteful system of keeping more men employed than was strictly necessary as a means of compensating for high absenteeism; this, critics argued, was a major cause of overcrowding. Others blamed the men who tried to bring rural symbols of status (polygynous households) into the city despite space constraints. Even the men who boarded in the camps during the week and returned home on the weekends came under attack. Their wives remained in the rural areas and a new type of household was created that preserved women's domestic roles without women; boy 'servants', were brought in. It was not uncommon for several men to share one boy. Officials considered adolescent boys that were unsupervised during the day to be a dangerous stratum. Thus African constructions of family life clashed with colonial notions of urban order. The war increased the pressure on the state to act. It did so by attempting to reconstruct the African urban household.

By the outbreak of the Second World War, concern with the radical potential of coal mining villages was eclipsed by the state's desire to control them. The consensus was that dangerous slums, not mining villages per se, were the true threat to the social order. The exigencies of war required that the labour force be stabilized, disciplined and controlled. However the severe conditions of the war incited further worker protest. The war proved to be a rude awakening to colonial policy makers in both London and Paris because it brought them face-to-face with their dependence on colonial production. After 1940, with most French Africans living under Vichy rule, Britain could not assume the loyalty of her African subjects. She had to create it. Propaganda, a necessary element of the productionist programmes of both French and British colonies, had the unintended result of empowering African workers and peasant producers by underscoring the importance of their labour for the survival of Empire.

Now policy makers realized the limitations on their control over labour recruitment, the labour process in the colonial workplace and the socio-political role of workers within the expanding and politically volatile urban classes. The contradictions were many and acute, the most fundamental being the economy's reliance on the cooperation of African labour and the state's refusal to give these same people a meaningful political role in their country. The wages of most African workers had not recovered from the Depression and, with inflation running at some 400 per cent, it was difficult for them to survive. This guaranteed unrest just when colonial resources were most crucial to the very survival of empire.

When Britain joined the war, West Africa became strategically important for her campaigns in the Middle East, and later in Asia. At Ghana's Takoradi port, British planes were offloaded, assembled and flown to the Middle East through Nigeria and the Sudan. From late 1940 until the attack on Pearl Harbor, under the Trans-African Air Base Programme, American and British airplanes were assembled in, and deployed from, newly constructed airstrips. More than 100 000 British, and several thousand American, troops came through such cities as Lagos, Kano, Ibadan and Enugu.[42]

West Africa was also an important part of imperial economic plans. Following the Japanese victories in South East Asia in 1942, West Africa became the main supplier of tin and tropical food products. These exports helped Britain to earn over £3000 million in sterling balances. Colonial exports also assisted in war preparation, both as supplies for British factories and for sale outside the sterling area in exchange for war materials from the United States.[43] Additionally, thousands of West African soldiers fought for Britain to reclaim Ethiopia from the Italians, and several hundred thousand participated in the Burma campaign from November 1943 to September 1945. With the Japanese victory in Asia, West Africa became important as a military transit stop in the defence of India and prosecution of the war in Burma.[44]

Like most of the colonies, Nigeria shouldered more than its fair share of wartime economic dislocations, and bore the brunt of both fiscal and general economic policy. Imperial economic policy was primarily concerned with control of the price of production in Britain through, firstly, the regulation of the price of colonial exports; secondly, the reduction of the amount of metropolitan and foreign imports to the colonies; and, finally, the accumulation of sterling and dollar reserves to finance the war effort.[45] All these policies had the impact of worsening the already beleaguered colonial working class. These elements of colonial wartime economic policy had drastic implications for the standards of living of colonial workers throughout the empire. Workers at the Enugu colliery were no different. The escalating costs of both essential and luxury goods were an obvious consequence of shortages, and inflation rates rose steeply.

But the war was a watershed event for African political activism. While the policies of colonial governments brought home the harsh reality of colonialism to broad sectors of the African population, it also gave colonial peoples the political and economic weapons to destroy it. African workers were at the centre of an imperial war strategy, supplying the military and civilian labour needed to prosecute the war and to conduct the productivity drives that delivered unprecedented amounts of colonial agricultural and mineral resources to the war effort. This experience, and the propaganda used to mobilize them, increased workers' awareness of their importance to the metropole and the potential that they had to bring the economy to its knees. Given this realization, colonial officials struggled to explain why Africans should make sacrifices for principles that gave freedom to European states when they were denied these freedoms in their own societies. These and other contradictions proved even more untenable in war than in peacetime. Within 15 years of the end of the war, virtually all of France's and Britain's African colonies were independent.

The war also emphasized the commonalities of coal mining, whether in the United Kingdom or Nigeria, and the men were even more interested in how their wages, conditions of service and living conditions compared with those of British and American miners. Increasingly, they behaved as part of an international brotherhood, a brotherhood of miners, acutely aware that they produced the primary energy resource needed in the waging of the war. Despite the special depredations of colonial racial polices, comparisons came easily with the information available in wireless broadcasts as governments worked hard to ensure the loyalties of their West African subjects. There were several strikes incited by wireless reports of gains by American or British miners. Enugu miners shaped their demands within an ideological context of 'similarity' not 'difference' from European coal miners. Gender ideologies reflected creative shifts in the ways that the men were appropriating British domestic discourses. This was the case with the male breadwinner norm (which hardly represented the realities of the African experience), which was used to frame demands for improved wages and conditions of work. This conscious manipulation of gender ideologies is but one challenge to the artificiality of the metropolitan/colonial dichotomy.[46]

Struggles over the 'Modern' African Family: the State Ventures into the African Bedroom

Both the war and workers' reaction to deteriorating living conditions forced colonialists to become more familiar with, and involved in, African society. They attempted to understand the conditions under which the social reproduction of the urban working class occurred and the intricacies of rural structures of power that directed labour systems and the production of strategic minerals and valuable tropical crops. In the workplace, policy makers solicited the assistance of a new breed of 'colonial labour experts' and the deployment of social scientists to calculate the quantitative dimensions of urban poverty, rural subsistence and the real costs of products that workers deemed essential to their lives.

The Colonial Office recognized that it was time to force colonial officials to establish more control over the workplace. The 'best' way of controlling Nigerian workers had not yet been determined so the Colonial Office fumbled along, experimenting by blending such British industrial relations institutions and processes as trade unions and collective bargaining with a large measure of colonial authoritarianism. Thus, while unions were legalized, their leaders were victimized. As a precondition for the introduction of collective bargaining, the Colonial Office ordered the Nigerian government to allow trade unions. It assumed that this would be the best way to prevent secretive organizations, such as the *Nzuko*, from operating outside the legislation governing labour relations. At the colliery, two unions were formed. One, the Colliery Surface Employees' Union, was dominated by clerks, and the other, the Enugu Colliery Workers' Union, included the underground workers. Elsewhere I have documented the stormy history of these unions and the attempts by both the Colonial Office and local officials to prevent them from being effective organizations of workers' activism. Nonetheless, the underground union, which came under the leadership of a radical nationalist leader, was very effective in using the state's own labour reforms against it.[47] Without these modifications the colonies would not be able to achieve productivity targets so necessary for the war. Additionally, the Colonial Office wanted the unions to understand the connection between 'disciplined' workers and stable working-class communities. As I have demonstrated elsewhere, Enugu miners had considerable control over the labour process.[48]

These productivity imperatives, and Britain's desperate need to increase the exploitation of its colonies, pulled African people into intimate contact with coercive state policies to expand colonial production. Intrusive administrative directives and social policies brought the state into their lives in more direct ways than had previously been the case. As these policies of social engineering took shape, the state challenged workers on the composition of the African family. While many of the men expressed their prominence as modern workers by heading polygynous families, the state attempted to restructure households by making residence in new housing contingent upon the adoption of the nuclear family model with a male breadwinner. Throughout West Africa, workers' struggles during the war focused on the definition of family and its entitlements as well as the boundaries of acceptable protest. Nonetheless, the inhabitants retained their identity as a militant mining community.

In the 1940s, the state renovated the colliery settlements as an example of the enlightened form of state labour policy. Inherent in the design was a challenge to the African family, an extended and often polygynous household, that often included distant relatives and non-kin. These households, which were a sign of male status within the rural idiom, were considered the cause of urban overcrowding by colonial authorities. Thus the hidden purpose of the new housing estates was to encourage nuclear family life which was considered the key to the creation of a 'responsible' working class community. Iva Valley's camps were modelled on a 'controlled' British mining village, with small cottages, adequate space for gardens, and social services. This experiment in social engineering reflected the state's wartime concern that deteriorating socioeconomic conditions would radicalize key groups of workers and push them into coalitions with the dangerous classes: those 'undisciplined' slum dwellers, whose contempt for industrial discipline, propensity for crime and subversive work habits gave the state increasing concern. A combination of the stable and, for the war effort, crucial urban working class with the city's lumpenproletariat was a worrying prospect, a recipe for urban disorder, political unrest and imperial vulnerability. It was important to fracture the working class by dividing casual workers from the more committed workers. Colonial labour officials believed that by encouraging Africans to form stable, nuclear families they could produce the type of efficient, disciplined worker that the industry needed. Moreover, by creating a model mining estate, in which they would regulate the occupancy rate of rooms, policy makers tried to encourage subtle changes in African home life. Once in the estates, African families were to be subjected to a number of other regulations to change their nutritional habits, expose them to modern medical care and provide recreational activities that would structure their leisure and working time.

For the first time officials studied African domestic arrangements and proposed policies to encourage a familiar stability based on the male breadwinner norm. Workers astutely appropriated this discourse to support their arguments for improved wages and conditions. In 1941, workers' demands resonated with the state's realization that undifferentiated poor conditions facilitated dangerous political alliances. In November 1942, the treasury approved a £104 000 loan for the construction of 461 semi-detached houses, providing 922 rooms.[49] From its inception, the project was envisioned as a model of state ownership and a demonstration to Africans of the correct, 'modern' way to construct houses. Construction began in February 1943. Orde Browne, like many in the Colonial Office, perceived the project as an 'object lesson on the possibilities of government ownership, both as a business proposition and in maintaining satisfactory conditions for working people'.[50] The plan was drawn up by a committee of all major government agencies involved in land use and social welfare, as well as labour officials and colliery welfare staff. But at no point did it include either the workers or their representatives in the unions.[51]

The estates were an excellent example of an 'imagined' British mining village, an illustration of the new fantasies colonialists entertained for the world of African workers: two-room cottages on plots 40 feet by 60 feet with verandas, electric lights and small gardens. The camps had wide roads, open spaces, bathrooms, social halls and chlorinated water. They were a demonstration of the superiority of British planning and of modern family living. Officials ignored the miners' own modern homes built with zinc roofs in the village. It was assumed that, once the communities saw how beautiful and modern these buildings were, they would modify their own construction in the villages.[52] Officials assumed that the estates would not become the rowdy type of British pit village that encouraged solidarity and militant worker unrest. They modelled the camps instead on the 'garden city' type of town planning: ordered houses, bright and airy with sufficient vegetation. In fact, the Udi Siding Camp was called Garden City.[53]

Female welfare officers were to supervise women and children, operate children's clinics and enter homes to teach nutrition and health. Initially, the wives of the staff welfare officer and deputy manager held these posts and also visited adjacent villages.[54] However, it was one thing to design the physical space that workers and their families occupied; it was much more difficult to make them occupy them according to the abstract standards developed by remote colonial labour experts. The men merely reminded the officials of 'the substantial quota the colliery workers were contributing towards the war effort'.[55]

The central conflict centred on occupancy rates. Only three people could occupy each room in the largest camp, Garden City, thus making the ideal polygynous household an impossibility. At a meeting with the Senior Resident and Acting Manager, Augustine Ude of the Colliery Workers' Union complained that, under the occupancy limits of three people per room in Garden City, residents would have to leave part of their family 'behind or throw them away'. The men also objected to the intrusions of welfare officers. The unions resented these intrusions and asked that the woman welfare officer stop 'supervising and molesting their wives when they went to work'.[56]

When their complaints went unanswered they organized a boycott of the estates and fined people who agreed to occupy the estates. The boycott embarrassed local planners and attracted the unwanted attention of the Colonial Secretary. When only one hundred people consented to live in the houses, the Colonial Secretary complained that 'no one had apparently attempted to find out the sort of accommodation which, while satisfactory from the health point of view, would be acceptable to the Africans'.[57]

Conclusion

Today the Iva Valley camps are but a sad reflection of these early days. The mining industry is no longer the central source of fuel for the railways. In shipping too, coal was eclipsed by petroleum in an expression of yet another industry marred by worker and community struggles. The order of the wartime camps was captured in a set of photographs in the Colonial Records Project, Rhodes House, Oxford University. They clearly exemplified the arrogance of colonial planning, the power to implement a fantasy and the assumption that African workers themselves could not possibly imagine, let alone create, a modern settlement. These neat cottages, broad streets and open communal spaces are today overgrown and dilapidated. They contrast with the villages whose development projects the miners both conceptualized and implemented. These communities are vibrant, 'modern' and well maintained. These, more than the Colonial Office's arrogant fantasy, are examples of the men's conceptions of 'modern' Africa. But the camps still have many men and women with valuable memories of the period when this fierce little community was in the forefront of the labour struggles in colonial Nigeria. When the colonial army killed 22 coal miners engaged in a mine occupation the people of this little settlement bravely erected barriers on the road to prevent further attacks. The incident, called 'The Iva Valley Shooting' is today considered the birthdate of the Nigerian nationalist movement.

Notes

1   See Nigerian National Archives Enugu (hereafter NNAE), P.E.H. Hair, 'Enugu: an African city', mimeo (1954).
2   David Eltis, Stephen D. Behrendt, David Richardson and Herbert Klein (eds), *The Transatlantic Slave Trade: A Database on CD-ROM* (Cambridge, 1999).
3   G.I. Jones, 'Igbo land tenure', *Africa*, **19** (1949), 309–23. For a discussion of the importance of slavery in this region, see C. Brown, 'Testing the boundaries of marginality: twentieth-century slavery and emancipation struggles in Nkanu, Northern Igboland, 1920–1929', *Journal of African History*, **37** (1996), 51–80.
4   P.E.H. Hair, 'Enugu', 10.
5   A.E. Afigbo, 'South eastern Nigeria in the nineteenth century', in J.F.A. Ajayi and M. Crowder (eds), *History of West Africa*, vol.2 (2nd edn: New York, 1974), p.442.
6   Dunbar Moodie with Vivienne Ndatshe, *Going for Gold: Men, Mines and Migration* (Berkeley and Los Angeles, 1994), p.38.
7   T. Uzodinma Nwala, *Igbo Philosophy* (Lagos, 1985), p. 194; Victor Uchendu, *The Igbo of Southeast Nigeria* (New York, 1965); Herbert Cole and Chike Aniakor, *Igbo Arts: Community and Cosmos* (Los Angeles and Berkeley, 1984), p.24.
8   Interview with James Alo, Ngwo, Nigeria, 6 June 1975.

9 Kristin Mann, *Marrying Well: Marriage, Status and Social Change Among the Educated Elite in Colonial Lagos* (Cambridge, 1985).

10 Hair, 'Enugu', 129; interview with Gabriel Mbalemlu, Michael Nwakuache and Clement Egbogimba, Ugwu Alfred, 5 July 1975; NNAE, NIGCOAL 2/1/138, Local Authority, 'Census of Enugu'; interview with Mokoro Osakwe, 8 July 1975 and Clement Ude, 9 July 1975, Ugwu Aaron, Enugu.

11 Brown, 'Testing the boundaries'.

12 These chiefs even developed paramilitary groups to police the camps. See Carolyn A. Brown *'We Were All Slaves': African Miners, Culture and Resistance at the Enugu Government Colliery* (Portsmouth, NH, 2003), ch. 2.

13 Mahmood Mamdanni, *Citizen and Subject: Contemporary Africa and the Legacy of Late Colonialism* (Princeton, 1996).

14 This was the 'Udi Uprising' which occurred during the First World War. See Brown, *'We Were All Slaves'*, pp.84–6. The system of chiefs came to an end after a revolt of women usually called the Aba Women's Riot. As a result the chief system was terminated and village councils were appointed to govern villages. See A.E. Afigbo, *The Warrant Chiefs: Indirect Rule in Southeastern Nigeria, 1891– 1929* (London, 1972) and Robert Gailey, *The Road to Aba: A Study of British Administration Policy in Eastern Nigeria* (New York, 1970)

15 '*Nzuko*' is a generic Igbo term meaning any type of 'meeting'. The urban 'tribal' unions were also considered *Nzuko*.

16 C.H. Croasdale, 'Report on the Enugu Colliery', 1938, Personal Collection of P.E.H. Hair.

17 This roughly translates to 'self-help meetings' or mutual aid groups. Their history has not been documented. The only study that explicitly mentions them is Croasdale, 'Report'.

18 Interview with B.U. Anyasado, Mbieri, Owerri, Nigeria, 23 July 1975.

19 In 1935, Colonial Secretary Ormsby-Gore sent a circular instructing colonial officers to establish systems to monitor labour conditions (NNAE, UDDIST 3/1/14, 'Circular from the Secretary of State to the O.A.G'., 9 November 1935).

20 Lugard was the chief architect of the consolidation of those territories that in 1912 became Nigeria. His interest in Enugu was apparently centred upon its becoming an example of 'enlightened' colonial labour policy which, to him, meant teaching the 'natives' British cultural and political practices. For a discussion of Lugard's impact on the colliery, see Brown, *'We Were All Slaves'*.

21 Nonetheless British culture was transferred to the 'bush'. The missionary and anthropologist, the Revd G.T. Basden, noted on a visit to Udi in 1916 that it was 'the custom in Ngwo to sit at the fire as in England': Church Missionary Society Archives (hereafter CMSA), University of Birmingham, G3A/0 1913-1916, the Revd Basden, 'Report on Visit to the Udi District'.

22 Powell Duffryn Technical Services, 'First Report to the Under Secretary of State for the Colonies on the Government Colliery, Enugu', mimeo (London, 1948), 'Characteristics of the coal', D–138.

23 Interview with Eze Ozogwu, Amankwo, Udi, 2 June 1975.

24 CMSA GBA 3/0 1913, Basden, 'Visit to the Udi District'.

25 For a history of the role of the Westernized elite of Lagos in the early nationalist agitation, see James Coleman, *Nigeria: Background to Nationalism* (Los Angeles and Berkeley, 1958).

26   Peter Hall, *Cities of Tomorrow: An Intellectual History of Urban Planning and Design in the Twentieth Century* (Oxford, 1988).

27   Anthony D. King, *Urbanism, Colonialism and the World Economy: Cultural and Spatial Foundations of the World Urban System* (London, 1990), pp.53–4.

28   Public Records Office (hereafter PRO) MPGG/129, Station Plan No. 107, Government Station, Enugu Ngwo, 12 March 1919.

29   F. Lugard, *The Political Memorandum: Revision of Instructions of Political Officers on Subjects Chiefly Political and Administrative, 1912–1918* (London, 1970), pp.405–22.

30   Maynard W. Swanson, 'The Sanitation syndrome: bubonic plague and urban native policy in the Cape Colony, 1900–1909', *Journal of African History*, **18** (1977), 387–410.

31   Lugard, *Political Memorandum*, p.416; emphasis added.

32   See Frederick Cooper, 'Urban space, industrial time and wage labour in Africa', in Frederick Cooper (ed.), *The Struggle for the City: Migrant Labour, Capital and the State in Urban Africa* (Beverly Hills, 1983), pp.7–50.

33   The literature on mining and masculinity is extensive. For a sample, see June Nash, *We Eat the Mines and the Mines Eat Us: Dependency and Exploitation in the Bolivian Tin Mines* (New York, 1979); N. Dennis, F. Henriques and C. Slaughter, *Coal is Our Life* (London, 1956); Michael Burawoy, *Another Look at the African Mine Worker* (Lusaka, 1972); Moodie, *Going for Gold*; and Patrick Harries, *Work, Culture and Identity: Migrant Laborers in Mozambique and South Africa 1860–1910* (Portsmouth, NH,1994).

34   'The word "pit man" carried with it meanings of social bearing; other men were "colliers" compared to "pit men", and others again were "labourers" compared to "colliers" ' (Robert Colls, *The Pitmen of the Northern Coalfield: Work, Culture and Protest, 1790–1950* (Manchester, 1987), p.12.

35   This developed because of the labour shortage following extraordinary mortality rates in the 1918 influenza pandemic. The miner brought in a 'helper' hewer and tubmen to work in his production team. His was the only name on the payroll and the output of his work team was attributed to him. See C. Brown, *'We Were All Slaves'*, ch. 6.

36   T.M. Klubock, *Contested Communities: Class, Gender and Politics in Chile's El Teniente Copper Mine, 1904–195* (Durham, 1998), p.143.

37   This refers to men with considerable oratory skills. See Ifi Amadiume, *Male Daughters and Female Husbands: Gender and Sex in an African Society* (London, 1987).

38   Croasdale 'Report', 29–30.

39   PRO/CO 583/263, Colonial Office Press Section, 11 May 1943, 'Model Villages for African Miners'.

40   NAE/NIGCOAL 2/1/94, letter from Colliery Department, Iva Valley to the General Manager (hereafter GM), Railway, Lagos, 12 July 1937 and Colliery Manager (hereafter CM).

41   PRO/CO 583/216, Ormsby-Gore Minute, 26 April 1937.

42   Deborah W. Ray, 'Pan America Airways and the Trans-Africa Air Base Program of the Second World War', PhD diss., New York University, 1973, and her article, 'The Takoradi route: Roosevelt's pre-war venture beyond the western hemisphere',

Journal of American History, **62**(2) (1975), 340–58; Michael Crowder, 'The 1939–45 War and West Africa', in Ajayi and Crowder (eds), *History of West Africa*, p.599.

43   Michael Cowen and Nicholas Westcott, 'British imperial economic policy during the war', in David Killingray and Richard Rathbone (eds), *Africa and the Second World War* (London, 1986), p.44.

44   Crowder, 'The 1939–45 War', pp.600–605.

45   Cowen and Westcott, 'Imperial economic policy,' pp.20, 21.

46   See Lisa Lindsay, 'Domesticity and difference: male breadwinners, working women and colonial citizenship in the 1945 Nigerian general strike', *American Historical Review*, **104**(3) (1999), 783–812.

47   Brown, *'We Were All Slaves'*.

48   For the full discussion of how unions were targeted by the Colorial Office and the impact this had on their effectiveness, see Brown, *'We Were All Slaves'*, ch. 6 and Carolyn A. Brown, 'The dialectic of colonial labour control: class struggles in the Nigerian coal industry, 1941–1949', *Journal of Asian and African Studies*, **23** (1988), 32–59.

49   PRO/CO 583/263, 'Report of the committee appointed to review the question of housing accommodations for the employees of the Government Colliery, Enugu', 27 September 1941.

50   PRO/CO 583/261, Orde Browne Minute, 4 March 1943.

51   Members included the Resident, Onitsha, Colliery Manager William Leck and F.J.W. Skeates, the Colliery's new Staff Welfare Officer; PRO/CO 583/263, 'Report of the Committee'.

52   PRO/C0 583/263, General Wormal, 'New homes in old Africa', n.d.

53   According to an informant, it was better placed than others. People were encouraged to beautify their compounds, and competitions were held for the best-looking garden. The District Officer and the Colliery Manager awarded prizes. Interview with B.U. Anyasado, Mbieri, Owerri, 23 July 1975.

54   *Hansard*, 37th Parliament Debates 1942–43, 11–17 November, 1943, p.344.

55   They also asked that the Lady Welfare Officer be an unmarried woman. The current officer was the wife of the Staff Welfare Officer, Skeates; NNAE/UDDIST 3/1/104, 'Enugu Colliery Workers and the New Housing Scheme'.

56   'Enugu Colliery Workers and the New Housing Scheme'.

57   PRO/CO 583/261, 'Notes on points arising in discussions with the Secretary of State, Wednesday 27 October, 1943'.

CHAPTER NINE

Everyone Black? Ethnic, Class and Gender Identities at Street Level in a Belgian Mining Town, 1930–50

Leen Beyers

It is not coincidental that Lockwood and Blauner, who in the 1960s classified miners as the most cohesive segment of the working class, were British social scientists.[1] Lockwood and Blauner identified the nature of mining (which implies a need for cooperation between workers and a relative autonomy from technical and supervisory constraints, the geographical isolation of mining settlements and the predominance of the working class in mining towns) as the main cause of the pronounced working-class identity and the overlap between work and leisure relations in mining societies.[2] The same structural features are present in the Limburg coalfield, situated in the Flemish region in the north of Belgium, but in contrast to the British coalfields the population in this area was far more ethnically diverse. The low density of the population in the Limburg province and their reluctance to work in the mines led to the influx of migrants, originating principally from Poland before the Second World War, from Italy just after 1945, and from Turkey and Morocco in the 1960s. Following Lockwood, strong class cohesion and a low level of ethnic antagonism can be expected. However, as Martin Bulmer has argued, ethnic cleavages often prevent a cohesive occupational community from developing, notwithstanding class homogeneity and geographical isolation.[3] This chapter analyses how in Zwartberg, one of the Limburg mining towns, spatial isolation and class homogeneity affected informal inter-ethnic neighbourhood relations before 1950.

The cohesiveness Lockwood and Blauner referred to was a matter of shared class identity. Identity, the sense of belonging to a group of people, is intrinsically bound to the categorization of other people as different, as outsiders. Did Polish, Italian and Belgian inhabitants consider each other as insiders, because they were all miners? Or were ethnic distinctions more dominant, possibly reinforced by class identities (for instance when Poles were sometimes labelled 'lousy workers')?[4] To refine the question further, not only class, but also sex roles, may have offered ground for collective identification. In the period under review, women in Zwartberg generally did not work outside the home and made use of neighbourhood support in managing their household work. Hence distinctions between good and bad housewives were probably as alive in the neighbourhood as class-based identities.[5]

This chapter investigates the salience of class, gender and ethnic identities in informal street life, that is to say, chance meetings, chatting and the giving of support outside the sphere of churches, associations and pubs. The focus on these street-bound neighbour relations requires an insight into the opportunities for inter-ethnic interaction in the mining town. If the particular spatial and population structures have encouraged inhabitants to meet, to chat and to support each other exclusively within their own ethnic group, the development of strong inter-ethnic ties and cohesive social identities may have been retarded.[6]

In what follows, I first outline how the spatial structure and composition of the population in Zwartberg affected the interaction between Belgian (mainly Flemish) and non-Belgian (mainly Polish) neighbours. Secondly, the ethnic, class and gender identities behind these contacts are discussed. The Belgian working class in Zwartberg included both Flemish, from the surrounding Dutch-speaking countryside, and Walloons, from the French-speaking Wallonia region in the south of Belgium. While the Walloons were often experienced miners, trained, for example, in the pits of the Borinage, the Flemish workers were inexperienced. They occupied lower positions and lived closer to the immigrant working class. Therefore this chapter is mainly concerned with Flemish Belgians on the one hand and Poles on the other.[7]

The Spatial Print of Occupational Hierarchy

Zwartberg had 3204 inhabitants in 1933, of whom 58.1 per cent were non-Belgian.[8] The town was already well developed when coal production started in 1925. The construction projects in Zwartberg and the other Limburg mining towns followed the style of the English garden cities built by industrial entrepreneurs (Figure 9.1).[9] Characteristically, these neighbourhoods were located close to the mines, were full of green open spaces in the form of private gardens as well as public parks, and had distinct residential zones for working-class citizens, on the one hand, and for a petty bourgeoisie, on the other. It is possible to discern four categories of houses located in different districts of Zwartberg (see Figure 9.1). The miners' houses were located in the North East Side and the South Side. Lower middle-class inhabitants (made up of occupations such as clerks, teachers and a few higher status miners) inhabited the North West Side and the northern part of the South Side. Meanwhile, more senior clerks and some engineers could be found at the

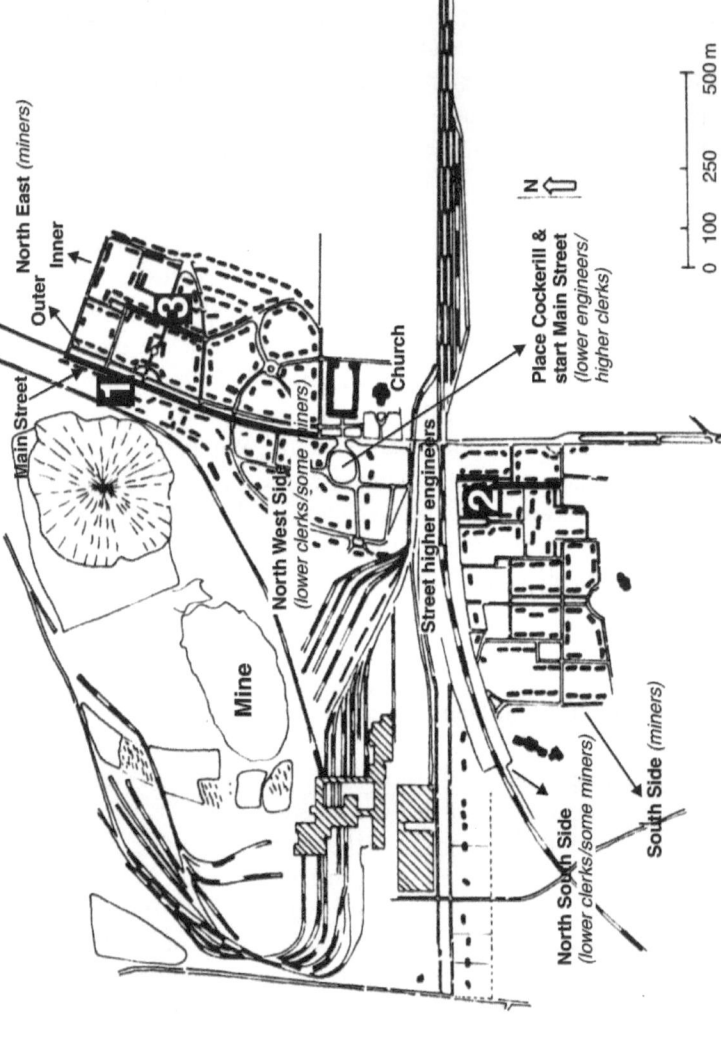

Source: Van Doorslaer, 'Zwarte monumenten in een groen Limburg. Naar een verdiende toekomst voor het mijnpatrimonium', *Monumenten en landschappen*, **9** (1990), 49 (map by L. Vandebergh).
Note: The numbers indicate the location of streets detailed in Figure 9.2.

Figure 9.1 The class-bound spacial structure of Zwartberg

Place Cockerill and at the beginning of Main Street. More important engineers and the mine director lived in a street close to the entrance of the mining site.

The spatial distribution already indicates that occupational position and status provided the main basis for the allocation of houses to miners' families.[10] The effect of this was that, from the 1930s to the 1950s, at least in the working-class areas, Belgians, Poles, Czechs or Italians were next-door neighbours. Clerks, teachers and engineers were generally Belgian and thus had far less chance to meet Polish or Czechoslovakian inhabitants. A 1933 report from the mine's housing service reveals, however, that national origin, albeit in a less decisive way than occupation, also determined the allocation of housing stock. The Inner North East Side was set aside for 'strangers' who had not yet proved their professional qualities. Belgians who were clerks, supervisors and section heads, or ordinary miners with a large family, had first call on the houses of the South Side. Nonetheless 'the good strangers' (*les bons éléments parmi les étrangers*) could also find accommodation there. The planning for the first half of Main Street was similar. The houses there were reserved for the Belgian engineers, clerks, supervisors and for (Belgian and non-Belgian) workers of long-standing good character.[11] This explains why hardly any Belgians lived in the Inner North East Side and only a few in the South Side in 1930, whereas the Outer North East Side (the Main Street and the street close to the school) comprised a sizeable population of Belgians (both clerks and miners).

As the figures show, by 1947 these distinctions were gradually becoming less pronounced (Figure 9.2). Belgians were found in the Inner North East, although they were still more numerous in the Outer North East. In the South Side they became predominant. The high number of Belgians, this time mostly Flemish, in 1947 was a result of the Second World War, when becoming a miner was seen as a more attractive alternative than forced labour in Germany, and of the postwar housing shortage and government policy to attract Belgians to the mines (the so-called 'battle for coal').[12] As for the decrease in the number of Poles and Czechoslovakians, a lot of them had left just before the outbreak of the Second World War. The return migration organized by communist activists after the war reinforced this trend.[13]

From this general picture of inter-ethnic meeting opportunities, we turn now to the character of Zwartberg's street life in the years before 1950. Inter-ethnic encounters were primarily composed of run-of-the-mill encounters in shops and in some streets in the working-class districts. Moreover Zwartberg was built in such a way that, in a few places, the backyards of working-class and middle-class inhabitants were close enough to allow social contact between inhabitants of different class and ethnic backgrounds. Did members of these different groups interact in these bridging areas and how intense was

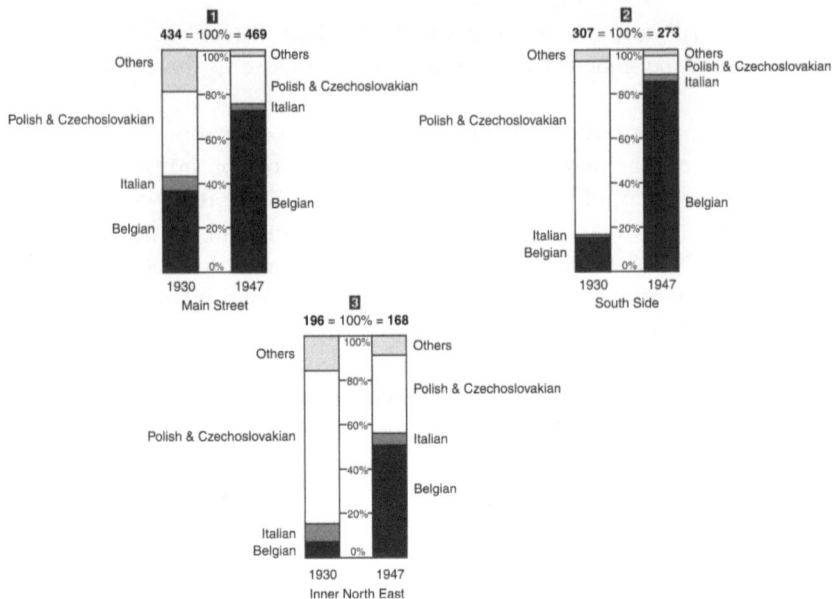

Source: Genk, Population census, 1930–47.
Note: These data cover the population of the Main Street and of two other typical streets, one in the South Side and one in the Inner North East. Families and lodgers staying with families are taken into account but not the lodgers staying in the six boarding houses on the Main Street, because their date of residence is untraceable. The figures at the top of each bar chart represent the total number of inhabitants.

Figure 9.2   The ethnic composition of Zwartberg

the social contact? Were the encounters meaningful exchanges of support and gossip, or were they merely acts of observation?

Shops

Inter-ethnic interaction was at its highest in Zwartberg's shops. The main reason for this was the town's geographical isolation. There were few shops as it was, and most working-class people (and, indeed, most middle-class inhabitants) were unable to travel regularly the 18 kilometres to Hasselt, the nearest shopping centre. Before 1950, the main shops were the official shop owned by the mining company, the so-called *Centrale d'Achat et de Vente* (CAV), and some privately owned shops located on the outskirts of the garden city. One of the biggest of these private shops was Pieters's, just below the South Side. The shopkeepers were a Flemish couple, although it was the wife who kept the shop while her husband was out working as a builder. The shop's customers came from the South Side for food, and from the whole of Zwartberg for textiles. Most of the customers were women. Belgian as well as Polish working-class families frequented the establishment. The wives of engineers and clerks also frequented Pieters's, although they also shopped in other places to find more luxurious products than the basic foodstuffs on offer there.[14]

The isolated nature of Zwartberg meant that the shopkeepers depended completely on the mining town's own citizens (including the 50 per cent that was of non-Belgian origin) for their business. Pieters tried hard to attract the customers from all ethnic backgrounds. Firstly, for immigrants who frequently had little in the way of savings, and whose pay was often poor and handed out fortnightly, the credit facilities open to them at the shop were especially welcome. In contrast, the official mining shop, located in the North West Side, did not extend credit as easily and did, if necessary, dock workers' pay in order to meet any outstanding debts.[15] Secondly, Pieters's eldest daughter learnt some Polish, while later on her sons picked up some Italian. Other shopkeepers and traders in the Limburg coalfield did the same.[16] At Pieters's, newly arrived immigrants with little or no knowledge of Flemish could interact to some degree at least with the shopkeeper and her family. Immigrant women confided their family problems to Pieters and many turned to her when they needed help understanding basic Flemish. Furthermore, as was the case in all Zwartberg's shops, inter-ethnic interaction was also high among the clients themselves, because most shopping took place on, or immediately after, the fortnightly payment of wages.[17]

Ethnically-mixed Miners' Streets

In the miners' streets the degree of meaningful inter-ethnic interaction was heavily determined by the level of ethnic segregation. In the 1940s, in the South Side, there was a large enough concentration of Flemish and Walloon Belgians to enable the development of a strong group identity.[18] While some Belgian inhabitants exchanged greetings with their non-Belgian neighbours, there was a clear social distance. Not only was contact with members of other ethnic groups avoided, antipathy could, on occasion, be openly expressed. Richeta, an Italian woman, settled in the South Side in 1940. She did not rely on her Belgian neighbours for help, but looked instead to a fellow Italian living two streets away. The two women frequently gathered coal together at the coal belt.[19] Similarly, in the 1930s in the Inner North East Side, Polish immigrants established a strong support network amongst themselves.[20] Only for the Main Street, where there was no clear pattern of residential segregation, can one find examples of sympathetic chats and reciprocal support (for example, babysitting and the sharing of milk and such like) between Poles and Belgians.[21]

It cannot be taken for granted that people in ethnically mixed areas automatically develop inter-ethnic contacts. Ethnic prejudices were sometimes too entrenched. However, two dimensions of neighbourhood life did stimulate inter-ethnic contact. As research has clearly shown, neighbours tend to combine friendly involvement with detachment; their interactions frequently stand mid-way between anonymity and intimacy, on a level which can be called familiarity. If neighbours get too close, it is possible for one to gossip about the other and damage reputations in the process. There are especially good reasons for next-door neighbours to maintain friendly relations, because it is hard for neighbours in such close proximity to avoid each other in the event of a falling out.[22] Richeta's experiences illustrate this point. She discovered that the children of a Flemish neighbour who lived at the far end of her street had stolen carrots from her garden. Richeta told the children's parents that they should ask if they wanted carrots and not damage her vegetable garden. This elicited the blunt reply: 'Plant your carrots in Italy.'[23] The breakdown in their relationship was relatively easy to negotiate on a day-to-day basis, as the two women rarely saw each other. However, in contrast, Richeta's next-door neighbour maintained friendly terms with her Italian neighbour.

The ability of immigrants to pick up the language of the host society was important in this respect. Although Richeta told me she preferred to behave 'discreetly', not looking for trouble, the fact that she spoke some French gave her the power to quarrel effectively with her neighbours if need be.[24] French was indeed the main language in the town up to 1950, when the French-speaking Walloons occupied the chief positions in the mine. As a consequence, Italians made their way by using a mixture of French and Italian. As for the Polish, they narrowed the language gap by using German or French, which they had picked up while working in the coalfields of the Ruhr, Wallonia and the north of France. Poles who migrated directly from Poland learnt this broken German or French from their countrymen.[25] In the Main Street, the different ethnic groups were relatively well mixed. Many Flemish inhabitants were the next-door neighbours of non-Belgians and hence they typically behaved in a friendly manner towards their immigrant neighbours.

In Zwartberg, as in many localities before 1950, the ties of mutual dependence between neighbours (and the need for a good reputation) appear to have reduced the inclination to quarrel. Before the era of the car, inhabitants of the isolated garden city often depended on each other for social support, leisure contacts and even marriage partners.[26] The second component of informal neighbouring which could enhance positive inter-ethnic interaction was the need for reciprocal support (for example, child care). The absence of a dominant Belgian group could leave Belgians in some parts of the Main Street or in the Inner North East rather isolated. Richeta must also have felt isolated as one of the few Italians in the South Side, and there was little chance that her Belgian neighbours would provide the support and sociability she required. Belgians were the most powerful ethnic group and when they appealed to their non-Belgian neighbours, they did so from a dominant position. This was the case of Angele, for example. Angele and her husband were Belgian dairy traders who had settled in the Main Street just after the Second World War. Angele's husband came from a mining town near to Zwartberg, but her family lived in another province and she seems to have felt lonely in Zwartberg.[27] The shop gave Angele the opportunity to get to know a Polish girl, Anna, who lived in a house on the opposite side of the street and who was always sent shopping there by her mother. After a short time, Anna's mother took on the role of childminder to Angele's one-year-old daughter. Every evening, when the shop closed, Anna returned the infant to her parents. The Belgian girl stayed with the Polish family until she was four and had to attend school. The support of Angele's neighbours was unpaid and repaid through small acts of consideration and kindliness.[28]

Backyard Neighbours of Higher and Lower Classes

Despite class-based internal divisions, the garden city of Zwartberg was a discrete urban unit. Its spatial boundaries were flexible enough to both constrain and facilitate interaction with 'other' people. Through chatting over their garden fences, the Belgian people with higher professional positions interacted with working-class people and thus possibly with Poles, Czechs or Italians. The northern part of the South Side, the start of the Main Street and the North West Side, were the settings for this 'backyard neighbouring' par excellence. In the South Side (and also in the North East Side) small footpaths ran between the backyards.[29]

Informal neighbouring between backyard neighbours of higher and lower classes in Zwartberg consisted of simple observation, the exchange of greetings, conversations and the giving of support, mainly of the non-reciprocal type (for example, the sharing of useful information, the donation of clothes, networking between employers and job seekers or the employment of a neighbour as a cleaner in one's house). The class difference between the actors was the main reason that so much of this help was of a non-reciprocal nature. It was simply not possible for working-class families to return this kind of support.

While in a few cases backyard neighbouring represented inter-ethnic familiarity, it also indirectly affected the inter-ethnic meeting opportunities. With a wealthier class of citizen (often with significant incomes at their disposal) living in close proximity to working-class inhabitants, there was a constant demand for cleaners, clothes washers and general helpers, all of whom could expect to get paid for their efforts. To the extent that members of the working class could supplement their income in such ways, their reliance upon neighbours from other backgrounds could be lessened. The example of Mart, a Flemish miner's wife, from the South Side, illustrates this.[30] Since her husband was often ill and unable to go to work, she supplemented the family income by doing the washing in the house of her neighbour, a Belgian clerk, and cleaning the house of a Walloon (which implies a higher status) family in the North Side. She also wallpapered the house of a teacher in the South Side. The networks Mart used were based on her backyard neighbouring as well as those developed in her local church. In other words, Mart was not fully dependent on her working-class neighbours, even when her husband's income did not fully cover the costs of her household. Zwartberg differs in this sense from London's working-class quarters before the First World War. As Ellen Ross has demonstrated, the reciprocal character of women's neighbour support was crucial there. Significantly, the neighbourhoods she describes seem to have been far more homogeneous in class terms than the Limburg mining towns.[31]

Charity and the Demarcation of Class and Ethnicity

Under certain circumstances shops, backyards and streets in Zwartberg were scenes of inter-ethnic interaction. We will now take a closer look at the identities that were to the fore and which worked to divide and unite the ethnically mixed neighbourhood.

In the case of non-reciprocal help from backyard neighbours, the two parties were often members of different social classes but the same ethnic groups. This, too, could result in difficulties for the potential benefactor and the recipient. For example, the Nijs family accepted an offer by the section head's wife to provide a dress for Holy Communion. Yet, when the section head's wife told Mrs Nijs that her husband was considering Mr Nijs for promotion, the offer of help from a manager to a minor was bluntly rejected. That this offer was refused, while the Holy Communion dress was accepted, is perhaps not surprising. The hierarchical difference between miner and section head was much more obvious in this case. Other examples of the social distance between donor and recipient breaking down can be found in proposals for adoption. A Russian miner's wife who became a widow just after the Second World War, was heavily dependent on caregivers. As her daughter remembers: 'And then you had, we called them "the barons" at that time. Now, I know it was the houses of the engineers, [but] my mother called them "the barons". I was allowed to go there to pick berries. And they also brought things for us sometimes ... They even wanted to adopt my youngest sister.' Other stories of this type can be cited. In all such cases, the proposal for adoption was the last stage in a series of gifts; it was considered a stage too far by the overwhelming majority of those on the receiving end of such 'kindness'.[32]

While being economically fruitful, non-reciprocal support between neighbours hardly contributed to inter-ethnic cohesion. The givers may have acted out of an inter-ethnic solidarity, but the offers they made must have left the recipients with the impression that class and ethnic barriers were unbridgeable, as the epithet 'barons' indicates. A class hierarchy could also be just as pronounced between shopkeepers and the non-Belgian miners' population. A lot depended on the shopkeeper's class and gender identities. The bigger the shops, the less likely were exchanges such as those between Anna and Angele.[33]

Ethnic Prejudices of the Flemish

Class distinction was central to non-reciprocal backyard neighbouring. Owing to the nature of residential segregation in Zwartberg, class difference was as good as absent in the miners' quarters; instead, ethnicity emerged as the main means of emphasizing difference. The more numerous and cohesive the Belgian group was, as in the South Side in the 1940s, the more chance there was that their initial ethnic prejudices were maintained and strengthened. However, the attitudes of Belgians towards their Polish, Italian or Czech neighbours changed drastically in the 1950s, when Italians entered the garden city. In this situation, the already established ethnically mixed group quickly designated the incomers from Italy as 'the outsiders'.[34]

Where did the ethnic prejudices of the Flemish in the 1920s and 1930s come from? Firstly, in Limburg, mining work was considered demeaning and the miners rough people, partly because there was no tradition of mining in Flanders.[35] Flemish and Limburg miners were not eager to exchange their villages for the mining town. They only did so after a while, when the commuting facilities appeared unsatisfactory. Often they considered a move as a step down the social ladder.[36] 'The people in the past, if one went to work [into the mines], said: "don't go to live in the city, because there resides the rabble"', attested a Flemish miner's wife. She settled close to the North Side on private ground, but even that felt like a demeaning move to her. If she had not been pregnant before getting married, she admitted, she would never have left her home village.[37] Louis's father started in the mines of Zwartberg in 1936, after he had been fired from his job as a baker's aide. Louis describes how his family experienced their move to the mining town, distinguishing his family from the 'rougher' miners:

> In the beginning it was very difficult and hard. It's a whole other mentality. We came from the city and in fact, miners, not directly the miners' families, but the language and everything. ... It was rougher than what we were used to before. Not that I found this personally hard all the time. ... some of these people also wanted to get up to a higher level. ... Some felt themselves commonplace and behaved accordingly in their contacts with other people. Miners spitting or telling dirty jokes ... [38]

During the whole inter-war period, trades unions, politicians and employers discussed the extent to which immigrant workers were threatening 'Belgian' jobs.[39] This discussion may have been reflected at the neighbourhood level, but the negative perception the Flemish in Zwartberg held of the immigrant population was mainly influenced by the formers' perception of the miners as being rough.[40] Louis's mother evaluated a Polish coal-supplier of the garden city in a way that highlights how the same distinction between the 'rough' and the 'educated' was ever present. This Polish man lived in a small house on a large estate nearby, where, when he was not delivering coal (his main job), he worked the land. One day, after delivering the coal at Louis's house, he asked Henri's mother in polite German if he could go to the toilet: 'Gnädige Frau, darf ich mal Pipi machen?' Louis's mother was pleasantly surprised by his politeness. 'You can see, that's one who lives on a large estate, who has had a good education,' she said.[41] Louis's school career also shows how this family associated the mining town and its immigrant population with a bad education. Louis was 11 when he arrived in the garden city and, although he should have attended the last year of the elementary school, he went straight away to a secondary school outside the mining town.[42] Louis explains hesitantly why this happened:

> First of all, my father did not know who was there [in the school ...] for us they were strangers. We didn't call them that; we called them 'compels'. This was not meant so badly. In Germany, the miners were called 'Kumpels. ... I had no bad contacts with them. They were not looked down upon in a racist way. They were strangers, that's all.[43]

As often occurs when people come to talk about their ethnic prejudices, Louis plays them down. There is no doubt, however, that the work 'compel' was like an insult to the Polish. As one of them explained: 'Nobody should have dared to call me that; I would have beat them up. "Kumpel" in German means "mate" or "workmate", but Polish people took it as "dirty Pole". "Kumpel"? We couldn't stand that.'[44]

Careful and Clean: Gender and Inter-ethnic Cohesion

We might expect that the common experience and mutual dependence of miners' wives would weaken the divisive force of ethnicity in informal neighbourhood life. However this is a far too simplified and romantic view of the interaction women had in Zwartberg, as the ethnic bonding in the South Side reveals. Nevertheless, thus far, two clear cases have been discovered of women's collaboration leading to changes in the way ethnic difference was perceived. The two cases of Lisa and Angele show how this could happen, albeit not on a regular basis. The example of Angele, whose Polish neighbour was a babysitter for her daughter, was mentioned earlier. The other example concerns Lisa, a Belgian upper working-class woman of the North West Side. One day in 1943, Lisa fell ill and was hospitalized. Her Polish backyard neighbour, Zofja, a miner's wife, looked after Lisa's five-year-old son during her absence, bathing him and putting him to bed. Just like Angele's neighbours, Zofja was not paid for this support.

Lisa and Angele appealed for support from their Polish neighbours, first of all because their Belgian network did not offer a solution at those particular moments in their lives. Angele could look beyond her own ethnic group for support, thanks to the fact that she kept a shop visited by a wide range of people. She got to know Anna through a process of regular contact, conversation and, finally, active support. As a shopkeeper, Angele was in a perfect position to observe her neighbours, who were also her customers. Another shopkeeper told me she observed her customers in the same way to decide whom she would allow to pay on credit.[45] Secondly, Angele asked Anna about her background ('You are not Flemish, are you? What are you then?'). Only after that did she ask her to go for a walk with her daughter. This simple question illustrates that a boundary had to be crossed.[46] Finally, Amelie and her daughter seem to have appreciated their Polish neighbours for the conscientious manner in which they cared for their children, ethnicity being of little consequence to this gender-bound identification.[47] As a result of the circular structure of the garden city and the limited mobility of its inhabitants before 1950, neighbours frequently saw each other out and about. Simple observation became an important way of identifying with neighbours.

On the street where Lisa and her family lived there could also be found a section head, teachers, clerks, the mine director's chauffeur and the mining shop.[48] Lisa's husband oversaw the painters who were employed by the mining company to decorate the interiors of the engineers' and higher clerks' houses. Lisa was familiar with the section head's wife and one of the clerks' wives as well. On occasion, these women asked Lisa for favours. Two miners' spouses, Zofja and a Belgian woman, would meet at Lisa's house for chats. They were the closest neighbours of Lisa.[49] However, the section head's wife and the clerk's wife, both of whom knew Lisa, did not chat with the other two miners' wives, although they lived almost as close to them as Lisa did. They felt themselves to be 'too good'. As Lisa's son put it: 'We were the only working-class family in the street.'[50] The snobbery displayed by the clerks' wives encouraged Rose to have closer contacts with the miners' wives. Although in the homogeneous miners' streets class was not usually a strong incentive for inter-ethnic cohesion, it was in this case where class prejudices were overtly present. Lisa's gender identity may have also played a part in her preference for Zofja. Lisa herself was a very clean and tidy housewife and she used to observe and evaluate others by her standards. However, she did not look down upon the miners' wives. On the contrary, she admired many of them who worked so hard to keep their husbands' clothes clean.[51]

Identity shifts such as those Angele and Lisa experienced were less likely in the South Side and may also have been isolated cases in the North Side. Two explanations for the lack of impact that gender identities had on ethnic prejudices can be given. Firstly, Belgian miners' wives were, certainly in the South Side in the 1940s, numerous enough to build their own exclusive networks. They had no need to reach out to the few non-Belgian neighbours in their street. Angele and Lisa, however, at those particular moments in their lives, found themselves in a different position. Secondly, the supply of non-reciprocal support made the inter-ethnic group of working-class women less mutually dependent than the stories of Lisa and Angele suggest. Finally, gender operated now and then as a cohesive factor in Zwartberg, but equally it had the power to reinforce ethnic lines of division. Richeta, the Italian woman, for instance, classified all Flemish neighbours who expressed their racism openly to her as mean women. They were mean not only because of their racism, but also because of their failure as wives, mothers and home-makers: she accused them of 'not caring for the children', 'not cleaning', 'always going out [to pubs]'. Richeta described the woman whose children had stolen her carrots as one who did not respect her husband. When her man left for the nightshift she used to hope 'that a stone falls on him', Richeta reported.[52]

Everyone Black?

Lockwood and Blauner assumed that geographical isolation and working-class dominance enabled cohesion in mining towns in the sense of a shared class identity. This chapter has raised the question whether these factors really were such vehicles of cohesion. We have examined how this worked on the streets, where neighbours – mainly women – observed each other, chatted with one another and gave support.

The spatial isolation and the internal circular structure of Zwartberg clearly facilitated inter-ethnic interaction. First of all, the isolated nature of

Zwartberg and the other Limburg mining towns caused people of different ethnic (and class) backgrounds to frequent more or less the same shops. The spatial isolation made shopkeepers fully dependent on this ethnically mixed clientèle, which explains their eagerness to speak at least some Polish, German or Italian. The rather dense and circular structure of the garden city encouraged inter-ethnic familiarity both between people of higher and lower class, and in the miners' areas themselves. It stimulated, first of all, simple observation. When it came to chatting and offering support, people were more selective, preferring their own ethnic group if it was numerous enough.

When did these inter-ethnic contacts lead to inter-ethnic cohesion? For reciprocal help it is clear both parties had to identify with one another to a certain extent, before daring to ask for help. However reciprocal inter-ethnic help seems to have been rare, partly because the supplies of non-reciprocal support made working-class women less mutually dependent. The effect on inter-ethnic cohesion of meeting and exchanging greetings was ambiguous. On the one hand, the Flemish mining population with strong ethnic prejudices merely used the information they got from day-to-day street encounters to categorize the non-Flemish inhabitants as outsiders. Louis's mother, who praised the Polish coal-deliverer for his politeness, by implication classified all other Poles as impolite. On the other hand, fleeting contacts in the streets could constitute the first phase in a process of transcending ethnic prejudices, as in the case of Angele and Lisa. This occurred in situations where the Belgian group, owing to class difference or spatial disparity, was not cohesive enough to maintain shared norms regarding who was 'in' and who was 'out'. Collective identities around class and gender could then come to the fore.

Interestingly, in contrast to the conclusions of Lockwood and Blauner, class homogeneity was not a strong incentive for inter-ethnic working-class networks and identities, except when the class distinction was experienced strongly, as in the case of Lisa. Some current inhabitants of Limburg, where the last mine closed in 1992, claim, not without pride, its ethnic tolerance by stating that 'in the mine everyone was black'. However we can be sure of one thing: class did not render everyone black in the mining town. Neither did class difference promote inter-ethnic cohesion. Miners' women readily profited from the non-reciprocal support of their better off backyard neighbours, while their own identity was clearly distinguished from that of their benefactors. Shopkeepers, engineers, clerks and miners on some occasions united in the face of a common threat, as when they protested against the pit closure in Zwartberg in 1966. Further research needs to be carried out to discover whether class and ethnic distinctions were less pronounced in informal neighbourhood life during and after this general strike.

As far as gender is concerned, Belgian and non-Belgian miners' wives had a lot in common. Most notable of all was a clear labour division between husband and wife, the husband bringing home the pay, the wife washing the dirty miners' clothes, cleaning the house, shopping and feeding the children, husband and lodgers. In Zwartberg, the supply of non-reciprocal support made income shortages less common a problem, but still the care of children, the fear of mining accidents, or of coping with alcoholic husbands, were common enough experiences.[53] However, not all of these issues were openly discussed by women, let alone women from different ethnic groups. As Melanie Tebbutt has shown, keeping up appearances implied that the worst problems were not made public by asking for support.[54] Even socially accepted problems were not often a shared concern that united non-Belgian and Belgian neighbours in Zwartberg before 1950. At least in the South Side, ethnicity proved to be a stronger focus of identification for Flemish women than the similarity of their own and their Polish or Italian neighbours' domestic troubles. Gendered prejudices – for example regarding the cleanliness of neighbours' houses or their abilities as mothers – could easily deepen ethnic distinctions in the miners' streets of Zwartberg in ways that class was unable to do. While gender could have a positive impact on inter-ethnic cohesion, it was only decisive when combined with collective class identities, as the cases of Lisa and Angele indicate and the history of Zwartberg after 1950 confirms. Such complexities reveal the extent to which historians must be careful before assuming, like Lockwood and Blauner, that coalfield communities were homogeneous entities. Quite clearly, ethnicity and gender mattered just as much as class did in the creation of identities and the structuring of social relationships.

Notes

1 The PhD research this article is based on is funded by Fonds voor Wetenschappelijk Onderzoek Vlaanderen. Thanks to Professor Patrick Pasture and the editors for comments on earlier drafts of this text. Some of the interpretation in this chapter have been refined. See Leen Beyers, *Iedereenzwart? Het samenleven van nieuwtomers en gevestigden in de mijncité Zwartberg, 1930–1980*, (Aksant Press: forthcoming).

2 David Lockwood, 'Sources of variation in working-class images of society', in Martin Bulmer (ed.), *Working-Class Images of Society* (London, 1975), pp.16–21.

3 Martin Bulmer, 'Sociological models of the mining community', *The Sociological Review*, **23**(1) (1975), 83–4.

4 Richard Jenkins, *Social Identity* (London, 1996), pp.90–103. People of course hold several social identities at once. Class and gender identities should thus not be seen as replacing ethnic identities, but rather as creating unity in diversity, certainly in the case of first and second generation migrants.

5   Tine De Rijck, *De ereburgers. Boek I: Mijnwerkers aan het woord* (Berchem, 2000), pp.349–51.

6   Talja Blokland-Potters, *Wat stadsbewoners bindt. Sociale relaties in een achterstandswijk* (Kampen, 1998), pp.288–9.

7   On the Walloons : De Rijck, *De ereburgers*, p.21. Not discussed in this chapter are the contacts these Belgians and Poles had with immigrants arriving in the period 1945–50, who were mainly Italian.

8   Charbonnages Des Liégeois, Service de la Régie, Règlement et rapport d'exploitation, 21/4/1933, p.6.

9   De Rijck, *De ereburgers*, pp.205, 212–13.

10 Ibid., pp.219–28.

11 Service de la Régie, Règlement, pp.5–6. For the workers in the Main Street, no nationality is specified in this report, but the population data confirm that both Belgian and non-Belgian workers were allocated there.

12 Luc Minten, 'Op zoek naar mijnwerkers in een agrarische provincie', in Luc Minten, Ludo Raskin *et al.* (eds), *Een eeuw steenkool in Limburg* (Tielt, 1992), p.116; Luc Minten, 'Van Mijnwerkersstatuut naar vreemdelingenstatuut', in Minten and Raskin, *Een eeuw steenkool in Limburg*, pp.165–6; De Rijck, *De ereburgers*, pp.206–7, 261.

13 Population Census Genk 1930–47; Idesbald Goddeeris, 'De verleiding van de legitimiteit. Poolse Exilpolitik in België, 1830–1870 en 1945–1980', unpublished PhD, Catholic University Louvain (2001), 332–9; private interviews. Owing to the housing shortage the Italians from the new migration wave (starting in 1946) were accommodated in 'prefab' houses and former army barracks on the outskirts of the mining town.

14 Interview with Fons.

15 Interviews with Irena, Paul, Janek, Victoria and Stanis. All interviewees have been given fictitious names. On the typical phenomenon of buying on credit in the Limburg mining towns, see De Rijck, *De ereburgers*, pp.317–19.

16 De Rijck, *De ereburgers*, 322; interviews with Stanis and Irena.

17 About Pieters: interviews with Fons and Gina; about shopping on payment day: De Rijck, *De ereburgers*, p.317.

18 Interviews with Flor, Victoria and Louis.

19 Interview with Richeta.

20 Interviews with Stanis, Janek and Lud.

21 Interviews with Anna and Irena.

22 On the neighbour role, see Blokland-Potters, *Wat stadsbewoners*, pp.157–79; Hugo Lis and Catharina Soly, 'Neighbourhood social change in West European cities from the sixteenth to the nineteenth centuries', *International Review of Social History*, **38**(1) (1993) 1–30; Martin Bulmer, *Neighbours: The Work of Philip Abrams* (Cambridge, 1986), pp.18–19, 27–33.

23 Quotation from an interview with Richeta.

24 For example, when a Belgian neighbour accused her falsely of having caused the cockroach plague in their street in 1946: interview with Richeta.

25 On language usage in the Limburg mining towns, see De Rijck, *De ereburgers*, p.335. On the migration patterns of the Poles in Limburg, see Pien (A.P.) Versteegh, *De onvermijdelijke afkomst? De opname van Polen in het Duits,*

Belgisch en Nederlands mijnbedrijf in de periode 1920–1930 (Hilversum, 1994), pp.149–50.

26 On the difference between neighbouring in our present, highly mobile society and in local-based societies, see Blokland-Potters, *Wat stadsbewoners*, pp.151–99.

27 Population Census Genk 1930–47; interview with Anna.

28 For example, Anna from time to time shared a meal with Angele's household: interview with Anna.

29 Map of Zwartberg garden city, 1965, Archives, Catholic Church Zwartberg; interview with Victoria.

30 Also other miners' wives looking for an extra income did jobs in the homes of clerks, better-off workers, teachers and engineers, but not in their own neighbours' houses: interviews with Maria, Jaak, Victoria and Anna.

31 Ellen Ross, 'Survival networks: women's neighbourhood sharing in London before World War One', *History Workshop*, **15** (1983), 5–6.

32 Interviews with Masha, Marian and Bentha.

33 Interview with Marian.

34 Norbert Elias, 'Introduction: a theoretical essay on the established and the outsiders', in Norbert Elias and J.L. Scotson (eds), *The Established and the Outsiders: A Sociological Enquiry into Community Problems* (London, 1994), pp.15–52. On the attitudes towards the Italians in the 1950s, see private interviews.

35 Jos Perry, 'Van vader op zoon', in Wiel Kusters and Jos Perry (eds), *Versteende wouden. Mijnen en mijnwerkers in woord en beeld* (Amsterdam, 1999), pp.85–8.

36 There were, for instance, no cycle paths for bikes, and trains often did not arrive on time. See De Rijck, *De ereburgers*, pp.277–80; interviews with Victoria and Flor.

37 Quotation from an interview with Joke.

38 Quotation from an interview with Louis.

39 Griet Van Meulder, *De ereburgers. Boek II: Een sociale geschiedenis van de Limburgse mijnen (1917–1985)* (Berchem, 2000), pp.558–62; Guy Vanthemsche, *Le chômage en Belgique de 1929 à 1940. Son histoire, son actualité* (Brussels, 1994), pp.173–6.

40 Interviews with Elza, Joke, Stanis and Victoria; see also De Rijck, *De ereburgers*, pp.280–81.

41 Quotation from an interview with Louis.

42 This choice of school was exceptional for Belgians at that time in Zwartberg and rather illustrates the strength of the initial prejudice. Part of the company's policy was to provide high-quality schools so as to keep the Belgian and higher-class employees tied to the mine. See private interviews. On the mine's school policy, see N.V. Cockerill-Ougrée, *Steenkolenmijn Zwartberg, Zwartberg 1907–1957* (Brussels, 1957).

43 Quotation from an interview with Louis.

44 Quotation from an interview with Stanis and Janek. Also to Italians 'compel' sounded denigrating; see Fred Vanhinsberg, *Waarom die Italianen?* (Berchem, 1996), pp.51–2.

45 Interview with Elza.

46 Quotation from an interview with Anna.

47 Interview with Anna. Amelie's daughter called Anna's mother 'mother', long after the years of babysitting.

48 Interview with Paul; Address List Zwartberg 31-12-1948, Archives Municipality Genk.

49 Population Census Genk 1930–47 ; Address List Zwartberg 31-12-1948.

50 Interview with Paul. This is an overstatement as the painter and also the director's driver had an intermediary position between workers and clerks. However this might not have prevented Lisa and her son from feeling themselves to be working class.

51 Interview with Paul.

52 Quotations from an interview with Richeta.

53 See De Rijck, *De ereburgers*, pp.341–51. The situation of miners' women in the adjoining Dutch coalfield was comparable; see Carla Wijers, 'In de schaduw van het kruis. Vrouwenlevens in de Limburgse mijnstreek, 1910–1970', in Hélène Vossen, Bien Kruijtzer *et al.* (eds), *Vrouwen tussen grenzen. Limburgse vrouwen in de 19de en 20ste eeuw* (Roermond, 1990), pp.47–9.

54 Interview with Richeta; Melanie Tebbutt, *Women's Talk: The Social History of 'Gossip' in Working-Class Neighbourhoods, 1880–1960* (London, 1997), pp.86–9. See also Blokland-Potters, *Wat stadsbewoners*, p.161.

CHAPTER TEN

Outsiders: Trade Union Responses to Polish and Italian Coal Miners in Two British Coalfields, 1945–54

Stephen Catterall and Keith Gildart

The study of coal miners, their trade unions and communities has often simplified the role of immigration and its impact on social and political developments. From the classic institutional studies by Robin Page Arnot, through to work that owed much to the growth of social history from the 1960s onwards, there is only a limited descriptive account of immigration. This aspect of mining history has now started to attract historians, especially those critical of the heroic, solidaristic and progressive portraits of writers who have moulded the story of miners to fit a particular Marxist version of labour history. The defeat of the miners' strike of 1984–5 and the subsequent disappearance of collieries from the industrial landscape has led to a period of reflection and revision. This process has been under way for some time in studies of coal miners in the United States and is slowly receiving the attention of British historians of the industry in Britain. In this chapter we focus on two British coalfields and assess trade union reaction to the employment of foreign labour. What emerges is a disjointed picture of trade union sectionalism, community suspicion and a clash between trade union officials who were keen to see management/union partnerships succeed and the more conservative views of the rank-and-file miner.

Foreign workers were brought into the British coal industry because of a 'manpower crisis' during the immediate postwar period.[1] These workers were mainly Polish but also included a variety of other European nationalities chiefly from Eastern Europe recruited from the ranks of 'displaced persons' (refugees) known as European Voluntary Workers (EVWs).[2] Hostility from rank and file miners arose from the perceived 'threat' that these workers posed as a result of prodigious output performance and the mining skills they brought or through domestic and social tensions.[3] Later, during the early 1950s, as the manpower 'crisis' deepened, Italians joined these workers. Nonetheless some union leaders were prepared to take a less tough line on the introduction of such workers. There was a feeling that in the aftermath of hostilities the National Union of Mineworkers (NUM) should give the

National Coal Board (NCB) support over its claim that the industry was facing a grave 'manpower crisis'. NUM area leaders such as Edwin Hall in Lancashire and Ted Jones in North Wales accepted that, in the chaos of demobilization, the advent of public ownership and through the challenges of reverting from war to peacetime conditions, there was a need to have the requisite manpower necessary to meet the demands placed upon the industry.[4]

Yet at both the national and area level of the union, there was some disquiet over the introduction of foreign labour. During the debate on Polish miners at the NUM national conference in Bridlington in 1946, the president, Will Lawther, had joked that 'we hope that we are not reaching the stage where an industry is so bereft that within a short period of time the only Britishers left in it will be the gaffers'. Lawther was concerned that the industry could well be adversely affected by a policy of employing foreign workers, especially in areas where recruitment of local labour was already in difficulty. Reconstruction was already diverting many young men in mining communities away from the collieries. When questioned on his internationalism by a delegate from the Midlands, Lawther replied that it 'is not simply a question of whether you sing the Internationale or the Red Flag, we have got an obligation to our membership'. The NUM vice president, James Bowman, reiterated this concern:

> Before there is any question of bringing foreign labour into our collieries we would prefer vastly as an Executive Committee that there should be an entire survey of the manpower problem at the collieries ... Let us see how many there are on haulage. Younger men at the collieries are prepared to be trained and upgraded for higher work. If that is done we say confidently that there are thousands of men who are on the surface who would be willing to go down and do these jobs.[5]

The placement of Polish workers was only achieved through agreements reached at national level, which allowed for specific undertakings to be made to the NUM on deployment. These included placement only after acceptance by the union branch concerned; agreement that in the event of redundancy the Poles would be 'first to go'; and finally the safeguarding of local agreements to ensure the preferential job grading of British mineworkers.[6] This created tensions between some coalfield leaders and their members. It was a particular problem in North Wales.

The experience of Polish and Italian miners in North Wales points to trade union concerns about job security, but also exhibits the importance of a particular local identity in initially working to exclude outsiders from the culture of the colliery and the local community. During the Second World War, the coalfield was one of the smallest in Britain. The eight collieries were spread across a wide geographical area in the counties of Flintshire and

Denbighshire.[7] The coalfield was characterized by a divided workforce in terms of union affiliation and the resilience of particular village identities. Nonetheless, Ted Jones, the coalfield union leader, had been successful in consolidating membership by the end of the war.

The first wave of European workers that entered the coalfield were Polish miners. This became a major concern for the North Wales miners and they expressed this at various meetings with management and through the forums of their local union lodges. At a meeting of the Area Executive Committee in Wrexham in 1947, the delegate from Llay Main Colliery asked for clarification on the position of the union concerning the employment of Polish labour. Ted Jones maintained that it would be a matter for each lodge to decide, although he was personally sympathetic to the view of the NCB. Nonetheless he insisted that the traditional union policy of 'last in, first out' would apply. Each Polish miner would have to join the union and be aware of the fact that in cases of redundancy they would be the first to lose their jobs.[8] Jones was merely reiterating the decision that had been reached between the NCB and the NUM in February 1947 at the national level. Both Ebby Edwards for the NCB, and the communist Arthur Horner for the NUM, were signatories to an agreement that discriminated against Polish miners. They concluded:

> Poles employed must be done so with the agreement of the Lodge. If Poles do not join the union they will be dismissed. In the event of redundancy the Poles would be the first to go. Also, they would be dismissed if another British miner was seeking re-employment in the pit.[9]

Union concerns about job security emerged in the context of instability for the coalfield. The NCB did not include North Wales in its plans for reconstruction and there was a clear indication that no new pits would be sunk. NCB officials estimated that the mining workforce would eventually fall from 8500 to 6700 between 1951 and 1965.[10]

Developments at Llay Main Colliery indicate that, although the foreign worker became a member of the union, he still carried little weight in terms of seniority within the organization. None of the Polish workers became officials of the union and the records show that there was little concern for problems of settlement and assimilation. There was an acceptance by lodge officials of the diminished status of Polish miners and this was reflected in the policies adopted. In 1948, the NCB Labour Officer informed Ted Jones that it was the intention of the NCB to introduce 100 such workers at Black Park and Ifton collieries. Facing pressure from the Black Park lodge, Jones utilized the power that the union had recently gained through public ownership. He stressed that management had little input into this area of industrial relations. Under pressure from union members, Jones reiterated the policy that it would be up to each lodge to decide whether to accept or reject Polish workers. After a series of tense meetings, it was agreed that the Poles would be accepted, but only after the NCB agreed to a number of safeguards relating to protection of local workers through the seniority system of job allocation.[11] The Poles that entered Black Park did not stay long and felt uncomfortable living in the confines of the villages. Initially they had been housed in temporary camps close to the colliery. Two years later, the NCB pointed to this as one of the reasons for closure, as miners were unwilling to transfer to this particular pit.[12] The union in North Wales and in other coalfields were responsible for screening the Polish workers with regard to their suitability. Horner complained that this was a very slow process. In North Wales it took Ted Jones three days to interview 103 potential miners. The delay no doubt allowed gossip to spread about job insecurity and deepened suspicions about the potential effects that the outsiders would have on the colliery villages.[13]

The aversion to the influx of outsiders was not only confined to the more isolated pits. Similar concerns were raised at Bersham Colliery, a mine close to the busy town of Wrexham. In 1951, the miners at a lodge meeting suggested that, in order to guard against the influx of skilled foreign labour, younger local miners should be trained as quickly as possible, ensuring that the local community was considered when jobs became available.[14] Nonetheless the union was in a difficult position. Throughout the 1950s, the NCB had to intensify publicity drives through the media to attract labour, as many young miners quickly left for jobs in other industrial sectors. Although some Polish workers found difficulty adjusting, many settled quickly and became accepted members of the workforce. Union concerns slowly evaporated, but were rekindled again when the second phase of European immigration was under way.

Opposition to the recruitment of foreign workers intensified during the early 1950s, when it was proposed to recruit Italian workers into British collieries. Italians had a tradition of working in the Belgian and French coal industries. A large number had been working in Belgian and French mines since the end of the Second World War, assisting the coal industries of these nations during the onset of peace. As the labour situation in both Belgium and France began to stabilize, so the Italians' contracts were not renewed and many returned to their homes, mainly in the poorer regions of southern Italy, to face the prospect of unemployment. Meanwhile, in Britain, the government pressed for coal production as a priority to underpin peacetime industrial expansion. This in turn put more pressure on the NCB to find the labour to maintain production. The NCB thus demanded that the government seek new sources of labour.[15]

While the NCB was pleased with the prospect of increased labour, from wherever it came, they had grave reservations about the reaction of mineworkers to the deployment of Italian workers. The NCB anticipated hostility on a number of fronts. It felt that the government, in its panic to ensure that production targets were met, might be tempted into rushing the Italians into British collieries with little consultation and minimal training. Similarly it held fears over operational and safety concerns. While the majority of the Italians had some mining experience, the NCB was worried that they lacked the necessary level of training and underground work experience for placement in British pits. These anxieties were compounded by the fact that the majority of the Italians had a poor command of English. The proposed English language training was believed by the NCB to be perfunctory and wholly inadequate to cope with demand.[16]

Certainly at both the official and 'rank-and-file' levels some of the objections raised were based on popular prejudice. Tired of trying to obtain the support of miners for the placement of Italian labour, an exasperated Ministry of Fuel and Power believed that the British workforce thought the Italians were too soft for a tough job like coal mining. They observed that miners have a 'feeling of contempt for their [the Italians'] historical association with ice-cream and organ grinding and also more latterly their rather poor performance in North Africa'.[17] Undoubtedly there was a hangover from wartime prejudice underpinning much of this sentiment. The Italians were not to be trusted because of their alleged record on the battlefield during the Second World War. This was underscored by a view which questioned the ability of the Italians to become involved in mining, where mutual dependency was paramount, based on their alleged martial weakness and political capriciousness.[18] The great problem with this sort of prejudice was that official frustration with the reluctance of mineworkers to accept the Italians was exacerbated by the persistence and prevalence of this base dialogue within the British 'establishment' itself. For example, Victor Raikes MP caused a storm in mining communities when he suggested that miners were reluctant to accept the Italians, not because they might lose their jobs, but because they might lose their wives to the allegedly amorous Italians.[19] Such interventions only served to fuel prejudice and further undermine official attempts to win over mineworkers.

Another problem was that of timing. The early 1950s witnessed a slackening in the pace of postwar economic expansion, with a sharp recession in 1952–3. This raised anxieties over job security, coming on top of the first moves toward rationalization of the coal industry under public ownership.[20] The government and NCB were thus in a dilemma over deployment of Italians. The government wanted to push ahead with a major deployment of up to 15 000 Italian workers in order to maintain levels of production. The government also came under intense diplomatic pressure from Rome as deployment was halted because of growing opposition.[21]

Hopes of overcoming opposition to the placement of Italians rested on winning NUM support. The NUM at national level, and in a majority of its areas, had accepted the deployment in principle, because leading officials supported the NCB's priorities on issues of output and 'manpower'. However branch union officials acting under rank-and-file pressure spearheaded opposition.[22] Resistance was intensifying as deployment proceeded, making life difficult for leading NUM officials. How they reacted to such recalcitrance was the key to settling the 'Italian labour question'.

With opposition increasing as deployment proceeded, the government, including Prime Minister Clement Attlee, made a vain attempt to make the position of the NUM less difficult by appealing to miners' sense of international solidarity.[23] Ultimately, all efforts at persuasion failed. Industrial action in the Yorkshire coalfield, with the threat of strike action spreading to other coalfields, was the final mortal blow to any hopes of further Italian deployment.[24] Although senior NUM officials in the coalfields had endorsed Italian deployment, they were forced to retreat under rank-and-file pressure. With the prerogative of acceptance passed to branch level, rejection 'on the ground' was now complete. The Lancashire and North Wales coalfields provide two arenas where this rejection of Italian workers can be analysed in some detail.

In North Wales, the view of the miners was unequivocal, although, again, Ted Jones promoted a more conciliatory stance on the question. At an Area Executive meeting in December 1950, reports from the lodges revealed an overwhelming rejection of NCB plans to introduce Italian workers into North Wales collieries.[25] At Ifton Colliery, local miners criticized the area union for not taking a more adverse attitude to the importation of such workers. Peter Morris, of the lodge committee, informed the NCB Labour Officer that, although the branch had voted 10–2 in favour of the employment of Italians, an amendment was moved at a subsequent General Meeting which reversed the policy. As a result the lodge could no longer sanction their employment.[26] It is clear that the more enlightened members of the various lodge committees, supported by Jones, were sympathetic to the plight of the Italian miners. This was reflected in the concerns of the NCB, who pressed the national union to take away the power of lodges and union members to decide on whether Italians would be accepted. The NCB felt that a more conducive environment would be the monthly consultative committee meeting between management and union officials at each pit. Nonetheless, rank-and-file members held firm and argued that this decision should rest with the lodges. At a meeting of the NUM National Executive in November 1951, there was concern that the removal of this veto would lead to further industrial relations problems.

If attempts were made to endorse the employment of these [Italian] men against the will of the workers actually employed in the pits, the committee believe that this would result in stoppages involving loss of coal production far in excess of any production which might be expected to accrue as a result of the introduction of such foreign labour.[27]

Between late 1951 and 1952, Lancashire was only one of two British coalfields which stood as beacons of enlightenment in this sad fiasco, but the reality was more complex.[28] The key figure in Lancashire was the NUM Area General Secretary, Edwin Hall. Initially, when deployment of the Italians had been proposed during 1950–51, he had been the first British miners' leader to offer Lancashire as a destination for Italian workers, for which he received many plaudits from both the British and Italian governments. However he faced stiff opposition from within his own executive because of fears over rank-and-file opposition. Hall, in typically forthright manner, had prevailed upon his executive to reverse a decision to reject the Italians.[29] Crucially, although Hall claimed that any final decision on deployment lay with NUM branches in Lancashire,[30] as was the case in other coalfields, he consistently ignored rank-and-file concerns. Despite his public claims to the contrary, there was never any question of Hall agreeing to Lancashire miners having any influence on the decision to accept the Italians or not. Not surprisingly, forcing a decision on union members in this manner was a mistake.

By January 1952, with opposition to deployment hardening, the NUM Lancashire Area voted afresh on the 'Italian question'. Lancashire miners rejected further deployments of Italian labour, though those already 'placed' in Lancashire collieries were allowed to stay because a major deployment had already been undertaken.[31] Thus, while Lancashire was only one of two British coalfields which made a positive endorsement of Italians already working in the coalfield, the decision to reject more was in direct opposition to Hall, who argued that Lancashire should accept further 'placements'.

Given that Hall faced intense opposition from within his executive and from the rank and file, what made him such an implacable devotee of the Italian deployment? In one sense Hall welcomed appeals from the British government, the labour movement and NCB based on international fraternity, though that was not the main reason. The overwhelming reason for Hall moving so decisively was his developing relationship with the NCB North Western Area (NCBNWA). The central figure in the region was Laurence Plover, NCBNWA Industrial Relations Director, former Lancashire NUM official and executive member now sitting on the opposite side since the industry passed into public ownership. Plover was a former colleague of Hall, who, together with Arthur Horner, NUM General Secretary, persuaded Hall that he should assist the Regional Board by taking more Italians. Moreover, such was Plover's influence, he even prevailed on Hall to take more Italians than originally anticipated either directly from overseas or from other British coalfields, where they were festering in camps, having been rejected. Hall thus became the only British miners' leader to make a positive response to this request.[32]

Throughout the first half of 1952, Hall was involved in several unsuccessful attempts to convince Lancashire miners to reverse their decision to reject more Italians. He won approval for another vote in February 1952 within his executive to agree to a request from Laurence Plover to take Italians 'still in the pipeline', which his executive duly rejected.[33] It was widely believed that Hall was becoming much too pliant on the 'Italian question', particularly as Lancashire was being asked to take ever-greater numbers of Italians as the only coalfield accepting them.[34]

Another important element in obtaining Hall's agreement to cooperate so earnestly was that these events came soon after one the of biggest strikes in the coalfield's history, involving a demand for parity on concessionary coal for Lancashire miners. The issues which gave rise to the dispute in 1949 were not fully resolved until 1952.[35] Accordingly, at this time, Hall felt the need to find an accommodation with the Regional Board after one of the most damaging episodes during the early years of public ownership. Above all, Hall believed that the dispute had been a serious impediment to developing a cooperative relationship with the NCB in his quest to build a modern nationalized coal industry. He saw in the 'Italian question' opportunities by which he was able to offer cooperation with the NCB on the 'manpower' issue in order to maintain its commitment to meet production targets as directed by the government.

Although Hall stood at the centre of glowing tributes to his enlightened leadership, it would be an understatement to say that the position he adopted on Italian labour was unpopular with Lancashire miners. Hall agreed to take the Italians against persistent objections from his own executive. Likewise, although Italian miners already placed in Lancashire were allowed to stay, continued acceptance was only achieved by Hall after he made significant concessions involving a series of guarantees to his executive as a means of assuaging rank-and-file anger. The 'guarantees' placed severe restrictions on Italian workers. Consequently, while Hall ignored rank-and-file sentiments as a way of demonstrating that he was in control of 'his lads', as he liked to call miners, he had, in reality, to make these concessions in order to win them over, just to retain those Italians already placed in Lancashire. At the same time, Lancashire miners were basking in the limelight of press attention as one of only two British coalfields to have responded 'positively' to the 'Italian question'. Journalists who made a beeline for Lancashire inaccurately reported that Italians had been accepted in Lancashire because Lancashire miners were intrinsically capable of working well with anyone. They were, according to press reports, more 'cordial' and 'cosmopolitan' than miners in other British coalfields. In addition to this unhealthy and misleading interest in British regional mining typologies, it was reported that acceptance of

Italians had been secured because 'placement' had been undertaken 'carefully' in a 'human' and 'personal' way with 'agreement' aimed at 'removing resentment or suspicion' as a result of the combined efforts of the Regional Board and Lancashire NUM.[36] This was another piece of press mythology, probably devised to shame other coalfields into accepting the Italians.

The reality of 'placement' was very different indeed. Acceptance by the rank and file and union branch officials was only achieved subject to agreement that in the event of redundancies or redeployment the Italians would be the first to go. Moreover, irrespective of whether the Italians had previous mining experience or had completed training in British collieries, they were not initially allowed to undertake 'face' or highest paid underground work although they were given the right to be upgraded according to experience and merit. The majority of Italians placed in Lancashire collieries were put to work on the surface, undertaking the most basic lower-grade labouring jobs. Those who did go underground were at best confined to haulage, road making and labouring duties.

Despite the press paying tribute to the 'real Lancashire welcome' the Italians had received, the reality was again rather different.[37] The majority were initially housed in a renovated camp at Garswood Park near St Helens. The British authorities missed the ironic point that this was a former Italian prisoner-of-war camp. Although hostel accommodation was not unusual for miners moving to a new coalfield, the isolation of the Italians was no coincidence. They were initially segregated from the rest of the mining community and transported in groups into and out of the pits where they worked. They made a pitiful sight during the early stages of their deployment in Lancashire. They were naturally glad of a job of sorts. However they had to endure the most basic facilities in a camp from where they were bussed each working day to collieries some miles away. They had to suffer the privations of loneliness in an alien country with its attendant inclement weather, indigestible food and official stuffiness, knowing all the while that their mere presence was resented by local miners.[38]

The nature of Italian deployment in Lancashire in this manner was not only agreed by Edwin Hall to minimize upset to 'the lads', but was also very much connected to British coal industry politics of the early 1950s. The great lie about deployment in Lancashire was that it was undertaken in order to plug gaps in manpower in the region. However the nature and pattern of deployment indicated that it had less in common with a putative 'manpower crisis', but had more to do with a 'production crisis'. The Regional Coal Board claimed that the Italians were in Lancashire in order help the manpower situation. Actually they were being used to help the Regional Board bolster output at selected collieries as a way of ensuring that government production priorities were met. For this reason the so-called 'careful deployment' of Italians was not so much 'careful' as deliberate. If deployment had been solely about 'manpower', the Italians would have gone to areas of the coalfield which were experiencing the most chronic labour shortages, such as the Manchester district.[39] In fact the majority of the deployment was in St Helens district, which was why the Regional Board was so keen on using Garswood Camp. The Italians were mainly placed in the expanding collieries of St Helens district such as Bold, Cronton and Sutton Manor, where the production drive of the early 1950s was being mounted before the focus shifted to new and reconstructed collieries elsewhere in the coalfield. The Italians were employed in jobs at these pits which were ordinarily undertaken by 'green labour' (school leavers, youths and those new to the industry), where there was a recruitment shortfall. Working in these low-grade jobs, the Italians were able to release more experienced local mineworkers for 'face' and ancillary tasks to ensure that the production drive was maintained. Hall thus agreed to use the Italians effectively as 'substitute juveniles' in order to ensure that production priorities were not derailed during the early 1950s while the nature of deployment allowed Hall to minimize opposition from within his own executive and from the rank and file. The hapless Italians had thus become unwitting pawns in coal industry politics.

Because of the difficulties encountered with Italian labour throughout British coalfields, the government's proposed target of 15 000 Italian workers fell well short. In fact, only 1268 permanent placements were ever achieved. In Lancashire, numbers reached 700 but fell to below 500 through 'wastage' and 'transfer'.[40] Throughout the 1950s, those Italians who stayed gradually gained a degree of acceptance in Lancashire collieries and began to assimilate into the local community. However resentment towards the Italians remained. This occasionally developed into tension where particular circumstances highlighted the nature of Italian deployment. For example, it became evident that some colliery managers were using the Italians to replace so-called 'light work' cases (disabled miners).[41] This hostility was still fresh enough in the late 1950s to create flashpoints in circumstances where the rank and file were faced with a large-scale programme of colliery closures.

The response of the NUM to the introduction of European migrant workers into the nationalized coal industry raises particular issues for labour historians. The Lancashire and North Wales coalfields provide case studies in which the tensions between miners over the issue of foreign labour led to delicate negotiation. Edwin Hall in Lancashire and Ted Jones in North Wales had to balance their commitment to the demands of the management/union partnership against the concerns of a rank and file that were increasingly concerned about pay, conditions and the workings of the recently nationalized industry. Perhaps more importantly, the success of nationalization depended on consensual industrial relations and the effective working of the newly established consultative machinery. Hall and Jones were totally committed to the new settlement and were keen to avoid the consensual environment being derailed by the issue of foreign labour. Jones carried this pragmatism forward when he was elected vice president of the NUM in 1954.

The construction of migrant miners as 'outsiders' by sections of the NUM membership had only a limited impact on relations in the workplace. After initial problems of adjustment, the very nature of mining ensured that all mineworkers were dependent on each other. Polish and Italian miners retained their cultural identities outside the workplace, but underground their primary concerns relating to pay and conditions were shared by the rest of the workforce. Popular stereotypes might have endured in terms of jovial banter, but in times of crisis there was a distinct lack of overt racism. By the end of the 1960s, most British pits could boast of a cosmopolitan workforce who enriched the underground environment and the culture of the colliery villages. Nonetheless, as this chapter has highlighted, even the most enlightened sections of organized labour could take reactionary measures when faced with economic uncertainty and threats to job security.

Notes

1   G.B. Baldwin, *Beyond Nationalization: The Labor Problems of British Coal* (Cambridge, Massachusetts, 1955), pp.194–9.
2   PRO Kew, File No POWE 37/233: Ministry of Fuel and Power: Employment of Italian Miners: 1950–1951: NCB HQ to Ministry of Fuel and Power 25 October 1950; *Pit Prop*, February 1949. The total deployment of Polish and EVW labour was 17 872 but owing to considerable 'wastage' of labour, as many left for jobs in other sectors, the total was actually 12 375 in work in all coalfield areas by October 1950. The EVWs mainly consisted of Ukrainians, ethnic Germans from the Sudetenland and Silesia, and Latvians and Lithuanians.
3   PRO Kew, File No POWE 37/233: Ministry of Fuel and Power: Report on Foreign Labour in Mining Communities 2 October 1951. Baldwin, *Beyond Nationalization*, pp.194–9.
4   *Coal*, February 1951.
5   National Union of Mineworkers (NUM), Annual Report, 1946.
6   Baldwin, *Beyond Nationalization*, pp.194–9.
7   For details of the coalfield, see G.G. Lerry, *The Collieries of Denbighshire: Past and Present* (Wrexham, 1968).
8   NUM North Wales, Minutes of Area Executive Committee, 8 August 1947.
9   NUM, Annual Report, 1947.
10 *Wrexham Leader*, 17 November 1950.
11 NUM North Wales, Minutes of Area Executive Committee, 5 April 1948.
12 NCB, Report on Ifton and Black Park collieries, 28 December 1949.
13 NUM, Annual Report, 1947.
14 NUM North Wales, Bersham Colliery Lodge Minutes, 20 November 1951.
15 PRO, File No POWE 37/233: UK Delegation, Organization for European Economic Co-operation, to Foreign Office 8 October 1950; Memo: Ministry of

Fuel and Power Manpower Working Party: Recruits from Overseas 15 November 1950; Minutes of Minister of Fuel and Power's meeting with the Mining Group of MPs 1 December 1950; Ministry of Fuel and Power: Brief to Prime Minister 2 January 1951 & Final Report on the Employment of Italian Labour: NUM Branch objections 20 June 1951; Baldwin, *Beyond Nationalization*, pp.194–5.

16 PRO, File No POWE 37/233: Ministry of Fuel and Power Note: NCB: Italians with mining experience, prior to Minister of Fuel and Power's Meeting with Mining MPs 1 December 1950; *Daily Telegraph*, 10 January 1952. The Italians were given 10 weeks' basic English tuition at Maltby in South Yorkshire prior to 'placement'.

17 PRO, File No POWE 37/233: Ministry of Fuel and Power: Report on Foreign Labour in Mining Communities, 2 October 1951.

18 PRO, File No POWE 37/234: Ministry of Fuel and Power: Employment of Italian Miners: 1951–1952: Various correspondence between Cabinet Office, Foreign Office, Ministry of Fuel and Power and National Coal Board 1952 including discussion of popular prejudice against Italians February 1952. The Ministry of Fuel and Power considered showing a propaganda film to all British mineworkers, entitled *To Whom Honour is Due*. This documentary film shot in 1946 depicted the assistance escaped British POWs had been given by Italian civilians and partisans during the Second World War. The idea was ditched since the 'informed' view within the Ministry was that it amounted to a 'rather moving' but 'crude bit of filming'.

19 PRO, File No POWE 37/234: Brief to Ministry of Fuel and Power, 3 December 1951; Ministry of Fuel and Power: Notes of discussion on how to handle reaction to a speech by Victor Raikes MP (Liverpool Garston) Hansard vol. 495.

20 NUM Lancashire Area, Annual Conference Report, 1952.

21 PRO, File No POWE 37/233: Ministry of Fuel and Power: Minutes of Meeting with Mining with Mining MPs 1 December 1950. PRO, File No 37/233: Foreign Office to Ministry of Fuel and Power: Notes of conversation with Italian Chargé d'Affaires Signor Theobaldi 25 January 1952.

22 PRO File No POWE 37/233: Ministry of Fuel and Power: Memo: Manpower Working Party: Recruits from Overseas 15 November 1950. PRO File No POWE 37/234: Ministry of Fuel and Power: Notes on Parliament: Parliamentary Question 757/1: 'How Many Lodges Refused to take Italians?' 27 November 1951; Arthur Horner to Coal Board 15 February 1952. In the Durham, Northumberland, Warwickshire, Staffordshire and South Wales coalfields, 108 out of the 153 NUM lodges approached had refused to accept Italian labour.

23 PRO File No POWE 37/233: Minutes of Meeting between Minister of Fuel and Power and Mining MPs 1 December 1950; Ministry of Fuel and Power: Brief for the Prime Minister 2 January 1951.

24 PRO, File No POWE 37/234: Ministry of Fuel and Power: Press Cuttings and Comment: *Yorkshire Post*, 20 March 1952: Strike at Bullcroft colliery: Comments by Arthur Horner.

25 NUM North Wales, Area Executive Committee Minutes, 4 December 1950.

26 Letter from Peter Morris, Ifton lodge, to T. Rogers, NCB Area Officer, 7 February 1952.

27 NUM, Annual Report, 1951.

28 *Colliery Guardian*, 3 January 1952; the other coalfield was Cannock. In the Cannock coalfield NUM branches were accepting Italians until a vote by the Midlands Area of the NUM decided to reject the placement of more Italians but allowed those already 'placed' to stay.

29 PRO, File No POWE 37/233: Ministry of Fuel and Power: Italian embassy London to Ministry of Fuel and Power: Press Statement by Ruggero Orlando (Press Agency) following a peak time broadcast on *Radio Italiana*; Note to Ministry of Fuel and Power, 13 January 1951.

30 *Daily Telegraph*, 10 January 1952.

31 NUM Lancashire Area, Annual Conference Report, 1952.

32 PRO, File No POWE 37/233: Final Report on Employment of Italians: NUM Branch objections: North Western Division 20 June 1951.

33 NUM Lancashire Area, Monthly Conference Minutes, 14 February 1952.

34 NUM Lancashire Area, Monthly Conference Minutes, 14 February 1952.

35 NUM Lancashire Area, Special Conference Report: Concessionary Coal, 7 February 1952; NUM Lancashire Area, Executive Committee Meeting Minutes, 22 September 1952.

36 *Daily Telegraph*, 10 January 1952.

37 Ibid.

38 PRO, File No POWE 37/233: Ministry of Fuel and Power to NCB HQ after correspondence with Ministry of Works regarding the conversion of Garswood Camp, Lancashire, 24 February 1951; NCB NW Division to NCB HQ regarding Garswood Camp and reception of Italian labour, 30 November 1951.

39 PRO, File Nos POWE 37/239 and 240: Ministry of Fuel and Power: Quarterly Reports on Manpower Recruitment and Wastage by Age Groups: Analysis and Report on Figures for the Period 1949–1954, 17 December 1954.

40 PRO, File No POWE 37/234: NCB HQ to Ministry of Fuel and Power, 22 February 1950; Notes of meeting at NCB HQ between representatives of the Ministry of Labour and National Service, Ministry of Fuel and Power and National Coal Board, 3 December 1951.

41 NUM Lancashire Area, Monthly Conference Minutes, 8 May 1954.

CHAPTER ELEVEN

The Struggle for Polish Autonomy and the Question of Integration in the Ruhr and Northeastern Pennsylvania, 1880–1914

Brian McCook

Beginning in the late 1870s, thousands of Polish labourers began migrating to the Ruhr basin in Germany and the anthracite fields of northeastern Pennsylvania to work in the growing coal industries of both regions.[1] By the eve of the First World War, well over 400 000 Poles lived in the Ruhr, while approximately 160 000 had settled in anthracite Pennsylvania. Initially, the Polish experience in both regions was strikingly similar. By the 1880s, employers began hiring Poles in large numbers, believing that they were cheaper and more docile labourers, and employed them in secondary, unskilled positions within the mining industry. The influx of Poles raised native working and middle class hostility to these 'foreigners', who were viewed as wage depressors and held to be culturally inferior. In both areas, the word 'Polack' was a widely used term of derision. Local and national governments on both sides of the Atlantic also added to this enmity, openly fretting about the moral effects the 'Slavic invasion' would have on their respective societies.

Despite the initially hostile environment, the Polish communities in each region persevered and ultimately thrived. In the workplace, Poles eventually were able to move up the occupational ladder and attain the status of full miners. With this gradual transformation, native workers increasingly accepted Poles as equals, particularly in the wake of significant labour struggles in both regions after the turn of the century. Acceptance of these migrant workers in the larger community, however, remained limited. As a consequence, Poles turned inward and built their own vibrant ethnic communities, supported by the pillars of faith, family and ethnic associations. The stability each respective ethnic community lent was significant, and provided a basis upon which Poles could interact and ultimately integrate, at least partially, into the community around them.

Among historians who have examined Poles in each individual region, the level of actual integration has aroused considerable debate and caused two general schools of thought to emerge. One school argues that, well into the 1920s, Poles were primarily not interested in integrating, but instead in maintaining their ethnic sub-culture. Polish contact with native society, when it occurred, was purely functional and based upon shared economic interests.[2] Another school presents an opposing view, holding that, after the turn of the century, an increasingly high degree of interaction between Poles and natives existed. While rejecting 'melting pot' myths, these studies argue that an 'ethno-class' consciousness emerged among Polish workers that combined, as one historian noted, 'both what united them with and separated them from their fellow miners'.[3]

Although the direction of these latter studies is intriguing, determining the degree of Polish adaptation to each society remains elusive, particularly because not enough focus has been given to the realm that influenced Polish life the most, namely religion and the Polish relationship with the Catholic Church. While most labour historians acknowledge the role of religion in working-class life, a tendency continues to exist that views religion and religious struggle as a means to an end rather than as an end in itself. In essence, the religious realm acts as an incubator, nurturing workers and providing them with a sense of stability, until they are able to evolve and develop broader and more mature, more 'modern' if you will, class and/or national identities.[4] Historians of the Poles in the Ruhr and in Pennsylvania are often no different from others in this regard.[5]

In examining certain working-class groups such as immigrant labour, however, religion not only serves as a base from which class and national consciousness can later develop, but also remains an important factor structuring identity and worker action. Consequently, this chapter first describes the development of Polish Catholicism in each region and then investigates the causes and consequences of Polish struggles with the Catholic Church over religious matters.

Polish Catholics and the Catholic Church

In the early stages of Polish migration, the Catholic Church provided a symbol of continuity and became the most visible unifying element for thousands of disparate Poles in both regions. The relationship of Poles with Catholic authorities in this period was comparatively warm when compared to the initial reception Poles encountered in the workplace. The prejudices of Protestant elites in both regions towards Catholics, as well as the growing attraction of socialism to workers, encouraged Catholic officials to emphasize a common confessional identity.[6] Moreover, Poles were permitted a degree of ethnic autonomy through the support given by the Church to the development of Polish ethnic associations, the employment of Polish or Polish-speaking priests and, in the United States, by some Poles being allowed to form their own ethnic parishes.[7]

By the early 1890s, however, the relationship between Poles and the Church was rapidly deteriorating in both regions. The causes of this trend vary. In both areas a chronic shortage of Polish speaking priests caused resentment, particularly as thousands of new Poles arrived in each region with the global economic upswing in the mid-1890s.[8] Further tension was caused by the fact that, throughout the Ruhr and in many parts of Pennsylvania, Catholics of different ethnic backgrounds had to share local parishes.[9] The most significant factor driving a wedge between the Church and its Polish parishioners, however, was the fear within the Church hierarchy that Poles could undermine their authority. The period of the late 1880s and 1890s was one in which both the German and American hierarchies attempted to reassert their institutional power. Polish activities were viewed as a direct threat to this process.

In Germany, the wounds of the *Kulturkampf* were still fresh and the increasingly conservative, predominately German, hierarchy was under intense pressure from the Prussian government to rein in Polish activities and cooperate in the policy of *Germanization*.[10] In the Ruhr, the transformation in Polish–Church relations began with the removal of the popular Polish priest Francis Liss from his position at the behest of the Minister–President of Westphalia, Heinrich Konrad von Studt, who was concerned by the growth of Polish–Catholic Associations that Liss founded throughout the Ruhr.[11] After Liss's removal, German priests were pressured to keep a close watch on the activities of Polish members in the Polish Catholic Associations and prohibit any activity that could be deemed political or nationalistic in spirit. This, in turn, led to constant conflicts between Poles and local priests over the activities and ultimate control of such groups.[12] In 1904, the Church hierarchy limited Polish spiritual care further. While allowing Poles to continue to hear confession and occasional masses in Polish, the hierarchy banned the singing of Polish songs in mass without permission, prohibited baptismal, marriage and funeral services conducted in Polish and ordered that Polish children only be instructed for the sacrament of communion in German.[13]

Meanwhile, in Pennsylvania, tensions were also rapidly rising over the level of autonomy that should be accorded to Poles. In the United States, the Church was emerging from a period of internal power struggles that saw the Irish rise to dominate the American hierarchy.[14] Having established themselves and their traditions as representing the Catholic Church in America, this Irish hierarchy was increasingly disinterested in granting wide latitude to the traditions of other ethnic groups. In this regard, the Poles, who bound national identity and Catholicism as closely together as the Irish, were seen as especially threatening.[15]

In particular, conflict raged over the right of Poles to form their own ethnic parishes and the level of control that local priests should be allowed to exercise. Although by 1890 a few Polish ethnic churches existed in the anthracite fields of Pennsylvania, the hierarchy, including the local Bishop O'Hara of Scranton, increasingly viewed such parishes as impediments to their broader goal of using the Church to *Americanize* immigrants.[16] While grudgingly allowing ethnic parishes to emerge, for example, the hierarchy required those ethnic groups who formed such parishes to sign over the property of the parish, which working-class Poles built with their hard-earned dollars, to the local, usually Irish, bishop.[17] Polish indignation was further inflamed by the fact that, within each ethnic parish, the local priest controlled how the financial resources of the parish would be spent and dominated Polish–Catholic associations.[18]

The anti-Polish attitude displayed by the Catholic Church generated common reactions among Poles in both regions. The Polish press in the Ruhr and in Pennsylvania, led by nationalistic agitators, condemned the respective policies of the German and Irish bishops.[19] Anecdotal evidence further suggests that the conflicts Poles had with the clergy did cause a decline in Church attendance.[20] More significantly, both regions witnessed the rapid growth of secular Polish associations independent of priestly interference.[21] Despite these similarities, however, there was ultimately a significant difference in the way Poles in each area responded to the challenges put forth by the Church.

In the Ruhr, the vast majority of Poles attempted to reform the Catholic Church from within. While the growing strength of socialism attracted a few converts among the Poles, and a particularly small movement of Poles wanted to break away from the German Catholic Church altogether and convert *en masse* to the Armenian Catholic Church,[22] most Poles remained loyal Catholics because their confessional identity was so intricately bound to their ethnic identity as Poles.[23] Consequently, Poles first engaged in grass-roots campaigns through the Polish–Catholic associations, drafting petitions to the German bishops and even to the Pope demanding ethnic Polish priests, an increase in the number of Polish language masses and the lifting of the 1904 restrictions.[24] The hierarchy, however, generally ignored these official Polish entreaties for improved spiritual care until the years after the First World War.[25]

Poles nevertheless persevered and achieved significantly more success in influencing the Catholic Church at the local level, particularly through their efforts to capture seats of local parish executive committees and councils. These two bodies aided the local priest in the administration of a given parish and provided the parishioners with a means of voicing their concerns and influencing local parish policy. At the turn of the century, despite their growing population, Polish representation on parish councils and executive committees was insignificant. By 1912, however, this picture shifted dramatically and, in some areas of the Ruhr, Poles were able to attain representation commensurate with their numbers in the Catholic population. Whereas, for example, Poles had no executive committee or council representation in the city of Dortmund in 1904, by 1912 Poles controlled 12.5 per cent of the executive committee and 15.8 per cent of the council seats, which paralleled closely the actual percentage of Polish Catholics in the city.[26]

In Pennsylvania, the majority of Poles also remained loyal to the Catholic Church and pursued similar tactics of appealing to bishops for an increased number of ethnic parishes and priests while attempting to exercise greater local control through parish councils. By the turn of the century, Poles did achieve a significant level of success in attaining more autonomy. For example, while 13 ethnically Polish parishes in the Diocese of Scranton existed by 1896, Poles formed 28 new ethnic parishes in the two decades thereafter.[27] However, the success of Poles in achieving these gains was due in no small part to the willingness of a minority of Poles formally to break with the Roman Catholic Church altogether and form the independent Polish National Catholic Church (PNCC), a religious institution with ties to the Old Catholic movement in Europe.[28]

Led by the charismatic Polish priest Francis Hodur, the PNCC emerged in Scranton in the period between 1896 and 1900. The immediate impetus for the formation of the PNCC rested in a conflict between working-class Poles and their Silesian-born parish priest over the control of parish finances at the Scared Hearts of Jesus and Mary Church in Scranton.[29] Growing dissatisfaction with Catholic Church policies among Polish communities throughout northeastern Pennsylvania, however, soon caused the independent movement to spread. Five national parishes appeared in the anthracite region by 1900, with other national parishes being created across the country.[30] After the First World War, the PNCC even created parishes in reconstituted Poland.[31]

The appearance of the PNCC caused a significant rift to open up in the Polish community in Pennsylvania. Nevertheless, a Polish minority was willing to endure ostracism and embrace the PNCC because the precepts of this church appealed to their working-class values.[32] While retaining much of the outward pomp and ceremony associated with the Catholic Church, the PNCC rejected the concept of Hell and eternal damnation, rejected papal infallibility, abolished celibacy, made confession personal rather than auricular, held services in Polish and also lowered the parish dues.[33]

Most importantly, under the leadership of Hodur, the PNCC articulated an ideology that was deeply influenced by peasant populist and socialist thought and could in many aspects be compared to more modern 'liberation theology' movements.[34] The National Church actively sided with workers during strikes, even going so far as being willing to mortgage its properties to aid striking workers, and saw in working-class Poles the vehicle through which the Polish nation could be spiritually and literally reborn. Hodur once declared:

> our heroes arose from the huts of farmers and the artisans, or as children of landless tenants, they will sweat and shed blood not for kings, not in the name of the Pope, nor for privileges and rights for special individuals and classes, as our titled ancestors had at one time ... but will go into battle for their own freedom and the rights of man ... the battlefield of the future is the tribune for the assembly of workers, the factory bench, the stand of grain, giving bread to the people, the schools developing minds and warming the heart, the fine arts and literature and the Church.[35]

The proletarian character of Hodur's movement is also clearly seen in the symbol chosen to represent the PNCC, namely an open book lying upon a cross, with the words Truth, Work, Struggle appearing underneath.

Causes and Consequences of Polish Activism

In describing the paths Poles in the Ruhr and Pennsylvania took in their drive for greater religious autonomy, two key questions arise. First, from where did this striving for autonomy in religious matters come? Second, what were the consequences of the different trajectories taken by Poles in each region?

The answer to the first question is complicated and reflects pre-industrial traditions brought from the homeland, combined with new outlooks Poles obtained through interaction with their host societies. During the nineteenth century, the Catholic religion in partitioned Poland was highly politicized. The Archbishop of Gniezno-Poznan was considered the official interrex of Poland, while the Virgin Mary was considered her figurative Queen. For many Poles, the Catholic faith served as the primary unifying element for the Polish nation, and devotion to the Church was proof of fidelity to the national struggle.[36]

Loyalty to the Church did not, however, necessarily mean subordination to the earthly representatives of the faith and actual religious practice among the peasantry differed from the formal form propagated by the Polish Catholic Church hierarchy. In peasant circles, devotional figures that were anthropo-centric, that is icons that were seen as 'of the people', were preferred to those that transmitted institutional power and authority. The most notable example of this is the widespread adoration of the figure of Mary, in contrast to the respectful though more detached worship of God himself, who was seen as a strict father. Moreover, in their daily forms of worship, peasants expected their religious figures, both iconic and real, to serve them in material ways.

Going to mass was less about worship itself and more about attaining favours from saints or the priest to improve their condition.[37]

As Poles migrated to the industrial centres of the West, the conditions they encountered provided fertile ground upon which their populist attitudes regarding the role of religion could further develop. As previously mentioned, within the religions. Poles increasingly felt oppressed by a foreign Church that sought to *Germanize* or *Americanize* them. Local priests were often viewed as being self-interested, snobbish and overbearing, an image not aided by the actions of certain priests that received significant attention in the local Polish press.[38] Overall, Poles on both sides of the Atlantic increasingly felt the need to reclaim what they believed was their Church from the corruption of the hierarchy and priests.

Polish willingness to confront the Catholic Church was also influenced by the socialist working-class milieu of both regions and specific developments emerging from the realm of work. The difficult, dangerous conditions under which Poles laboured encouraged Poles to form worker associations in which broader themes of social justice were discussed. Strike-actions and the experiences they gained encouraged Poles to mobilize and challenge various forms of authority. It should come as no surprise that the Polish agitation for parish council seats in Germany grew in the aftermath of the failed 1899 Herne strike, in which Poles were resoundingly condemned by all elements of German society, including those institutions connected with the Catholic Church.[39] Similarly, the 'Lattimer Massacre', in which 19 unarmed Polish and Hungarian workers were shot during a strike in 1897, served as a catalyst for the growing PNCC movement.[40] The situation was exacerbated by the article published in the semi-official newsletter of the Scranton Diocese which noted that the Poles involved brought 'terror to the hearts and the hearths of the better elements of the community' and were 'the scum of Europe who were carried here and dumped on our shores'.[41]

Finally, Polish agitation over religious matters was also spurred by the general political weakness of Poles. While Poles in Germany automatically had the right to vote since they were Prussian citizens, they could generally only influence politics indirectly by acting as a swing vote in elections. In Pennsylvania, the political aspirations of Poles were even weaker because many of the Polish immigrants were not citizens. In religious matters, however, Poles could have a significant impact on developments in both regions and as a result Polish agitation was channelled in this direction.

Although the similarity in conditions caused a tremendous general upsurge in Polish activism in religious matters, the directions Poles took in each region did differ. Two key factors are primarily responsible for this divergence. First, unlike the situation in Pennsylvania, where the right to religious self-determination was enshrined in the constitution, Poles in the Ruhr could not realistically have built their own ethnic parishes, let alone an independent national church, owing to the policies of the Prussian government.[42] Consequently Poles in the Ruhr needed to attempt to reform the Church exclusively from within. Second, the period in Pennsylvania when the PNCC was emerging, 1896–1902, was also a period marked by high levels of Polish worker radicalism, radicalism encouraged not only by the Lattimer massacre in 1897, but also by the lengthy strikes of 1900 and 1902 and the general growth of socialism in the region.[43] As a result, there was a willingness of some Poles to embrace the more radical religious ideology propagated by priests such as Hodur.

Despite the differences in the paths Poles in each region chose to take in ensuring their ethnic autonomy, the consequences of Polish religious agitation were markedly similar: in the long run, Polish agitation led to greater integration into their host societies, albeit in a form that could be described as 'negative' since it was the actions of the Catholic Church that forced Poles to become active in order to defend their beliefs. This assertion may initially seem contradictory, since Polish success in attaining parish offices, forming ethnic churches or even founding an independent church was based first and foremost on appeals to Poles' national consciousness. Nevertheless, in the process of agitating for and attaining greater ethnic autonomy, Poles in both regions forged stronger ties to their local communities and became stakeholders in the local society.

In the Ruhr, for example, the growing political strength of Poles in church elections caused Germans to reach pre-election compromises with Poles over the allotment of parish council seats rather than risk losing more seats in elections. Such agreements highlight the fact that inter-ethnic cooperation was possible and that ethnic relations, despite increasing at the national level, were stabilizing in various areas across the Ruhr. Moreover, in the aftermath of the First World War, when two-thirds of the prewar Polish population left the Ruhr, anecdotal evidence suggests that those Poles who were active on parish councils and in parish associations were more likely to remain.[44]

Similarly, in Pennsylvania, the founding of an ethnic parish represented a sizeable material investment by Poles and was in essence a declaration of intent to stay in their adopted community. Those who went further and joined the PNCC integrated over the long term, in many respects faster than those who did not. In this regard, three points are pertinent. First, despite arguments put forth by PNCC leaders that this institution was the true church of the Polish nation, joining this independent movement represented a clear break with the traditions of the ethnic community and the Polish minority that joined was subject to the ostracism of the majority. Second, the PNCC did not have the same resources as Polish–Catholic churches to preserve their ethnic identity long term. While most Polish–Catholic parishes had their own parochial schools, the vast majority of national parishes did not and, as a result, second generation children went to English-speaking public schools. Finally, the formation of the PNCC helped to settle for its members the ethnic animosity stirred by religion. While Catholic Poles continued to fight the Irish well into the 1930s and 1940s in the Diocese of Scranton, National Catholic Poles were making ecumenical outreach attempts to other American religious groups, most notably to the Episcopalians.[45]

Conclusion

In analysing working-class immigrants and their integration patterns, both in the period before the First World War and in the present, historians must carefully examine the continuing power of religion to inform worker outlook and action. Nowhere can this influence be better seen than among Polish coal miners in the Ruhr valley and northeastern Pennsylvania. In both regions, pre-migration traditions combined with post-migration experiences to transform religion into a powerful vehicle that provided the best means through which Poles could directly empower themselves.

The methods by which Poles defended their interests differed according to the environment in which they lived, with the overwhelming majority of Poles in Germany adopting a 'reformist' approach, while in Pennsylvania both Polish reform and schismatic movements existed. Despite these different paths, Polish mobilization in each region prior to the First World War led to similar results. That is, a significant segment of each respective migrant community was able to break out of an ethnically homogeneous sub-culture and engage with the larger local society in which they lived. Further, because this religious struggle was simultaneously linked to continuing efforts to achieve greater equality in the workplace, a broad, multifaceted ethnic working-class consciousness was able to emerge that in the long term aided the integration of many Poles into their adopted societies.

Notes

1 For supporting the research and writing of this article, I wish to thank the Alexander von Humbolt Foundation, the German Historical Institute, the Kosciuszko Foundation and the Institute for International Studies at UC-Berkeley.

2 For examples of such interpretations analysing Poles in the Ruhr valley of Germany, see Hans-Ulrich Wehler, 'Die Polen im Ruhrgebiet bis 1918', in Hans-Ulrich Wehler (ed.), *Moderne deutsche Sozialgeschichte* (Cologne, 1966); Krystyna Murzynowska, *Die polnische Erwerbsauswanderer im Ruhrgebiet während der Jahre 1880–1914* (Dortmund, 1979); Christoph Klessmann, *Polnische Bergarbeiter im Ruhrgebiet: Soziale Integration und nationale Subkultur einer*

Minderheit in der deutschen Industriegesellschaft (Göttingen, 1978). For examples
of such interpretations analysing Poles in northeastern Pennsylvania, see Michael
A. Barendse, *Social Expectations and Perception: The Case of the Slavic
Anthracite Workers*, Pennsylvania State University Studies, no. 47 (University
Park, PA, 1981); Perry K. Blatz, *Democratic Miners: Work and Labor Relations in
the Anthracite Coal Industry, 1875–1925* (Albany, NY, 1994).

3   John L. Kulczycki, *The Foreign Worker and the German Labor Movement:
Xenophobia and Solidarity in the Coal Fields of the Ruhr, 1871–1914* (Providence,
RI, 1994), p.262. For similar arguments regarding Poles in the Ruhr, see Richard
Charles Murphy, *Guestworkers in the German Reich: A Polish Community in
Wilhelmine Germany* (Boulder, CO, 1983); Valentina-Maria Stefanski, *Zum
Prozess der Emanzipation und Integration von Aussenseitern: Polnische Arbeitsmi-
granten im Ruhrgebiet* (Dortmund, 1984); John Kulczycki, *The Polish Coal
Miners Union and the German Labor Movement in the Ruhr, 1902–1934* (Oxford,
1997); Susanne Peters-Schildingen, *'Schmelztiegel' Ruhrgebiet: Die Geschichte der
Zuwanderung am Beispiel Herne bis 1945* (Essen, 1997). For such arguments
regarding Poles in northeastern Pennsylvania, see Victor Greene, *The Slavic
Community on Strike: Immigrant Labor in Pennsylvania Anthracite* (Notre
Dame, IN, 1968); Harold W. Aurand, *From the Molly Maguires to the United
Mine Workers: Social Ecology of an Industrial Union, 1869–97* (Philadelphia,
1971).

4   A classic example of this is found in E.P. Thompson's *The Making of the English
Working Class*, where religion acts as a vehicle through which workers are able
eventually to attain a broader class-consciousness and then recedes in importance.
To be fair, Thompson clearly points out that his history pertains to the English
experience only. However many labour narratives, particularly those focused on
labour movements in western Europe and the United States, often give only
perfunctory treatment of the role of religious identity within the working class,
viewing it as more of a sensibility than an active, need-oriented construct
influencing working-class mobilization.

5   While most histories written on Poles in both the Ruhr and Pennsylvania
acknowledge the role that priests and Church-supported Polish–Catholic Working Associations played in encouraging Poles to join trade unions, little in-depth
work has been undertaken on such figures or organizations and their continued
importance in the Polish workers' lives. One important source for primary
documents on Polish–Church relations in Germany is Hans Jurgen Brandt, *Die
Polen und die Kirche im Ruhrgebiet, 1871–1919* (Münster, 1987).

6   Of particular importance in helping to forge Catholic bonds across ethnic lines
was the *Kulturkampf* in Germany and the attacks of American Protestant circles,
represented in the period of the 1880s and 1890s most clearly by the emergence of
the American Protective Association (APA). The APA was an organization
created to counter the supposed threat that Catholic immigrants, viewed as agents
of Rome, posed to the American Republic. In Germany, Klessmann cites a telling
example of a German Catholic Centre party official calling Poles in 1885 'our
Catholic brothers', while in 1891 in Pennsylvania, the Irish Catholic brother of
the future Bishop of Scranton called Poles 'lovers of liberty and respectable
people'. Native Catholic pronouncements such as these would be all but unheard
of within another five years. See *Tremonia* (Dortmund), 1 June 1885, cited in

Klessmann, *Polnische Bergarbeiter im Ruhrgebiet*, pp.57–8; *Diocesan Record* (Scranton, PA), 16 May 1891.

7   By 1900, over one hundred Polish–Catholic organizations in Germany existed while, in Pennsylvania, there were 15 ethnically Polish Roman Catholic parishes. See Kulczycki, *Foreign Worker*, pp.40–47; *Scranton Diocesan Directory* (1990). While, in Pennsylvania, the government did not concern itself with religious matters, in Germany the initial support given Poles by the Church raised the ire of the state. In 1896, the Minister–President of Westphalia, Heinrich Konrad von Studt, warned in a report to the Prussian Interior Minister that the Church and Catholic Press were far too lenient with Poles and warned of the dangers of allowing Polish priests to act as political leaders. See Hauptstaatsarchiv Düsseldorf (HSTAD), Regierung Düsseldorf (RD) Praes. 867, Report of the Ober-Präsident der Provinz Westfalen, 31 October 1896.

8   By 1890, there was only one full-time Polish priest in the Ruhr, while in Pennsylvania there were fewer than 10. By 1903, the ratio of Poles to Polish-speaking priests was estimated at ten thousand to one in the Ruhr, while in Pennsylvania Poles often had to resort to hiring travelling clerics of dubious qualifications to meet their spiritual needs. See Klessmann, *Polnische Bergarbeiter im Ruhrgebiet*, p.139; Gallagher, *A Century of History: the Diocese of Scranton 1868–1968* (Scranton, PA, 1968), p.157.

9   In Pennsylvania, the tensions of sharing parish facilities caused extensive ethnic factionalism and led to the formation of many distinct national parishes. By 1899, 40 ethnic parishes existed, which exceeded the number of territorial parishes. The largest number of ethnic parishes belonged to Poles. See Gallagher, *A Century of History*, p.152. In Germany, Poles did not have the luxury of creating distinct ethnic parishes and as a result tensions in certain parishes often remained high throughout the period under study. For in-depth detail, see Brian McCook, 'Divided hearts: the struggle between national identity and confessional loyalty among Polish–Catholics in the Ruhr, 1904–1914', *Polish Review*, **47**(1) (2002), 67–96.

10 As a result of the *Kulturkampf*, there was a significant decline in the numbers of Poles in the upper echelons of the Prussian Catholic Church hierarchy. By 1900, there was only one Polish bishop in the hierarchy, the Archbishop of Gniezno-Poznan Stablewski, while subordinate positions throughout eastern Prussia were increasingly given to German priests. For a further elaboration, see Lech Trzeciakowski, 'The Prussian state and the Catholic Church in Prussian Poland, 1871–1914', *Slavic Review*, **26**(4) (1967), 618–37.

11 HSTAD, RD Praes. 867, Report of the Ober-Präsident der Provinz Westfalen, 31 October 1896.

12 Brandt, *Die Polen und die Kirche*, pp.15–16; Klessmann, *Polnische Bergarbeiter im Ruhrgebiet*, pp.93, 96; McCook, 'Divided hearts', 78–84.

13 HSTAD-Landesamt (LA) Essen 101, Report of the RP Düsseldorf to the Landräte ..., 25 August 1904. These various prohibitions unfolded over the course of 1904.

14 Hieronim Kubiak, *The Polish National Catholic Church in the United States of America from 1897 to 1980* (Cracow, 1983), pp.56, 62; Jay P. Dolan, *The American Catholic Experience: A History from Colonial Times to the Present* (New York, 1985), p.302. Dolan notes that, by 1900, two-thirds of the Catholic bishops in the United States were of Irish descent.

15   Joseph A. Wytrwal, *The Poles in America* (Minneapolis, MI, 1971), p.49. According to Wytrwal, pre-eminent American Catholic bishops such as John Ireland, James Gibbons and John Spalding 'regarded the use of the Polish language for church services, school instructions and the press as un-American and also un-Catholic'.

16   For a broader discussion of the idea of using the Church to promote the *Americanization* of immigrants, see Dolan, *American Catholic Experience*, pp.294–320; Kubiak, *Polish National Catholic Church*, pp.56–66. For a detailed description of the battle over ethnic parishes in northeastern Pennsylvania and Bishop O'Hara's involvement in them, see Gallagher, *A Century of History*, pp.163–9; James Earley, *Envisioning Faith: The Pictorial History of the Diocese of Scranton* (Devon, PA, 1994), pp.109–25.

17   Continuous conflicts over control of local parish property led American bishops to decide in the Third Synod of Baltimore in 1884 that such property be signed over to the local bishops in exchange for lay participation in local parish administration. The 1884 decision, however, never fully resolved the issue and many non-Irish Catholics continued to press for greater control over local parishes. For more detail see Dolan, *American Catholic Experience*, p.180; Kubiak, *Polish National Catholic Church*, pp.58–9.

18   Gallagher, *A Century of History*, pp.163–4.

19   Staatsarchiv Münster (STAM), Oberpräsidium (OP) 6396 – 'Stand der Polenbewegung', 22 April 1912.

20   STAM OP 6396 – 'Stand der Polenbewegung', 22 April 1912; Brandt, *Die Polen und die Kirche*, p.24; Gallagher, *A Century of History*, p.163.

21   According to STAM OP 5758 – 'Zahlenmaessige Angabe über das Polentum im rheinish-westfälische Industriebezirk-1912', Report of the Regierungspräsident Münster to the Ober-Präsident der Provinz Westfalen, 27 March 1913, over half of the total 1038 Polish associations in Germany were secular by 1912. In Pennsylvania, statistics are unavailable regarding the total number of Polish associations; however secular organizations such as local branches of the Polish National Alliance, many of whose leaders in northeastern Pennsylvania were socialists, were widespread. For further information on the PNA in northeastern Pennsylvania, see Joseph Wieczerzak, 'Bishop Francis Hodur and the socialists: associations and disassociations', *Polish American Studies*, **15**(2) (1983). Both regions also experienced a rapid growth in the politically active *Sokols* (Falcons), which were gymnastic/paramilitary societies.

22   STAM OP 6396 – 'Stand der Polenbewegung', 15 April 1911.

23   A Pole in a Polish–Catholic Association meeting in Castrop in 1905 best captured this sentiment when he declared, 'We Poles and Catholics assembled in this meeting ... resolve, as true sons of the Catholic Church, never to vacillate in our Faith in the fundamental tenets of the Catholic Church, even though we are increasingly oppressed in political and religious respects by the government and German clergy; just the same, however, we will never depart from the demand for equality in the Church' (*Wiarus Polski*, Bochum, 2 July 1905).

24   The campaign to appeal to the Pope lasted from 1905 to 1907 and ended because of the lack of necessary funds to support a delegation as well as disillusionment with the stance of the Pope in 1907 over the Polish Schools strikes in Poznan. See HSTAD RD-902, Report of the Polizeiverwaltung Essen, 25 February 1907.

25 Historisches Archiv des Erzbistums Köln – Generalia Tit. XX 25, vol. II, Letter from the Archbishop of Cologne to the Polish-Catholics of Oberhausen, 21 March 1919; Letter from the Generalvikariat in Köln to the RP Duesseldorf, 4 October 1919; cited in Brandt, *Die Polen und die Kirche*, pp.328, 330.

26 McCook, 'Divided hearts', 90–92.

27 *Scranton Diocesan Directory*, 1990.

28 In addition to the monograph by Kubiak, other key works on the history of the PNCC include Ksiega Pamiatkowa, '*33*' (Scranton, PA, 1930); Paul Fox, *The Polish National Catholic Church* (Scranton, PA, 1957); Stephen Wlodarski, *The Origin and Growth of the Polish National Catholic Church* (Scranton, PA, 1974).

29 *Scranton Republican*, 11 August 1896; *Scranton Republican*, 7 September 1896; Kubiak, *Polish National Catholic Church*, pp.102–7.

30 PNCC Archives, List of PNCC parishes with founding dates. For a detailed description of each parish, see Pamiatkowa, '*33*'. The independent movement among Poles did not emerge only in the Scranton area. In the 1880s and 1890s, independent Polish parishes led by renegade Polish clergy also developed, most notably, in Chicago and Buffalo. The PNCC, however, integrated these other movements into their own by the First World War.

31 Ksiega Pamiatkowa '*33*', pp.565–86; Pennsylvania State Archives, MG-215, 40th Anniversary Book of the Good Shepherd Church in Plymouth, PA, 1938. By the late 1930s, there were over 70 PNCC parishes in Poland, parishes that continue to exist today.

32 Those who stayed loyal to the Catholic Church derogatorily called those who joined the PNCC 'kickers' well into the mid-twentieth century. Conversely, PNCC members called Polish Roman Catholics 'suckers'.

33 Much of the ideology and precepts of the Church developed over the first three decades after the PNCC's founding. The basic tenets of the Church are laid forth in the '11 Principles of the Polish National Catholic Church', first published on the 25th anniversary of the PNCC in 1923. See *Po Drodze Zycia* (Scranton, PA, 1923).

34 In fact, as a seminarian in Cracow, Hodur was strongly influenced by the activities of the peasant populist priest Stanislaus Stojalowski, who in the late nineteenth century was a driving force in the organization of a peasant party in the Austrian partition. See Wlodarski, *Origin and Growth*, pp.40–41; *Straz* (Scranton, PA), 16 April 1898.

35 *Prace i pisma ksiedza biskupa Franciszka Hodura* (Scranton, PA,1939) p.69; cited in Kubiak, *Polish National Catholic Church*, p.134.

36 Trzeciakowski, 'The Prussian state and Catholic Church', 618–37.

37 Barbara Strassberg, 'The origins of the Polish National Catholic Church: the Polish national factor reconsidered', *PNCC Studies*, 7(1) (1986), 29–31.

38 For example, in Pennsylvania, reports circulated in the press that, in the early 1890s, the Irish priest Michael Hoban, later archbishop of Scranton, withheld sacraments and threatened to excommunicate Poles who failed to meet assigned contributions. See *Straz* (Scranton, PA), 12 June 1897. In the Ruhr, a priest in Hamborn engaged in a two-year legal battle with the Polish–Catholic associations in his parish, at one point using the German police to intervene and remove property that the priest claimed was stolen from him. (See HSTAD RD-897, Report of the Oberbürgermeister Hamborn to the RPD, 28 August 1913 and 28 August 1914.)

39   For an excellent discussion of the Polish strike in Herne, see Peters-Schildingen, *'Schmelztiegel' Ruhrgebiet*, pp.76–94; Klessmann, *Polnische Bergarbeiter im Ruhrgebiet*, pp. 75–9.

40   Earley, *Envisioning Faith*, p.134. The most thorough account of the Lattimer massacre itself can be found in Michael Novak, *The Guns of Lattimer* (New York, 1978).

41   *Diocesan Record* (Scranton, PA), 11 September 1897.

42   After the end of the *Kulturkampf*, Poles were increasingly subject to the oppressive *Germanization* policies of the Prussian state, including after the turn of the century laws attempting to restrict Polish associational activity and land ownership. For an excellent general overview of the political relationship between Poles and Germans see William Hagen, *Germans, Poles, and Jews: The Nationality Conflict in the Prussian East 1772–1914* (Chicago, 1980); Richard Blanke, *Prussian Poland in the German Empire, 1871–1900* (New York, 1981).

43   In addition to the works of Aurand, Greene, Barendse and Blatz, see also Donald Miller and Richard Sharpless, *The Kingdom of Coal: Work, Enterprise, and Ethnic Communities in the Mine Fields* (Easton, PA, 1998), for a detailed account of the 1900 and 1902 strikes.

44   There exists significant difficulty in determining the exact nature of the out-migration from the Ruhr in the early 1920s because precise statistical data for the Ruhr are unavailable from both German and Polish government sources. In individual towns, some information on the types of out-migrants can be ascertained by examining local city address books. For example, in the city of Gelsenkirchen, of the 16 Poles who held seats on various Church councils in 1910, ten, or 63 per cent, were still living in Gelsenkirchen in 1924–5, well after the massive out-migration of the early 1920s. See *Addressbuch Gelsenkirchen*, 1910, 1920, 1924–5. Similarly, 12 Polish associations, three of them connected to the Church, existed in the district of Osterfeld (city of Oberhausen) prior to 1921. By 1927, only four associations remained, including the same three religious associations that existed before 1921. See *Addressbuch Osterfeld*, 1921 and 1927.

45   The decision to enter into communion with the Episcopal Church USA occurred in 1946. This inter-communion ended in 1974 over conflicts pertaining to the Episcopal Church's ordination of women as well as the belief that the Episcopal Church's attitudes regarding homosexuality were too liberal. See Kubiak, *Polish National Catholic Church*, pp.140, 196–7. The larger issue of whether the formation of the PNCC led to greater or less integration in American society is the subject of continuing debates. Kubiak, for example, claims that 'the process of integrating the members of the PNCC into the American nation did not proceed any slower or faster than in other communities of Polonia'; ibid., p.187. For a more detailed discussion of the differing positions, see Eugene Obidinski, 'Ethnic conflict to ethnic conservation: changing functions of Polish National Congregations', *PNCC Studies*, **7**(1) (1986).

CHAPTER TWELVE

Networking among Welsh Coal Miners in Nineteenth-century America

Ronald L. Lewis

The main reason for British emigration to America in the nineteenth century was the hope of an improved standard of living. The tide of emigration ebbed and flowed, therefore, in response to the cycles of the British and American economies. Most British emigrants were skilled industrial workmen who resumed their old trades in America. Like their English and Scottish counterparts, the Welsh did not come merely to find work, as did so many unskilled Europeans of peasant background; they moved instead from skilled jobs in British mines and mills to comparable jobs in America. The United States Industrial Commission of 1907–10 reported that, of the bituminous coal miners, 88 per cent of the Welsh had been miners in Britain. A similar pattern emerged in the iron and steel industry where 72 per cent of the Welsh had prior experience.[1] This chapter explores the critical role of emigrant Welsh managers in the development of ethnic Welsh coalmining communities in nineteenth-century America.

During the nineteenth century, British coal production doubled and redoubled until, by 1860, the number of miners reached 300 000, and by 1914 their numbers exceeded 500 000. American coal operators depended on this vast pool of experienced labour to establish the industry in the United States. America's rise as an industrial power is measured by the dramatic expansion of production and labour in the basic industries. That growth was indeed dramatic in the coal industry. The US census enumerated only 6800 mine workers in 1840, but their numbers grew to 36 500 by 1860, doubled during the Civil War decade, surpassed 127 000 in 1900 and peaked at over 650 000 in 1920.[2] Eighteenth- and early nineteenth-century Welsh immigrants in the USA were primarily farmers seeking better land. By far the greatest influx, however, came with the migration of industrial workers between the 1840s and 1900 in search of better jobs. In 1850, there were only 30 000 foreign-born Welsh in the USA, but that number more than trebled by 1890 to reach a historic peak of 100 079. Their actual numbers were undoubtedly higher, perhaps double the official figure, because census takers often lumped the Welsh and English together, and the major Welsh coal-producing county of

Monmouthshire was administratively part of England during this period. The most striking feature of the nineteenth-century migration was its concentration in specific industrial locations.[3]

Welsh miners came first to the anthracite fields of eastern Pennsylvania, but they were soon followed by others bound for the bituminous fields of western Pennsylvania, Ohio, Illinois and beyond the Mississippi River as the mining frontier advanced westward during the nineteenth century. Known for their opposition to slavery, their support for the Republican Party and loyalty to craft unionism, Welsh immigrants found little to attract them to the southern states, which were noted for their oppressive labour systems. Moreover, the limitless coal region of Central Appalachia was developed after Welsh immigration began to decline. Therefore the vast majority of Welsh miners settled in the anthracite fields of Pennsylvania, and the bituminous fields of Ohio, and then scattered more thinly throughout the mining regions further west.

Welsh industrial immigrants, like the British generally, were the expert workers of the industrial revolution who transferred their knowledge and skills to the infant coal industry of nineteenth-century America. They moved from field to field, coal town to coal town, and mine to mine, making decisions about when and where to go based on knowledge passed along through networks of personal and professional relationships that bound Welsh occupational communities together.

Ethnic networks are a general feature of American immigration history. Transitional institutions, such as clubs, societies and fraternal orders, newspapers and churches not only helped to fill the void immigrants felt when torn from their own culture and transplanted in America, but they also gave form and meaning to ethnic communities. In these enclaves members of the group found necessary services, sympathetic people who understood their language and culture, and acquired vital information about work opportunities, which further bound them together as an expatriated community. As the Welsh adapted their institutions to American life, they continued to maintain communications with kith and kin back in Wales and with other émigrés in America through complex formal and informal networks linking fellow travellers within the diaspora and back to the homeland.

Various modes of communication, both institutional and individual, formal and informal, linked the Welsh. Labour leaders in Wales often cooperated with their American counterparts by assisting in the formation of emigration societies to support miners who wanted to emigrate, and maintained contact with and occasionally visited the emigrants in the USA. Moreover mining news on both sides of the Atlantic was reprinted in labour journals, providing an important two-way flow of information regarding the state of the coal trade, employment and mining legislation in Britain and America.[4]

One of the most important information networks, however, was personal correspondence with other émigrés, and with family and friends back in Wales. The importance of personal correspondence in Welsh emigration has been ably demonstrated by Anne Knowles's study of the Welsh in the Ohio counties of Jackson and Gallia, and by Alan Conway's edition of Welsh émigré correspondence to friends and relatives back home,[5] but contemporary observers also understood the significance of personal correspondence in the movement to America. The *Merthyr Times* was distressed in the late 1860s 'to witness the large number of colliers and miners who are constantly leaving the iron and coal districts of South Wales from Merthyr, Aberdare, Pontypool, and other centres of population. It is an ordinary occurrence to witness the departure of 100 to 120 ... weekly'. The writer noted that 'the passage money of a large number of them has been paid by relatives and friends who left their native home years ago, and who have since so far prospered as to be able to render this assistance to their connexions. As usual a large majority of the emigrants are leaving for the United States'.[6] As Aled Jones and Bill Jones have demonstrated in their recently published study, *Welsh Reflections: Y Drych and America, 1851–2001*,[7] Welsh American newspapers were a vital source of news about Wales and Welsh America which helped to reinforce intra-group identity.

The ethnic press reinforced Welsh identity at a general level, but newspapers like the *Druid* also had more pragmatic information for loyal Welshmen. For example, in 1914, 'a well known Cymro' of Moundsville, West Virginia, posted notice that Gwelym Stephens had passed the examination for his first-class mine foreman certificate. The writer hoped that 'some Welshman in need of a first class foreman will soon meet him'.[8] Similarly, the Welsh language newspaper, *Y Drych*, published a letter from a group of Welsh miners who had opened a mine in Cadiz, Ohio, declaring their preference for hiring Welsh miners to work in the mine.[9]

How miners operated within this broader ethnic web of communications is an intriguing topic, but this chapter focuses on the key role played by mine owners and managers in initiating and sustaining Welsh mine communities. In his famous history of Calvinistic Methodism in America, the Reverend Daniel Jenkins Williams lamented that strikes, closings and new openings 'played havoc' with church organizations by having an 'unsettling' impact on Welsh communities. 'It was the habit of Welsh miners,' he observed, 'to follow their leader from one location to another. When a Welsh superintendent was transferred from one place to another, or found a more lucrative and inviting position elsewhere, many of those in his employ followed him to the new location to work.'[10] Clustering in occupational communities was controlled in large measure by the geographical location of the coal itself, of course, but it is clear that Welsh miners nearly everywhere in America saw more opportunity when one of their own was in charge of the operation. This belief was more important earlier in the industrial migration than later, but for most of the nineteenth century it remained a prominent feature of Welsh mining settlements. Eastern Pennsylvania, one the earliest American coalfields to be developed, reveals this pattern very clearly.

Just outside St. Clair, Schuylkill County, William and Thomas Johns leased coal lands in 1846 and built Johns Eagle Colliery, one of the most substantial of the early coal mines in the anthracite region, as evidenced by the commentary it elicited from contemporaries. The Johns brothers, and then their sons, were experienced miners from the coal valleys of South Wales who lived within 100 yards of the breaker, exposing them to the same noise and dirt that afflicted their employees. Even after they became millionaires they continued to live near their operation and those they employed. Nearly all of the 150 to 200 miners and bosses were Welshmen, Protestants and Republicans like the Johns. The first superintendent, John Reese, came to America about the same time as the Johns and served from 1846 to the time of his death in 1856. His son William also served as superintendent from 1871 to 1888.[11]

The Welsh were numerous throughout the middle and southern anthracite fields of eastern Pennsylvania as well, but the largest concentration was in the northern field of the Wyoming Valley. This coalfield lay primarily within Carbon, Luzerne and Lackawanna Counties, and included such well-known destinations for Welsh emigrants as Nanticoke, Plymouth, Kingston, Pittston, Olyphant and, most importantly, Wilkes-Barre, Scranton and Carbondale. Numerous Welsh managers were appointed with instructions similar to those given to John T. Griffiths of Wilkes-Barre, who was appointed general superintendent of the Consolidation Coal Company. Charged with the task of increasing the number of Welsh coal miners, he and his friend and fellow manager Lewis S. Jones succeeded in luring 'many Welsh labourers to the mines' to Wilkes-Barre and the surrounding mine towns.[12]

In fact, the largest concentration of Welsh in the entire USA was in the northern field encompassing two counties, Luzerne and Lackawanna, where a total of 16 286 foreign-born Welsh made their homes in 1900. At the centre of this population was the city of Scranton, which historian William D. Jones has described as the 'epicentre' of Welsh America.[13] The most influential Welsh mine superintendent in nineteenth-century America was probably Benjamin Hughes (1824–1900) of Scranton, the general superintendent of mining for the Delaware, Lackawanna and Western Rail Road Company (DL & W). According to Reese Hughes, one of the Welsh pioneers of nearby Carbondale, who was himself a prominent miner, the legendary general superintendent 'was more instrumental probably than any other man in aiding in this emigration of the Welsh miner by promptly finding employment for every one who arrived in that city'.[14] Benjamin Hughes was born near

Brynmawr, the son of a foreman at the Nantyglo Ironworks. He emigrated to the United States in 1848, and went to work in the mines first in Pottsville, but by 1850 had settled in Scranton, where he became foreman and then superintendent of the DL & W's Diamond Mine in the Hyde Park neighbourhood of Scranton. His reputation was such that the company appointed him general inside superintendent of mines, which by 1890 placed seven thousand men under his supervision. In addition, he was involved in numerous local businesses, and served on important professional boards, including the Pennsylvania Board of Examiners for mine inspectors.[15]

Hughes was noteworthy for his dedication to furthering the interests of Welsh Americans, particularly in mining. One authority, William D. Jones, declares that Benjamin Hughes was 'the father of the Welsh community in Scranton'. He actively recruited Welsh miners, and then provided them with jobs. He often wrote to his Welsh subordinates in the Welsh language to inform them of their failings in order to protect them from their enemies within management.[16] While he was obviously a good company man, Ben Hughes was a good *Welsh* company man. Patronage carried a price, however. Reese Hughes, himself a mine manager, described Benjamin Hughes as 'a remarkable man' who 'insisted on every one joining and attending church as soon as he gave them a job'. Nearly all the Welsh mining bosses of those days, he recollected, 'were good religious men and they would go into the mines to see the men who neglected [to attend] services the Sunday previous'. Sunday was a long day for those who deferred to their superiors because the old-time bosses 'believed in spending the whole day in church'.[17]

Welsh owners, bosses and miners migrated westward with the mining frontier during the last half of the nineteenth century, but these patriarchal characteristics were not so fully developed elsewhere in the American coalfields because the country was so large and the Welsh so few in number. Nevertheless many Welsh miners continued to find security in mine towns run by Welsh owners and managers in other American coal regions. Immigrants often headed for established Welsh mining centres like Scranton and Wilkes-Barre, and then moved about following news of job opportunities either to smaller regional mine towns or further west to the newly opening coalfields.

Ohio claimed the second-largest population of Welsh mining immigrants. The largest percentage of them were concentrated either in Trumbull and Mahoning Counties, in the Mahoning Valley coalfield in the northeastern part of the state near Youngstown, or in the Hocking and Ohio River valley mining towns of southeastern Ohio. These fields developed rapidly during the 1860s and 1870s, when Welsh managers placed the industry on a commercial footing, and then recruited skilled miners from the homeland. John Davis was hired by David Tod, the Civil War governor of Ohio, to develop the coal found on Tod's farm, which now lies within the city of Youngstown. During the 1830s and 1840s, Davis recruited miners from Wales, and personally led a 'substantial number' of emigrants to the Mahoning Valley. Thomas Davis and Morgan Reese were the first to cut the coal on Tod's Brier Hill farm. Another Welshman, William Philpot, erected the famous Eagle Furnace at Brier Hill, the first in this major iron and steel centre, and it too was managed by a Welshman, William Richards.[18] The pioneering mine in adjoining Trumbull County was the Cambria, opened and operated in the 1850s by Welshmen John Morris, W.T. Williams and John Lewis.[19]

The Mahoning River valley became a major Welsh destination during the last half of the nineteenth century, and by 1880 it was home to between 6000 and 8000 Welsh miners and their families. Because the 1890 census was destroyed by fire, and the 1880 census combined the English and Welsh into one category, it is difficult to say precisely how many residents were Welsh. A manual calculation of the Welsh-born residents in the Mahoning Valley gives a minimum of 5000. It is impossible to determine how many were coal miners from the census, but, from manual counts in strategic locations, two-thirds of the Welsh heads of household in Trumbull were miners in 1880.[20]

Even though it would be impossible to link directly the thousands dependent on mining with the number of Welsh mine owners and managers in Ohio, the positive correlation is unmistakable. In a mine-by-mine survey conducted by the Ohio inspector of mines in 1875, a total of 24 mine managers were listed for Trumbull County. Of the 20 whose nativity can be identified, 13 were Welsh born, and two others probably were Welsh. Of the 18 mine managers identified in Mahoning County, eight of the ten whose nativity can be determined were born in Wales, and another was married to a Welsh woman.[21]

On a smaller scale than the Pennsylvania fields, the Mahoning Valley was often the first home for many Welsh miners and managers. The large Welsh mining community was a welcome oasis for immigrants, and enabled them to acquire an economic foothold and get their bearings before moving on to other Ohio mining districts or further west where new opportunities beckoned. Even in the small, new mine communities created by Welsh owners and managers, however, the established pattern was perpetuated. Anthony Howells, for example, was born in Dowlais and worked underground until 1850 when he emigrated to the USA. He headed straight for the Brier Hill mine in Youngstown, which was still being managed by a Welsh superintendent who continued to employ a substantial Welsh workforce. After five years underground, Howells opened a provisioning business, and in 1870 moved to Massillon, Ohio, where he launched the Howells Coal Company. He hired another Welshman, Evan Evans, as his superintendent, who in turn recruited a primarily Welsh workforce. The town of Justus grew up around the mine and remains to this day occupied by many descendants of those early families.[22]

Few coal mining settlements revealed the community generative power of Welsh managers so clearly as nearby Thomastown, now incorporated into the city of Akron. The first commercial mines there were opened after the Civil War, but were particularly active in the 1880s.[23] Thomastown probably derived its name from Llewelyn Thomas, one of the original coal mine owners. According to the manuscript census for 1880, about 58 per cent of the population were born in Wales. That does not tell the whole story, however, for another 20 per cent were connected with the Welsh through marriage to a spouse either born in Wales or whose parents had been born in Wales. Of the total population of 622, at least 80 per cent were ethnically linked with the Welsh. Of the 193 employed outside the home, 147 (76.16 per cent) were coal miners.[24]

Thomastown certainly supports the thesis that Welsh miners gravitated to coal towns where Welsh owners and superintendents were in charge. Of the six men recorded in the 1880 manuscript census as bosses, three were 'coal operators' (owners), two of whom were born in Wales and also married to Welsh women, while the third was born in Ohio although his wife was born in Wales. Two superintendents were born in Wales, and one was an English (although he might have been born in Monmouthshire) 'stable boss' who was wed to a Welsh woman. In short, four of the six supervisors were definitely Welsh born, and the other two were at least linked to the group through marriage.[25]

Mine managers continued to generate Welsh communities on the mining frontier as it expanded westward during the last half of the nineteenth century. Experienced Welsh mine managers were hired to open new mines in Iowa, Missouri and Kansas, and as they took up new positions some of their former employees followed them. When the Mahoning Valley field in Ohio began to decline in the 1880s, a Welsh manager was hired from Trumbull County to open a new mine at Hiteman, Iowa. Among the nucleus of 12 Welsh mining families whose origins were known to the Reverend Owen Thomas, seven had moved there from the Mahoning Valley, three came directly from Wales, and two from Soddy, Tennessee, another mining centre established by Welsh managers.[26]

Even as far away as Black Diamond, Washington, the most distant coalfield on the mining frontier, being Welsh still meant something as late as 1900, when Albert Garrett's father brought his wife and two children to America. Black Diamond was managed by Morgan Morgan, a Welshman prominent in mine engineering circles. Morgan filled all of the supervisory and skilled positions with Welsh miners. According to Albert Garrett, his father found Morgan's son minding the office when he arrived. As he had been coached to do, Garrett declared specifically that 'he was a Welsh coal miner and would like to have work'. Young Morgan responded, 'We don't have any work for you. We can't hire you. We have all the men we need.' But his father overheard the conversation; emerging from his office he chastised his son, 'You hire that boy. He's a Welsh coal miner.'[27]

A map displaying the distribution of the Welsh population would show that the Welsh were not in abundance in all American coalfields, however, and it is no accident that very few Welsh miners located in the South. Not all mine managers with strong Welsh identities conformed to the general pattern outlined above, and Llewellyn W. Johns (1844–1912), probably the most influential mine manager in the Birmingham, Alabama, coal and iron district, was one such manager. Even though he had benefited so much from the Welsh ethnic network, when in the position to repay his ethnic obligation Johns rejected traditional cultural nepotism as a matter of good company policy.

Johns boasted that he was 'born atop of a coal mine' in Pontypridd, Wales. Economic need forced him to leave school at a tender age and enter the mines, and in 1863, at 18, he emigrated to the anthracite fields of Pennsylvania.[28] Johns's early years in America read like a wild west adventure. He seems to have been constantly on the move, working at a variety of jobs from coast to coast, prospecting for gold, building bridges, mining coal, cutting timber, even fighting Indians. He returned to the anthracite field from his first trip west in 1868, quickly found employment in the mines near Scranton, and was soon promoted to the position of mine engineer. He also worked for several family members of David Thomas, founder of the anthracite iron industry in the USA, including Major William R. Thomas, a nephew of Samuel Thomas (David's son). Major Thomas informed Johns that the Thomases were expanding their iron interests into the South, that he himself was moving to Rising Faun, Georgia, to build a furnace, and he urged Johns to seek his fortune in this developing region. The year 1877 found Johns back in the Birmingham coal and iron district working for James Thomas as the superintendent of the Helena coal mines. Johns benefited from a complex set of acquisitions and mergers of the fledgling coal and iron companies which occurred during the rapid development of the Birmingham coal and iron district, managing ever larger operations until he achieved the position of chief mining engineer for Republic Iron and Steel Company.[29]

A biographer writing of Llewellyn Johns in 1887 declared that 'Captain Johns' was 'one of the important factors' in the rise of the district as 'the iron and coal center of the south'.[30] However that may be, he did not extend the occupational advantages of the Welsh network to others of his nationality. In fact, even though Johns employed some Scots, English, Irish, Cornish and German miners, he refused to hire Welshmen at all. In an interview with a newspaper reporter in 1886, he explained: 'We have no Welsh at the mines. I am a Welshman, and I do not have them because being my countrymen, they would expect favors from me which they could not receive. They are among the best miners in the world,' he declared, but he still refused to hire them.[31]

Like Benjamin Hughes, his counterpart in Scranton, Johns retained a strong Welsh identity throughout his life in America. He made return visits to Wales on several occasions and remained in close contact with relatives who still lived there. His national pride was apparent in 1908, when he responded to a letter in the *Druid* inquiring about the identity of Evan James (Ieuan ap Iago), the composer of the Welsh national anthem, 'Hen Wlad fy Nhadau' ('Land of My Fathers'). Johns wrote that James was a resident of Pontypridd when Johns was growing up. On 'my last trip to Wales,' he wrote, 'I ... visited his resting place at Carmel where my dear old grandfather (also) lies'. Johns proudly proclaimed that all of his immediate family were 'personal friends' of Evan James.[32] In fact, the bard wrote a poem and gave Johns a photograph of himself when he emigrated to America which Johns prized to the end of his days. While on an inspection tour of the Panama Canal, Johns 'was called on for a song, and on the top of the greatest cut ever undertaken ... I arose and stepped on top of a boulder and took off my hat to the Atlantic and Pacific Oceans ... and sang, as I never sang before, that good old song "Hen Wlad"'. Johns signed the letter with his bardic name, 'Tragolwyn'.[33]

Llewellyn Johns also maintained close association with a small circle of Welsh-born friends who met periodically at his Birmingham home, 'The Elms'. The press reported that he was 'the best known and most popular Welsh–American in this part of the country'.[34] In 1912, the papers reported another meeting of 'The Welsh Boys' who spent an evening in song at the Johns home. Who made up this circle of Welsh managers remains uncertain, but along with some prominent Welsh ironmasters, such as Samuel Thomas and Giles Edwards, several Welsh mine superintendents undoubtedly belonged to that august body, including Richard Thomas, born in Baglan Parish, Glamorganshire, who operated coke ovens and coal mines at Coalburg, Alabama.[35] His son, Edward Thomas, was superintendent of Little Warrior Coal and Coke Company in Littleton, Alabama. Also there was John X. Thomas, born in Rhymney, who rose through the ranks from digger to general mine superintendent at Ensley, Alabama.[36]

George A. Davis was a member of this circle as well. A native of Dowlais, he immigrated to America and settled in St. Clair, Pennsylvania, and in 1888 moved to Alabama. Davis was the exception that proved the rule among Welsh mine managers in Alabama. He managed the Belle Sumter mines, and erected the only known Welsh church ever built in Alabama, where the Reverend D.M. Lewis preached in Welsh. He was unique in another way for, unlike Llewellyn Johns and the other 'Welsh Boys', Davis built the church to serve a Welsh mining community that he recruited. Little is known of this community, but, if his obituary is correct, 'wherever he lived there would always be found a colony of Welsh miners, for he was never known to turn away a countryman'.[37] The large mining complex managed by Davis was owned by another Welshman, David Roberts, who emigrated in 1873 from Anglesey, Wales.[38]

Even though Johns, and apparently most of the other 'Welsh boys', were proud of their heritage, the larger social and political imperatives of racial segregation they found in the American South overrode their ethnic impulses when it came to business. Johns and his compatriots continued to play the role of managerial patriarch, but, with the single exception of Davis, their dependants were African–Americans trapped in the nexus of segregation. It is clear that Johns and nearly all of the other Welsh managers in the Birmingham district accepted the South's racial norms and ideology. 'The majority of our men are negroes,' Johns explained to a reporter, 'who are the best workers and stand heat and cold better ... We could not do without negroes ... The work does not require brains and they do splendidly.'[39] By the late nineteenth century, thousands of convicts from the Alabama state prison were leased to coal mining companies. Pratt Mines, which was under Johns's direction, was the largest coal mining complex in the South, and also the largest employer of convict labour in Alabama, most of whom were displaced agricultural workers who found themselves toiling in a prison mine. One authority estimated that more than 50 per cent of the African–American coal miners in the Birmingham district learned their trade while employed as convicts.[40] When a US Senate investigating committee asked a black convict if an ex-convict could find employment in the mines, he responded: 'When he is released he will generally get employment immediately. He generally goes to Mr. Johns here and he will give him a job of some kind.'[41] The system functioned like a vocational labour school for black convicts who learned the craft at a cost of pennies a day to their employers. Upon release they provided a labour reserve which by state law and local custom required that they be paid less than a white workforce. Small wonder, then, that few British miners found the prospect of working in the South attractive.

How do we explain the different hiring strategies adopted by Welsh managers represented by Benjamin Hughes and Llewellyn Johns toward members of their own ethnicity? While Hughes fused professional management, ethnic progress and his own personal responsibility, Johns found success by separating them and elevating cost accounting and regional racial norms above personal and ethnic loyalties. Even though Welsh mine owners and managers played an important role in the South's coal industry, unlike their counterparts in the North they insulated themselves within a closed racial system rather than recruit their countrymen to open, dynamic communities. Consequently few Welsh miners either were recruited or moved to the Birmingham coal and iron district. In fact, the 1900 census recorded only 306 residents in the entire state of Alabama who had been born in Wales;

of the 252 (82.35 per cent of the total) who resided within the Birmingham district,[42] most worked for George A. Davis at Belle Sumter. But this represented a minuscule population for a coal region known as the 'Pittsburgh of the South', particularly when compared with the tens of thousands of Welsh living in the mine and mill regions of Pennsylvania and Ohio.

Even though Welsh industrial immigrants were overwhelmingly miners and iron mill hands, occupation alone does not fully explain why Welsh mining communities were concentrated in particular locations. Ethnic networks and institutions are important in understanding Welsh migration patterns, but within those ethnic networks mine managers played a critical role in determining where Welsh coal communities actually took root. Even a cursory comparison of the two polarities in ethnic managerial strategies represented by Hughes in Pennsylvania and Johns in Alabama dramatically illustrates this point.

Notes

1   Rowland Tappan Berthoff, *British Immigrants in Industrial America, 1770–1950* (Cambridge, 1953), pp.19, 23, 28–9.
2   Ibid., p.7; United States censuses of population for 1840, 1860, 1900 and 1920.
3   William D. Jones, *Wales in America: Scranton and the Welsh, 1860–1920* (Cardiff, 1993), pp.xvii–xviii.
4   Amy Zahl Gottlieb, 'The regulation of the coal mining industry in Illinois with special reference to the influence of British miners and British precedents, 1870–1911', PhD, University of London, 1975, 36–51; Bill Jones, ' "We will give you wings to fly": emigration societies in Merthyr Tydfil in 1868', *Merthyr Historian*, **13** (2001), 27–47. American labour newspapers of the late nineteenth and early twentieth centuries which were devoted to the interests of coal miners regularly carried international news about the industry, particularly from Britain. See especially *United Mine Workers Journal* and *National Labor Tribune*.
5   Anne Kelly Knowles, *Calvinists Incorporated: Welsh Immigrants on Ohio's Industrial Frontier* (Chicago, 1997); Alan Conway (ed.), *The Welsh in America: Letters from the Immigrants* (Minneapolis, 1961).
6   *Merthyr Times*, 25 November 1869, quoted in Amy Zahl Gottlieb, 'Immigration of British coal miners in the Civil War decade', *International Review of Social History*, **23** (1978), 369.
7   Aled Jones and Bill Jones, *Welsh Reflections: Y Drych and America, 1851–2001* (Llandysul, Wales, 2001).
8   *Druid*, 2 November 1914.
9   *Y Drych*, 26 May 1881, cited in Jones and Jones, *Welsh Reflections*, p.59.
10 Daniel Jenkins Williams, *One Hundred Years of Welsh Calvinistic Methodism in America* (Philadelphia, 1937), pp.113–14.
11 Anthony F.C. Wallace, *St. Clair: A Nineteenth-Century Coal Town's Experience with a Disaster-Prone Industry* (New York, 1981), pp.103–5.

12 Williams, *Calvinistic Methodism*, pp.102–13, quotation on p.106.

13 Jones, *Wales in America*, pp.xx, xxii.

14 *Druid*, 26 June 1913.

15 *Cambrian*, **15** (January 1895), 1–5; **20** (May 1900), 232–4; Jones, *Wales in America*, p.34.

16 Jones, *Wales in America*, p.35.

17 *Druid*, 26 June 1913.

18 Howard C. Aley, *A Heritage to Share: The Bicentennial History of Youngstown and Mahoning County, Ohio* (Youngstown, 1975), p.45.

19 *Sixth Annual Report of the State Inspector of Mines to the Governor of the State of Ohio for the Year 1880* (Columbus, 1881), pp.59–60.

20 Norman and Mary Lou Ulam (compilers), *1880 Census Index of Trumbull County, Ohio* (Warren, 1991). There is no similar index for Mahoning County for 1880.

21 Ibid; *Second Annual Report of the State Mine Inspector to the Governor of the State of Ohio for the Year 1875* (Columbus, 1876). The Ohio mine inspectors were assigned districts for which they were responsible. Their annual reports contained information on each mine, including the names of the managers. I have compiled a database of these mines and managers, and attempted to identify the nativity of each from the dicennial censuses for 1870 and 1880 and other sources. The results are uneven because 1875 falls between censuses, and not all managers were at the works in 1870 or 1880.

22 *Druid*, 26 June 1913; *Cambrian*, **17** (July 1897), 337–8.

23 Karl H. Gismer, *Akron and Summit County* (Akron, 1952), p.575 for quotation. See also Scott Dix Kenfield, *Akron and Summit County Ohio, 1825–1928*, vol. 1 (Chicago and Akron,1928), p.150; Samuel A. Lane, *Fifty Years and Over of Akron and Summit County* (Akron, 1892), p.986.

24 Compiled from the 1880 United States Census of Population, Summit County, Ohio, Village of Thomastown, microfilm manuscript schedules.

25 Ibid.

26 For examples of early Welsh mining communities in Iowa, Missouri and Kansas, see the Reverend R.D. Thomas, *Hanes Cymry America*, translated by Phillips G. Davies (Lanham, MD, 1983, reprint of 1872 edn), pp.213–90; Berthoff, *British Immigrants in Industrial America*, pp.50–51. *Druid*, 18, 25 April, and 16, 30 May 1912. The Reverend Owen Thomas himself had lived in the Mahoning valley and remembered some of the Mahoning Welsh who moved to Hiteman (*Druid*, 18, 25 April, and 16, 30 May 1912).

27 Diane and Cory Olson (eds), *Black Diamond: Mining the Memories: An Oral History of Life in a Company Town* (Black Diamond, Washington, 1988), pp.13–14.

28 John Witherspoon DuBose, *Jefferson County and Birmingham, Alabama: Historical and Biographical* (Birmingham, 1887), p.561.

29 Ibid., pp.562–5; Ethel Armes, *The Story of Coal and Iron in Alabama* (Birmingham, 1910), p.292.

30 DuBose, *Jefferson County and Birmingham*, p.564.

31 *Birmingham Age*, 5 February 1886.

32 *Druid*, 19 March 1908.

33 Ibid.

34 Ibid., 19 October 1911; 11 January 1912.

35 Cambrian, **26** (April, 1906), 180–81; **29** (May, 1909), 13.

36 *Druid*, 22 April 1909; 18 April 1912.

37 *Druid*, 1 January 1920 (quotation); *Y Drych*, 15 May 1893 (courtesy of Eirug Davies).

38 Armes, *Story of Coal and Iron in Alabama*, pp.330–36, 425.

39 *Birmingham Age*, 5 February 1886.

40 United States, 61 Cong., 2d Sess., Senate Document No. 633, *Reports of the Immigration Commission, Immigrants in Industries*, Part 2 (Washington, DC, 1912), p.218.

41 United States, 48 Cong., 2 Sess., Senate, *Report of the Committee of the Senate Upon the Relations Between Labor and Capital, and Testimony Taken by the Committee*, Vol. 4 (Washington, DC, 1885), p.434.

42 United States, Twelfth Census, *Census Reports: Population*, Part I, Vol. 1 (Washington, DC, 1901), p.736.

CHAPTER THIRTEEN

Gender and Ethnicity in Japan's Chikuho Coalfield

W. Donald Smith

Japan's elites have been successful in painting a picture for the world of their country as homogeneous and harmonious, leaving many Western observers with the impression that the history of working people in Japan has little in common with Western labour history. In reality, however, ethnicity, gender, region and status have divided the modern Japanese workforce in all too familiar ways, and struggle and conflict have been integral parts of labour relations. Japan is in many ways part of the mainstream of modern labour history, with one major exception: the role of women miners, who were integral to Japanese coal mining until the onset of the Depression, decades after they vanished from European mines.

Japan is often called a resource-poor country, but in fact it has several major coalfields that fuelled its industrial revolution and imperialist expansion into Asia. This chapter will focus on one of these fields, Chikuho in southwestern Japan. This 787 square kilometre area between the present-day cities of Fukuoka and Kita-Kyushu was important both as Japan's top-producing field for most of the modern period and for its diverse labour force. Most Japanese coalfields employed large numbers of either women (mostly Japanese) or Koreans (mostly men) between the two world wars, but Chikuho had large numbers of both. Chikuho and other coalfields in Fukuoka Prefecture employed 51.5 per cent of all male and 59.7 per cent of all female coal miners in Japan in 1930, along with 56.6 per cent of Korean men and 69.3 per cent of Korean women mine workers.[1] Women were important as coal haulers while Koreans, as a reserve army of labour, made it easier for collieries to keep the workforce divided and to delay mechanization.

Women and Men in the Chikuho Mines

Women played an important role in Chikuho long before the first modern mines opened in the late nineteenth century. For centuries, men, women and children had dug coal as a sideline to agriculture in this densely populated region, so modern mines hired both men and women as a matter of course. (In the sparsely populated northern island of Hokkaido, Japan's second most important coal mining region, by contrast, little mining was carried out until after the 1868 Meiji Restoration that put Japan on the path of capitalist industrialization. With no tradition of family-based mining, very few women ever became underground miners in Hokkaido.[2])

While women had thus worked in Chikuho mines from the beginning, it was not until the closing years of the nineteenth century that they became structurally indispensable to modern mining. With the Japanese industrial revolution picking up speed, enabling and spurring imperial expansion into Asia, large mines run by the *zaibatsu* conglomerates began installing winch engines to speed the hauling of coal out of their mines. These winches had to be operated continuously to maximize profits, but the workers who dug the coal were not provided with any new equipment to help them keep pace. To produce a steady supply of coal for the winches to haul up, the rapidly expanding mines needed large numbers of new full-time workers, and many of them were women. There was a surplus of female labour, and women could be paid less, so for management it made sense to use men only in jobs for which they were seen as indispensable, such as hewing, and to use women for other tasks, such as hauling the coal to a central loading point.[3]

Women mine workers were thus employed in large numbers in Chikuho because of pre-existing patterns of work, women's relatively low wages and the need to expand the labour force rapidly in a time of soaring production. Chikuho mines had an additional reason for hiring women, however. By putting women to work underground along with men, often their husbands, Chikuho mines found they could improve male retention rates. This was an important consideration in an era when monthly turnover in Chikuho approached 15 per cent.[4] The number of women mine workers continued to increase through the first two decades of the twentieth century, peaking in about 1920. In that year, the census counted 66 396 women working underground in Japan's coal mines, making up 26.6 per cent of the underground workforce. Another 28 474 women worked on the surface (30.6 per cent of surface workers), for an official (no doubt understated) total of 94 870 women mine workers (27.7 per cent of the total).[5]

Just what kind of work did women do? As of 1930, nearly two out of three (63 per cent) worked underground, most (54 per cent of the total) of them as haulers or helpers, who often worked in pairs with male hewers. The others hauled timber, did repairs and engaged in other manual tasks. Of the remaining third who worked on the surface, most (28 per cent of the total) were coal sorters, while the others hauled coal or did other manual tasks. Women, not surprisingly, were shut out of managerial and technical positions.[6] There was a clear link between marital status and place of employment, but not the link that students of European mining might expect.

While single women in Belgium, for example, tended to work underground and married women sorted coal,[7] the opposite pattern prevailed among Japanese in Chikuho (but not among Koreans, who tended to be married whether they worked in the pits or above ground).[8] And, in contrast to countries like the United States, where coal sorting, like other mine work, was generally a male preserve, the few males in prewar Japanese sorting plants were either supervisors or boys too young to go into the mines.[9] Women filled the gap, doing work that was dull and repetitive but comparatively safe.

Hauling coal underground was much harder work, of course. A hauler generally carried coal either in a bamboo carrier on her (or, in some cases, his) back or by pushing a basket on runners; either way, it was an exhausting task that often required the worker to crawl through narrow passageways, grasping a lantern with her teeth. 'A lot of women had miscarriages right down there in the pit,' one miner recalled. 'We had to climb up and down these steep inclines carrying things was why.' After inserting a straw sandal or a piece of newspaper to staunch the bleeding, 'we'd go on working as if nothing had happened'.[10] Conditions underground were frequently dangerous too. Many women were well aware of this before they took their first steps into a coal pit, because it was the death of a family member that compelled them to begin working underground. Miyata Sumiko's[11] mother, for example, went into the pit in 1918, the year after Sumiko's father and elder sister were killed in an explosion. Sumiko herself followed in 1922, when she was about 13.[12]

In addition to the danger and the arduous nature of life underground, women workers also found themselves subjected to sexual harrassment and assault. The style of dress common in prewar mines was often cited as an explanation for such incidents. It was hot and humid in the Chikuho mines, and exposed skin was said to make it easier for miners to notice the shower of small particles that often provided early warning of an impending roof collapse.[13] As a consequence, men wore only loincloths while women were commonly seen wearing only short skirts. Because women haulers usually worked with their husbands or other family members, this partial nudity was not a particular problem at the working face. When women hauling coal through dark mine shafts encountered other men, however, 'mistakes could easily occur', as a government report coyly put it.[14]

As difficult and dangerous as the work was, women could earn much more for their families by going underground, especially if they formed teams with their husbands, because this would ensure that the men actually went to work most days instead of drinking and gambling. The pay was also better than working-class women could hope to earn elsewhere. While women coal haulers were generally paid some 20 to 30 per cent less than workers in male-dominated specialisms underground, their earnings were some 70 per cent higher than those of female factory workers and double those of coal sorters.[15] The high wages available to many men above ground (almost as high as wages for underground work) suggest that gender was at least as important as the location of work in determining pay.

Women, like men, entered the mines out of economic necessity, and some naturally disliked the work, but their strenuous labour and high earnings also brought them higher status and more freedom within the family than most women had in prewar Japan. A daughter's earning power made her a valuable asset, so in at least one mining region a bridegroom would be asked to live and work with his bride's family for two or three years or pay a comparable sum in exchange for her hand in marriage.[16] Women were expected to do most of the housework, but they expected men to apply themselves as well, at least to wage labour. It was not uncommon for a woman to break off relations with a man who was not a good worker and take up with someone else.[17]

Women and the Labour Movement

Women mine workers stood up for themselves on the job as well as at home, displaying a potential for militancy that male labour leaders did not seem to recognize (and in fact impeded in some cases). This potential was sometimes demonstrated in informal actions in the pit. At one Chikuho mine, for example, a supervisor named Yamashita was giving young women a hard time for sneaking rides out of the mine on a coal car, so they decided to teach him a lesson. 'One of the older girls knocked Yamashita's light out of his hand, and with that we attacked him with our poles,' a co-worker recalled. 'We twisted his arms and beat him down on the tracks.'[18]

Women took part in a variety of larger-scale actions as well, staging mine strikes for the first time during the Rice Riots that swept Japan in July and August 1918. Female longshore workers began these riots in reaction to soaring food prices, but in mining areas they grew to encompass demands for higher wages and better conditions as well. In Chikuho, 'miners' wives' (many of them probably miners themselves) looted dry goods and clothing stores that had shown a discriminatory attitude toward miners, while women miners staged a brief strike at one pit. Women, including mine workers, also boiled rice to feed male strikers, as they would in many later walk-outs.[19] Women also displayed militancy in at least two defensive strikes in Chikuho during the Depression, but their role was limited by male leaders. There were 76 women among the 273 workers who occupied the Nittetsu Takao mine in 1931, but strike leaders sent all the women outside to form a food battalion. (Unusually, however, the strike committee gave some consideration to women's interests, demanding that women's pay be raised to 90 per cent of the male standard and that provisions be made for laid-off female workers.)[20]

The next year, women played a key role in a Korean strike at the Aso mines, which had already laid off all their women miners in advance of a ban due in 1933. Four hundred male workers from Korea passed out fliers throughout the 20-day defensive walk-out in an effort to put pressure on the company and win support from Japanese co-workers, but without success. Women protesters, however, most of them the wives of Korean miners, seem to have played a decisive role in the Aso company's decision to take part in talks that led to a limited victory for labour. The women staged sit-ins and barged into company offices, yelling out their grievances while 'scratching and clawing' at Aso officials, in the words of a Korean miner. They also took part in demonstrations, often with each woman carrying a baby on her back and holding another child by the hand. Company officials were clearly unsure how to deal with this 'unladylike' behaviour, while the police hesitated to beat or arrest mothers and their children. Strike leaders, for their part, seem to have been unwilling, or unsure how, to take full advantage of women's militancy and seeming immunity to arrest.[21]

Even after their exclusion from the mines, women played a key role in several post-Second World War strikes that linked the workplace and community. During a nationwide mine and electrical strike in 1952, for example, women blocked coal shipments at one mine and played a key role in opposing the formation of a pro-management second union at another.[22] In a defensive strike at the Mitsui mines in 1953, the housewives' militancy frightened company officials so much, by some accounts, that they insisted on holding negotiations at sea and then agreed to reinstate miners who had refused to accept early retirement.[23] And women were also key players in the 1960 strike at the Mitsui Miike colliery. This was a walk-out that began as a battle over lay-offs and control of the workplace but escalated into a contest over the future of Japanese labour, bringing modern Japan the closest it has ever been to all-out class warfare. Workers from around Japan supported the 16 000 strikers, while big business interests lined up behind Mitsui in its attempt to destroy the union. The walk-out ended in defeat for labour, but women provided the base of support that kept it going for nearly a year. They took part in strike strategy sessions, underground sit-ins and demonstrations, helping to bring attendance to 100 000 at one pithead rally.[24]

Women Mechanized out of Work

Just as large numbers of women were brought into modern mining in response to technological change, further mechanization led to their expulsion. When the Japanese government announced in 1928 that it would ban women from underground work as of 1933, it claimed it was acting to protect women, but the fact was that most mines no longer needed women to haul coal. The shift from room-and-pillar mining to the long-wall system made it practical for the first time in the 1920s to mechanize face transport, the hauling of coal from the working face to the nearest level shaft, where it could be loaded into a coal car for transport out of the mine. International pressure was a secondary factor behind the ban. In 1919, just as the number of women miners in Japan was nearing its peak, the International Labour Organization was set up. Japan, as a founding member of the ILO and as the only country other than India or the Soviet Union then with significant numbers of women underground miners, came under pressure to bring its labour regulations into greater conformity with international standards, including tightened restrictions on where women could work.

A ban on female underground mine work was first proposed in Japan in 1922. While this proposal allowed for a ten-year grace period,[25] industry opposition prevented its adoption. Just as in the concurrent debate on outlawing night work for women in the textile industry,[26] both proponents and opponents of a ban on female miners couched their arguments in terms of women workers' sexual morality. While middle-class reformers like Ishimoto Shidzue maintained that it was 'ridiculous to expect morality' in the mines, where the 'wives and daughters of miners went down in a half-naked condition, mingling with the naked men laborers',[27] coal operators turned this argument on its head. The Chikuho Coal Industry Association, for example, cited what it called the undesirable moral implications of 'regular miners' leaving women at home alone at night when the men went to work, as well as the drop in family income that would result from women leaving the mines.[28] The government finally decided in 1928, as we have seen, to ban most underground female labour from 1933, but women were allowed to continue removing coal in mostly worked-out areas and to work in thin seams. It seems unlikely that an exception would have been made for work in such dangerous and difficult places if the ban had actually been intended to protect women.

While the overall number of women working underground declined drastically, the number employed in seldom inspected smaller mines, where wages were lower and conditions worse, actually increased, at least between 1932 and 1934.[29] Some major mines set up workshops to employ former women miners, but the pay was low and the number of jobs limited. The majority were left unemployed. The ban also had consequences for the division of household labour, as historian Iwaya Saori has pointed out. Government labour officials expected that 'freeing' women to take complete charge of housework and child care would allow men to devote themselves more fully to paid labour, leading to increased efficiency on the job. While it seems unlikely that women's removal from the mines increased male efficiency, the abolition of day care centres quickly made child care women's unquestioned responsibility.[30]

When Japan shifted to a full-fledged war economy following its invasion of China in 1937, the government withdrew from the ILO and decided that the 'protection' of women was no longer necessary. From August 1939, women aged 25 or older were sent back into the mines.[31] While their numbers peaked at just 11 411 (4.7 per cent of underground workers) in March 1944, they were important to production because many had mining experience, a scarce commodity at wartime collieries, and some became low-level supervisors for the first (and last) time. With many women continuing to work above ground at mines, women coal mine workers hit a wartime peak in March 1945 of 65 613, 15.6 per cent of all mine workers.[32]

After Japan's defeat in August 1945, demobilized soldiers returned from Asia and the Pacific asking for their old jobs back, and most women quickly left the coal pits. The exclusion of women from underground work was then formalized by the US authorities who oversaw the 1945–52 Allied Occupation of Japan, and the last officially authorized women were gone by March 1947.[33] Some women continued to work illegally in small, rarely inspected mines as late as 1961,[34] but mining came to be seen for the first time in Japan as a 'naturally' male occupation. Women kept working as coal sorters, but they were phased out as mechanization transformed sorting from a labour-intensive, largely female occupation into a relatively capital intensive and almost entirely male occupation beginning in the 1950s.[35]

The War brings Koreans into Labour Force

Koreans became important to the Japanese coal industry during the First World War. That conflict, like Japan's wars with China (1894–5) and Russia (1904–5) before it, was a boon to Japanese coal production, but it also set in motion changes that would transform the ethnic make-up of the mine labour force. The war cut off British shipments of coal to Asia, creating new markets for Japanese coal, and drastically reduced shipments of other products from Britain as well, opening new markets for Japanese manufactured goods and, in turn, increasing domestic demand for coal. From late 1916, with little surplus labour left in the rural sector and many Japanese miners leaving for more attractive factory jobs, the booming Japanese coal industry turned to Korea, a Japanese colony since 1910. Korean workers allowed the industry to increase production without immediately changing its almost total dependence on manual extraction.

Even before the First World War, Chikuho had a diverse workforce, with workers stratified along lines of regional origin and status as well as sex. In addition to residents of Fukuoka Prefecture, many of them outcast *burakumin*, workers from across western Japan came to work in the Chikuho mines from late in the nineteenth century. In some ways, Koreans were just another group of migrant workers, but their colonial status made them easier to exploit. Their arrival in large numbers also made it easier for mine owners to encourage divisions among workers, playing upon the resentment Japanese miners felt over slights at the hands of farmers, shopkeepers and other 'good citizens'. Many Japanese assumed that miners must be criminals or outcast *burakumin* unsuited for any other occupation, and ascribed their poor living standards, heavy drinking and seeming lack of concern for sanitation to moral failings on the part of miners rather than to low pay, high stress and poor company facilities.

Japan sank into recession after the war, and the mines laid off thousands of Japanese miners. At the same time, however, collieries recruited more Koreans. Between 1923 and 1928, for instance, while the total number of workers at major Chikuho mines fell from 78 703 to 63 646, the number of Korean miners rose from 3943, or 5 per cent of the workforce, in 1923 to 5626, or 9 per cent, in 1928.[36] This phenomenon – the reverse of the last hired, first fired principle so familiar to minority workers – can be explained by two major factors. First, mines sought out Koreans even during the postwar recession because they were willing to work for less and in places where Japanese were reluctant to go.[37] Companies may also have seen Koreans as a way of creating further divisions within the workforce and (inaccurately) as docile and unlikely to strike.

Second, companies hired many Korean men in the 1920s to replace women, most of whom were Japanese. Many companies were beginning to mechanize face transport and lay off women coal haulers in anticipation of the approaching ban on female underground labour. The new mechanical face conveyors still had to be loaded by hand, so companies hired Koreans to perform the combined tasks of hewing and loading coal.[38] When women miners were laid off, their husbands often left with them to seek higher-paying work because of the difficulty of supporting a family on the pay of one miner, so Korean men filled the jobs vacated by many male Japanese workers as well.

The shift to long-wall mining, in addition to setting the stage for the expulsion of women from the mines, as we saw above, also facilitated the employment of large numbers of Koreans. Under the older room-and-pillar system it was much more difficult to employ workers fresh from Korea because of their minimal Japanese language skills and lack of mining experience. Each pair worked in a separate room with minimal supervision, so at least one member of each pair under the older system had to be fairly experienced. Koreans could have transported the coal removed by the hewers, but the language problem would have made it difficult for them to communicate with the Japanese hewers. The long-wall system, which was first employed in Japan before the First World War and then spread during the postwar recession, resolved these problems by replacing pairs of workers with larger work teams. Each team could be staffed by an interpreter, and inexperienced Koreans could be used in unskilled positions. However, during the Depression that hit Japanese mining in 1930, Koreans were laid off out of proportion to their numbers, especially at major companies like Mitsubishi that had begun to mechanize coal extraction in the late 1920s. Local companies like Aso actually increased their dependence on Koreans, but the majors sought to create an all-Japanese workforce until the 1939–45 wartime period of forced labour importation.

Wartime Labour

Japan could not have fought the Asia–Pacific War without Korean miners, who dug over half the coal that was by far the country's most important energy source. Beginning in 1939, Koreans took the places of called up Japanese workers as the collieries struggled to meet an ever-escalating demand for coal. At their February 1945 peak, Koreans comprised over half of hewers and about a third of the total coal workforce, compared to 57.6 per cent for long-term Japanese workers. The remainder of the workforce comprised Japanese on short-term work programmes (6.9 per cent), Chinese prisoners (1.8 per cent) and Allied Prisoners of War (POWs) (1.5 per cent), all of them relatively late additions to the wartime labour supply.[39]

Koreans were subject to much more coercive means of mobilization and to approximately 20 per cent higher death rates than Japanese. Japan brutally exploited Koreans (and treated Chinese and Allied POWs even worse) during the war, but conditions were abysmal for all mine workers, even Japanese.[40] The vast majority of wartime Korean workers were there against their will, but up to 10 per cent were prewar voluntary migrants.[41] Some of these long-term residents rose somewhat in the mine hierarchy, serving as interpreters or low-level supervisors, but at the cost of lasting hostility from fellow Koreans who felt they had sold out to the Japanese.

Korean Militancy

Korean and Chinese miners staged the first strikes in postwar Japan, beginning on the very day of Japan's surrender.[42] By one estimate, in the weeks after 15 August 1945, some 90 000 Korean and Chinese miners took part in 'uprisings' at 40 to 50 coal mines,[43] some of them put down by the US-led Occupation forces. While doing everything possible to speed the repatriation of Allied POWs, Occupation authorities sought to maintain coal production by keeping colonial workers at the mines until Korean and Chinese strikes and uprisings forced officials to send them home as well.[44]

Most colonial workers were gone by the end of 1945, but by then Japanese workers had recovered from the lethargy of defeat and the postwar mine labour movement had got off to an energetic and militant start, thanks in large part to the Koreans and Chinese. This willingness to stand up to management was nothing new for Korean workers in Japan. Most of the Koreans' actions, however, were sporadic and poorly planned (as were most Japanese labour actions). The Korean miners' potential contribution to the Japanese labour movement was further limited by working-class Japanese racism.

Within weeks of the arrival of the first large groups of Korean miners in Chikuho in 1917, they became involved in violent incidents unfairly dismissed by the local press as 'riots' or 'drunken brawls'. These incidents in fact show that Koreans were conscious of themselves as workers and as Koreans in conflict with Japanese capital, and that they were militant enough to put their jobs on the line to protest against poor working conditions and Japanese prejudice. Two-thirds of the incidents reported in the local press between 1917 and 1920 involved ethnic friction.[45] Just three of the incidents were sparked by overt prejudicial actions on the part of Japanese, but ethnic tension and mutual suspicion, amplified by the language barrier and the separation of the two groups in the workplace and in company housing, formed the background to all these clashes. The other third of the incidents, those not involving Japanese workers, demonstrate both divisions among Korean workers and spontaneous labour activism.

There are no reliable statistics from the 1910s and 1920s, but Home Ministry figures compiled between 1930 and 1942 show that Korean workers in Japan were on average almost four times as likely to take part in labour disputes as all workers, the vast majority of whom were Japanese. If we examine strikes alone, Koreans were more than twice as likely on average as Japanese to strike between 1932 and 1942.[46] Even after 1939, when voluntary Korean migrants were quickly outnumbered by wartime forced migrants, Koreans, especially those in Hokkaido, continued striking, even as Japanese strikes practically ceased to exist. The annual Home Ministry statistics stop in 1942, but Special Higher Police figures suggest that Koreans continued to stage more labour disputes than Japanese throughout the remainder of the war.[47]

As we saw in the discussion above of the 1932 Aso walk-out, in which not a single Japanese miner supported the Korean strikers, solidarity across ethnic lines was rare. None of the incidents that occurred during the period of the First World War involved Koreans and Japanese joining forces against management, but both groups did take part in some of the 1918 Rice Riots[48] which, in Chikuho, reportedly incorporated Korean demands for an end to discrimination and higher wages, Japanese miners' demands for better conditions, and a general demand for lower prices for rice.[49] Most of what cooperation there was took place on the left, especially after Korean communists in Japan joined the Japanese Communist Party in 1929 under the new Comintern policy of 'one country, one party' and the Federation of Korean Workers in Japan dissolved itself in 1930 into the JCP-affiliated National Council of Japanese Labour Unions (Zenkyo). By 1933, Koreans made up more than half of Zenkyo's members[50] and, as the ranks of Japanese leftists were thinned through arrests and defections, Koreans gradually assumed positions of authority.

Conclusion

Chikuho's last underground colliery closed in 1976, and the region has never recovered from the blow. The history of Chikuho, however, retains plentiful reserves of unmined comparative questions for historians. Why, for example, were women banned so much earlier in Western Europe than in Japan, and (at least ostensibly) for such different reasons? Comparative studies of groups such as Koreans in Chikuho, African–Americans in Appalachia and Poles in the Ruhr, meanwhile, could do much to sharpen our understanding of the relative importance of structural factors and local economic and political conditions in shaping the discrimination minority-group workers faced on the job, in housing and from majority-group workers. As the example of Chikuho shows, however, ethnicity and gender (along with class) must be examined in tandem for us to gain a nuanced understanding of occupational distribution and daily life in coal-mining communities.

Notes

1 Calculated from Naikaku Tokeikyoku, *Showa 5-nen kokusei chosa hokoku*, zenkoku no bu, vol. 2, shokugyo oyobi sangyo, 41–3, 225; and vol. 4, fukenhen, Fukuoka-ken, 118.

2 Japan was not alone in employing women miners, of course. As late as the 1930s, about 14 per cent of coal miners in India and over 22 per cent in the Soviet Union were women, according to International Labour Organization, *The World Coal-Mining Industry*, Studies and Reports Series B (Economic Conditions) 31(2) (Geneva, 1938), p.7.

3 Nishinarita Yutaka, 'The coal mining industry', in Nakamura Masanori (ed.), *Technology Change and Female Labour in Japan* (Tokyo, 1994), pp.60–62.

4 Noshomusho Kozankyoku, *Kofu taigu jirei* (1908), cited in Ogino Yoshihiro, *Chikuho tanko roshi kankeishi* (Fukuoka, 1993), p.21. Chikuho had a monthly average turnover of 14.9 per cent in 1906, compared with 7.1 per cent for Hokkaido.

5 Noshomusho Kozankyoku, *Honpo kogyo no susei* for 1920, in Nishinarita, 'The coal mining industry', p.87.

6   Most women mine workers were Japanese, but the small Korean female minority had a strikingly different occupational profile. For details, see W. Donald Smith, 'Sorting coal and pickling cabbage: Korean women in the Japanese mining industry', in Barbara Molony and Kathleen Uno (eds), *Gendering Modern Japanese History* (Cambridge, MA, in press).

7   Patricia Hilden, *Women, Work and Politics: Belgium, 1830–1914* (Oxford and New York, 1992), p.94.

8   Nishinarita, 'The coal mining industry', pp.69–70. This pattern did not hold true for Korean women in the Japanese mines, however.

9   Retired miner Nagaoka Iwao, for example, was a coal sorter as a boy at Mitsubishi Hojo mine, while his father was a coal sorting supervisor, the highest job available, according to Nagaoka, to outcast *burakumin*. (Interview, Hojo-machi, Tagawa-gun, Fukuoka Prefecture, 25 May 1995.)

10 Morisaki Kazue, 'Tough girls', *Concerned Theater Japan* 2, nos. 3 and 4 (1973), 172, translated excerpt from *Makkura* (Tokyo, 1970).

11 Japanese names are presented in the traditional East Asian order, with surnames first.

12 Miyata Sumiko (a pseudonym) herself survived the mines, later to do construction and factory work and be interviewed by Iwaya Saori in 1995. See Iwaya Saori, 'Work and life at a coal mine: the life history of a woman miner', in Wakita Haruko, Anne Bouchy and Ueno Chizuko (eds), *Gender and Japanese History*, vol. 2 (Osaka, 1999), pp.419–21.

13 Interview with Idegawa Yasuko, 16 May 1995, Kurate-machi, Fukuoka Prefecture.

14 Osaka Chiho Shokugyo Shokai Jimukyoku, *Chikuho tanzan rodo jijo* (Osaka, 1926), p.76.

15 At the Mitsubishi Shinnyu mine in June 1928, for example, coal sorters were paid an average 16.54 yen, compared to 26.67 yen for male miscellaneous workers, 28.32 yen for porters, 35.90 yen for operatives and 39.16 yen for drivers. The only workers paid less above ground than coal sorters were female miscellaneous workers, at 15.32 yen a month. Underground, by comparison, women 'hewers' (probably the haulers or helpers known as *atoyama*, literally '[to the] rear [of the] mine') averaged 33.20 yen a month, compared to 35.15 yen for porters, 36.39 yen for drivers, 38.75 yen for 'miners' other than hewers, 39.84 yen for miscellaneous workers, 41.36 yen for male hewers, and 44.70 yen for operatives. See Tomigashi Fumiya, 'Shinnyu tanko dai-6-ko jisshu hokoku', University of Tokyo, 1928, cited in Ogino, *Chikuho tanko*, p.304. For a comparison of average wages by gender in mining and factory labour, see Rodo Undo Shiryo Iinkai (ed.), *Nihon rodo undo shiryo*, vol. 10 (Tokyo, 1959), pp.284–5.

16 This report from 1918 is cited in Maekawa Masao (ed.), *Tankoshi: Nagasaki-ken sekitanshi nenpyo* (Fukuoka, 1990), pp.233–4.

17 Ichihara Hiroshi, *Tanko no rodo shakaishi: Nihon no dentoteki rodo shakai chitsujo to kanri* (Tokyo, 1997), pp.31–5.

18 Morisaki, 169.

19 Hayashi Eidai, *Yami o horu onnatachi* (Tokyo, 1990), pp.196–202. For an account in English of the Rice Riots in mining areas, see Michael Lewis, *Rioters and Citizens: Mass Protest in Imperial Japan* (Berkeley, 1990), pp.192–241.

20 Ogino, *Chikuho tanko*, pp.389, 392.

21 For details, see W. Donald Smith, 'The 1932 Aso coal strike: Korean–Japanese solidarity and conflict', *Korean Studies*, **20** (1996), 94–122.

22 Mitsui Kozan K.K., *Shiryo Miike Sogi* (Tokyo, 1963), p.123. For an English-language account of the 63-day strike, see John Price, *Japan Works: Power and Paradox in Post-war Industrial Relations* (Ithaca, 1997), pp.114–15.

23 Price, *Japan Works*, pp.115–17; Isayama Hiroshi, *Mitsui Miike* (Tokyo, 1960), p.186; Tanro: gekito ano hi ano toki hensan iinkai (ed.), *Tanro: gekito ano hi ano toki* (Tokyo, 1992), p.285.

24 Miike comprises a separate coalfield southwest of Chikuho. For an overview in English of the 1960 Miike strike, see Price, *Japan Works*, pp.191–218.

25 See Ogino, *Chikuho tanko*, p.280, for a summary of proposed amendments to Japanese mining regulations.

26 Barbara Molony, 'Activism Among Women in the Taisho Cotton Textile Industry', in Gail Lee Bernstein (ed.), *Recreating Japanese Women* (Berkeley and Los Angeles, 1991), p.229.

27 Ishimoto Shidzue, *Facing Two Ways: the Story of My Life* (Stanford, 1984, orig. published New York, 1935), p.161.

28 Shimura Kazushige (ed.), *Nogatashi-shi hokan: sekitan kogyo hen. Nogata sekitan kogyoshi* (Nogata, 1979), p.228.

29 Regine Mathias, 'Female labour in the Japanese coal mining industry', in Janet Hunter (ed.), *Japanese Women Working* (London and New York, 1993), p.116.

30 Iwaya, 'Work and Life', p.433.

31 Tanaka Naoki, *Kindai Nihon tanko rodoshi kenkyu* (Tokyo, 1984), p.600.

32 Shokosho Kozan Kyoku, *Honpo kogyo no susei*, various years, compiled in Tanaka, *Kindai Nihon tanko*, pp.160–61.

33 'The coalfields of Kyushu', p.21, Supreme Commander for the Allied Powers, General Headquarters, Record Group 331, Box 2494, Folder 1, 'Records of Allied Operational and Occupation Headquarters, World War II', Washington National Records Center, Suitland, MD.

34 Chikuho Sekitan Kogyo-shi Nenpyo Hensan Iinkai, p.589, cited in Iwaya Saori, 'Nihon kingendai no sekitan kogyo ni okeru josei rodosha', undated graduation thesis, Osaka University of Foreign Studies, p.14. Also see the account of Tachibana Shizue, a prewar mine worker who went back to work in a small mine at the age of 54 or 55 in 1956 or 1957, in Nakayama Akira, *Chikuho Zanzo: onna kofu wa ima* (Tokyo, 1983), p.55.

35 For a study of this phenomenon at two Hokkaido mines, see Hokkaidoritsu Rodo Kagaku Kenkyujo, *Tanko ni hataraku fujin no rodo jittai* (Sapporo, 1961).

36 Fukuoka Chiho Shokugyo Shokai Jimukyoku, *Chikuho chiho tanko rodosha shusshinchi chosa 1923*, cited in Chung Jin Sung, '1920-nendai no Chosenjin kofu no shiyo jokyo oyobi shiyo keihi: Chikuho chiho no Mitsubishi-kei tanko o chushin to shite', *Nihon shigaku shuroku*, **10** (1990), 31; and Fukuoka Chiho Shokugyo Shokai Jimukyoku, *Chikuho chiho tanko*, 6.

37 Government reports and observations by engineering students suggest that Koreans were cheaper to employ. For details, see W. Donald Smith, 'Ethnicity, class and gender in the mines: Korean workers in Japan's Chikuho coalfield, 1917–1945', PhD diss., University of Washington, 1999, 68–9.

38 For the example of Mitsubishi Shinnyu mine in 1923, see Okada Hideo, 'Shinnyu

tanko dai-rokko hokoku', 1929, University of Tokyo, 48, cited in Chung, '1920-nendai no Chosenjin', 32.

39 To be precise, there were officially 136 825 Korean coal miners in Japan in February 1945, 32.1 per cent of the coal workforce. Sekitan Toseikai Kinrobu, 'Showa 19-nendo zenkoku tanko romusha ido jokyo shirabe', in Nagasawa Shigeru (ed.), *Senjika Chosenjin Chugokujin Rengokoku horyo kyosei renko shiryoshu*, vol. 1 (Tokyo, 1992), p.39.

40 For a study of wartime Korean mine labour in Japan, see Smith, 'Ethnicity, class and gender', 223–334.

41 There were 13 315 long-term Korean workers in the Japanese mines in April 1943, calculated from Sekitan Toseikai Romubu, 'Tanko romu tokeihyo', in Nagasawa, *Senjika Chosenjin Chugokujin Rengokoku horyo kyosei renko shiryoshu*, p.317.

42 For details on early postwar strikes and Korean repatriation, see Smith, 'Ethnicity, class and gender', 335–80.

43 Ohara Shakai Mondai Kenkyujo, *Taiheiyo sensoka no rodo undo* (Tokyo, 1965), p.30.

44 For the Occupation's ordering of Chinese workers back to the mines at Bibai, Hokkaido, on pain of arrest or death, for example, see Smith, 'Ethnicity, class and gender', 339–40.

45 For press accounts of 22 incidents in and around Chikuho from 1 July 1917, to 24 March 1920, see Smith, 'Ethnicity, class and gender', 140–48, 480–86.

46 The discussion of strike and dispute proclivity is based on the following sources. The figure for all workers in Japan in 1930 was calculated from Naikaku Tokeikyoku, *Showa 5-nen kokusei chosa hokoku*, zenkoku no bu, vol. 2, shokugyo oyobi sangyo, cited in Rodo Undo Shiryo Iinkai, *Nihon rodo undo*, p.78. Because of the lack of annual figures after 1930 for all non-agricultural workers in Japan, the strike proclivity of all workers is calculated using the 1930 figure as a constant denominator. Figures for strikes and all labour disputes in Japan are from Naimusho Shakaikyoku, *Rodo undo gaikyo* and *Rodo undo nenpo*; and Koseisho Rodokyoku, *Rodo undo nenpo* and *Rodo jiho*, cited in Rodo Undo Shiryo Iinkai, *Nihon rodo undo*, pp.440–41. Figures for Korean workers and their involvement in strikes and other labour disputes are from Naimusho Keihokyoku, *Shakai undo no jokyo*, 1930–42 editions (Tokyo, various years).

47 In the first 11 months of 1944, for example, 10 838 Koreans in Japan were involved in 157 'labour incidents', while 11 340 workers of all ethnicities in Japan were involved in 322 'labour disputes' during the same period. For Koreans, see Naimusho Keihokyoku Hoanbu, *Tokko Geppo* (November 1944), 67, recalculated by John W. Dower and cited in his 'Sensational rumors, seditious graffiti, and the nightmares of the thought police', in Dower (ed.), *Japan in War and Peace: Selected Essays* (New York, 1993), p.118. For workers of all ethnicities, see *Tokko Geppo* (November 1944), 45, adapted and cited by Dower, *Japan in War and Peace*, p.116. The threshold for an event to be classified as a 'dispute' was undoubtedly higher than that for an 'incident' but, with some 1.9 million Koreans in Japan in 1944, compared to some 70 million Japanese, it seems likely that Koreans were considerably more militant in this as in other periods.

48 Koreans reportedly participated along with Japanese in the uprisings at the Chuo and Minechi mines in Chikuho and the Shinbaru mine in nearby Kasuya-gun, and a Korean was among those killed when the military opened fire in the Ube

coalfield in Yamaguchi Prefecture. See Yoshioka Yoshiaki, 'Shokuminchi Chosen ni okeru 1918-nen: kome sodo to Chosen', *Rekishi Hyoron*, **216** (August 1968), 36.

49   Hayashi Eidai, 'Kaisetsu', in Hayashi (ed.), *Senji gaikokujin kyosei renko kankei shiryoshu*, Vol. 2, Part 1 (Tokyo, 1991), p.11.

50   Naimusho Keihokyoku, *Shakai undo no jokyo*, 1933, p. 1549, cited in Richard H. Mitchell, *The Korean Minority in Japan* (Berkeley and Los Angeles, 1967), p.61.

CHAPTER FOURTEEN

Coal Mining, Foreign Workers and Mine Safety: Steps towards European Integration, 1946–85

René Leboutte

The Treaty of Paris (18 April 1951) was signed by 'the Six' (Belgium, France, Germany, Italy, Luxembourg and the Netherlands). As a consequence, the European Coal and Steel Community (ECSC) was established and, two years later, there was a common market for coal, iron and steel. This chapter examines the implications of this development for workers employed in the coal industry. By the mid-1950s, more than 1 533 000 people were employed in the two sectors of coal and iron- and steel-making in the ECSC. Among them, 170 800 (11 per cent) were 'foreign workers'. This figure not only included those who hailed from outside the ECSC, but also those who had migrated from one member state to another.[1] The years immediately following the Second World War saw huge movements of workers across and within the borders of the Six. Belgium and France and, to a lesser extent, Luxembourg, imported more labour than they exported; in contrast, Italy was primarily a point of departure.

The establishment of the ECSC ostensibly brought with it the prospect of improvements in working conditions and standards of living for workers employed in the steel and coal industries. Moreover, in theory, it allowed for greater mobility of labour between member states. However it took more than 15 years to remove fully the obstacles to the free movement of workers in the European Community.[2] Furthermore, as will be shown in this chapter, notwithstanding the establishment of the ECSC, the migration of foreign workers in the late 1940s and in the 1950s was still largely shaped by bilateral agreements between member countries. The ECSC did not modify the pattern of labour migration characterized by the recruitment, on a temporary basis, of unskilled foreign workers to meet a shortage of underground workers both in iron-ore and coal mines. Where the European Community did play a key role was in improving health and safety standards in coal mining in the face of a number of pit disasters, mainly in the Walloon coalfields during the 1950s; the scandalous working conditions and poor safety records of Belgian coal mines became a major problem that demanded a Community-wide solution.

New Opportunities, Old Receipts

In the immediate aftermath of the Second World War, Europe faced a critical shortage of coal and other raw materials that seriously threatened any hope of economic and social recovery. Most of the coal needed to facilitate rebuilding projects, to heat exhausted populations and to supply heavy industries and transportation systems had to be imported from the United States of America under the auspices of the European Coal Organization. Against the background of this general shortage, the situation in Belgium was significantly different. Coal mining was traditionally one of the most important industrial sectors (along with the iron and steel industries) in Wallonia. And, after the First World War, a new coalfield was opened in Kempen (Campine), a Flemish region close to the older coalfield of Liège. Whereas French and German coal mines had been damaged during the First World War, the Belgian coalfields were in a position not only to provide enough coal for Belgian economic recovery, but also to supply other countries.[3] The Belgian government took advantage of this opportunity, and launched a campaign for massive coal output (Table 14.1); the 'battle for coal', as Prime Minister Achille van Acker described it in 1945, was under way.

But at the very moment when the coal industry urgently needed manpower, Belgian miners were leaving the mines in their droves, particularly in the older Walloon coalfield, which had an appalling safety record and was frequently the scene of disasters and accidents. In response, the government introduced a carrot and stick policy. A 'miners' statute' (*statut des mineurs*) was established by a decree law of April 1945. It offered wage bonuses, additional days off, a complementary pension scheme and other benefits. Another decree imposed a 'civil mobilization' of all the workers employed in the coal industry since September 1944. Unemployed 'ex-miners' were now required by law to return

Table 14.1   **Coalminers and output of coal in Belgium, 1939, 1944–7**

Year	Workers		Output of coal	
	Thousands	Index	Millions tons	Index
1939	130.6	100	29.7	100
1944	97.5	75	13.5	45
1945	100.4	77	15.6	53
1946	133.1	102	22.7	76
1947	138.2	106	24.2	81

Source: Stenuit R., 'Royaume de Belgique. Mines de houille. Renseignements statistiques, années 1850 à 1947', in *Annales des mines de Belgique*, 1949, 1, p.52.

to their coal mines. Such measures failed to meet the necessary targets, however, and a campaign to recruit foreign workers was launched by the government and the *Fédéchar* (a federation of Belgian coal-mining companies, created in 1918). With the government's blessing, the *Fédéchar* started recruiting Italian workers in the inter-war period.[4] Between 1922 and 1930, more than 39 500 Italians came to work in Belgian coal mines. And the importance of the contribution of all foreign miners increased significantly between these dates. In 1922, 8 per cent of Belgium's miners were non-Belgian; in 1930, this figure had risen to 19 per cent.[5]

The migration policy was based on a short-term contract that gave enough flexibility to match the numbers of foreign workers with the available jobs and the volume of stocks of coal. In periods of labour shortages and increasing demand for coal, the Federation recruited foreign workers who could be disbanded quickly when the demand for coal decreased. The national authorities approved of the Federation's attempt to create a 'flexible' labour force that relied heavily on foreign workers. Indeed the government relaxed institutional entry barriers to any foreign workers who did not compete directly with the indigenous labour force. It was not until 1937 that the first regulation was introduced and coal-mining companies were obliged to pay for repatriating their redundant foreign workers.[6]

The same policy was pressed into service in the immediate aftermath of the Second World War when Van Acker's government retained German prisoners and set them to work underground. The first prisoners entered coal mines in May 1945. There were already 46 000 of them at work in December. By early 1946, German prisoners mined more than a third of the total output of Belgian collieries. Belgium retained 64 000 German prisoners in its mines until March 1947, while Belgian collaborators with the enemy (*inciviques*) were also sent into collieries. In addition, between 1947 and 1949, more than 23 000 displaced people coming mainly from Eastern Europe started working in Walloon mines (Table 14.2). However, those casual

Table 14.2   Refugees of war and displaced people recruited by Fédéchar, 1945–9

	Wallonia	Kempen (Campine)	Total
1945	2 369	0	2 369
1946	—	—	—
1947	12 999	6 202	19 201
1948	1 340	874	2 214
1949	1	0	1
Total	16 709	7 076	23 785

Source: Martens, *Les immigrés*, p.72.

Table 14.3   Employment, coal output, labour migration, 1947–61

	Stocks (thousand tons) December	Number of underground coal miners, 31 December		Foreign workers recruited by Fédéchar
		Belgians	Foreigners	
1947	448	54 188	60 146	40 640
1948	836	56 212	72 252	40 894
1949	1 812	57 803	58 515	865
1950	1 037	55 889	52 788	0
1951	224	52 081	67 689	27 389
1952	1 673	51 963	67 615	13 967
1953	3 073	53 008	62 216	3 590
1954	2 814	53 701	56 065	105
1955	370	49 917	64 535	15 279
1956	179	47 281	60 818	7 667
1957	1 412	45 958	70 931	17 048
1958	6 958	44 284	61 304	740
1959	7 495	39 163	51 771	—
1960	5 552	34 106	43 227	—
1961	1 773	29 833	36 626	—

Source: Martens, *Les immigrés*, pp.83, 88, 105.

workers did not entirely cover the shortage of manpower, while prisoners of war had to be freed after two years, that is to say, by May 1947.[7]

By 1947, foreigners accounted for more than a half of the underground workforce (Table 14.3). This situation did not change in the following decades. The Federation clearly recruited foreign workers according to the fluctuations of coal stocks, a policy dating from the inter-war period. Thus, in years when companies had difficulty getting rid of coal (such as 1949, 1950, 1954 and 1958), the numbers of foreign workers brought in by the Federation dropped markedly. Clearly foreign manpower was regarded as a reserve army.[8]

Waves of Italian Workers entering Belgian Coal Mines

By late 1945, Italy had already sent some of her sons to work in Belgian mines in exchange for coal. In June 1946, a bilateral Belgo-Italian agreement (*Protocole d'accord*) authorized the immigration of a further 50 000 workers for the coal industry in return for greater stocks of coal.[9] Some 75 000 Italians went to work in Belgian collieries between June 1946 and December 1949 (Table 14.4).

Table 14.4   **Italian workers recruited by the Belgian Federation of coal mining companies, 1945–9**

	Wallonia	Kempen (Campine)	Total
1946	19 915	891	20 806
1947	15 987	3 715	19 702
1948	27 095	8 634	35 729
1949	612	168	780

Source: Martens, *Les immigrés*, p.71.

The Belgo-Italian agreement was in line with the system for recruiting foreign labour based on bilateral 'treaties of trade and navigation' that mushroomed in Europe from the mid-nineteenth century onwards. Belgium and Italy signed a similar treaty in December 1882, by which Italians were allowed to work in Belgium. In 1938, this agreement was renewed by a treaty explicitly concerned with the labour force (*Traité d'établissement et de travail entre la Belgique et l'Italie*).[10] Both countries benefited from these arrangements; Belgium got its cheap workforce, while Italy got its guaranteed coal stocks. However the relationship between Belgium and Italy quickly deteriorated because of the poor safety devices in Walloon coal mines.[11] The number of accidents and deaths per thousand miners dramatically increased in the late 1940s (Table 14.5) as a result of the 'battle for coal'. With attention focused on increasing output as rapidly as possible, working

Table 14.5   **Number of accidents and fatalities in Belgian coal mines, 1935–47**

Year	Number of accidents	Number of dead	Per 1000 workers	
			Accidents	Dead
1935	177	125	1.47	1.04
1936	178	146	1.47	1.21
1937	202	135	1.61	1.08
1938	201	131	1.53	0.99
1939	181	149	1.39	1.14
1940	197	175	1.68	1.49
1941	275	204	2.19	1.63
1942	312	224	2.56	1.84
1943	292	202	2.39	1.65
1944	154	108	1.58	1.11
1945	159	112	1.58	1.11
1946	220	163	1.65	1.23
1947	190	128	1.38	0.93

Source: Stenuit, 'Mines de houille', p.52.

conditions were effectively ignored. Underground equipment, machinery and infrastructure were all out of date. Moreover neither the German prisoners nor the Italian workers and refugees had any training in underground mining.

On the basis of the *Protocole*, the Italian government repeatedly requested that the Belgian authorities take radical measures to improve the safety conditions in the mines and to ensure better standards of living generally. In 1951, the Italian Under-Secretary of State visited Belgium and strongly denounced the unacceptable conditions of housing and underground safety. The following year, Prime Minister Alcide De Gasperi, one of the 'founding fathers' of the ECSC, also visited Belgium and requested urgent improvements for his compatriots. Against such a background of tension, a disaster at the Charbonnage d'Ougrée-Marihaye near Liège occurred on 24 October 1953. Four days later, the Italian government decided to stop sending miners to Belgium and set up a committee of inquiry into safety conditions in the country's coal mines. As the committee was drafting its report, new accidents and ever worsening safety records (Table 14.6) provoked indignation throughout Italy,[12] although it can be argued that both Italians and Belgians were responsible for the poor conditions underground. As one commentator has observed, 'The employers, with their entrepreneurial egoism, and the emigrants, with their anxiety for a secure tomorrow that they can enjoy in their homeland, conspire in a paradoxical manner to maintain the inhuman wage system in the Belgian basins.' The incentives to boost production at all costs did little to encourage a culture in which health and safety issues were taken seriously.[13]

Table 14.6   Fatal accidents in Belgian coal mines, 1926–80

	Number of dead per 10 000 workers	Number of dead per million tons	Dead per million days of work
1926–30	11.9	6.8	3.88
1931–35	10.6	5.3	3.82
1936–40	11.8	5.0	3.98
1941–45	15.0	7.6	4.83
1946–50	10.5	5.3	3.52
1951–55	11.9	4.8	3.92
1956–60	12.1	5.1	4.67
1961–65	7.9	2.9	3.48
1966–70	7.7	2.1	2.91
1971–75	5.9	1.4	1.98
1976–80	5.2	1.2	1.91

Source: Leboutte, R., 'Mortalité par accident dans les mines de charbon en Belgique aux XIXe–XXe siècles', *Revue du Nord*, LXXIII (293) (1991) 703–36.

Trade Unions and Labour Migration

Belgian trade unions were alarmed at many of the developments in the mining industry during these years. Worried by the arrival of immigrant workers in Wallonia, the poor housing conditions in the mining districts, and the very poor safety record in the mines, trade union representatives met with government officials and coalowners in July 1948 to discuss the situation. Top of the agenda were the problems that were perceived to flow from the recruitment of foreign workers, issues surrounding the safety of miners and the general standard of living of those employed in the coal industry. Concerns at seeing Italian workers remaining outside any kind of labour organization had already led to some efforts being made to address this matter. The Catholic trade union (*Confédération des syndicats chrétiens*) had tried to gather them together and from June 1946 it had established regular contacts with the Italian trade union, *Associazione Christiana di Lavoratori Italiani*, in an effort to set up regional secretariats with Italian representatives in each of Belgium's coalfields. However, most young Italians, not to mention workers coming from Eastern Europe, had had little exposure to the idea of free labour organization.[14]

The European Coal and Steel Community and the Free Movement of Labour

When Robert Schuman delivered his declaration on 9 May 1950, the notion of a European 'pool' of coal miners and steel workers was an acknowledgment, in part, of the massive movements of labour that had occurred in the immediate aftermath of the war. Indeed, in 1950, miners employed in the Belgian coalfields already constituted a kind of 'European community'. During the negotiation of the Treaty of Paris, the Italian delegates insisted upon a free movement of workers as well as coal, iron and steel to solve the problem of acute unemployment in Italy. Belgium and Luxembourg were facing a shortage of manpower and, in theory, were interested in receiving foreign workers. But both countries wanted to limit the immigration to unskilled jobs in the mining and steel industries. The same was true of other members of the Six who were all reluctant to open their labour markets to large numbers of out-of-work Italians. Consequently, the Treaty of Paris did not create a free movement of labour within the ECSC notwithstanding Article 69 of Chapter 8 which stated:

> the Member States bind themselves to renounce any restriction, based on nationality, on the employment in the coal and steel industries of workers of recognized qualifications for positions in such industries possessing the nationality

of one of the Member States. This commitment shall be subject to the limitations imposed by the fundamental needs of health and public order.[15]

In March 1953, a committee of experts was appointed by the 'High Authority' of the ECSC and charged with the task of defining a 'skilled worker'. An intergovernmental conference took place in May 1954 to endorse their subsequent report. In December 1954, the Council of Ministers of the ECSC approved an agreement, which came into effect only in September 1957 after ratification by the Six. This agreement dealt with a limited category of people. To be eligible a worker had to prove that he was qualified in coal mining or steel making, and that he had already been employed for two years in one of these industries. If he fulfilled all the requirements, he had to apply to his government for a *carte de travail*. This 'labour card' allowed the worker free movement from one member state to another in response to an offer of a skilled job in either the coal-mining or steel sectors. So the free movement of workers painfully established in 1957 had nothing to do with the bulk of Italians recruited for iron-ore mines in Luxembourg and Lorraine and for coal mining in Belgium and France. Moreover Article 69 did not affect bilateral and multilateral agreements on labour recruitment established by the different member states. The *Protocole italo-belge relatif au recrutement de travailleurs italiens pour les mines belges*, signed in Rome in March 1954, and a similar agreement between Italy and Germany, signed in December 1955, both remained fully in force.[16] Nevertheless the concept of a 'common professional nationality' which began to emerge from the Treaty of Paris carried with it the right for the worker to choose his place of work, while the social aspects of labour migration started to receive more and more attention from the ECSC.[17]

Employment in Coal Mining and Steel Industry within the ECSC in 1955

Today, it is still difficult to get an overview of labour migration within the ECSC from the early 1950s onwards. However, a survey completed in 1955 gives us a broad impression. More than 170 000 foreign workers were employed by a member state different from their home country, that is to say 11 per cent of the total of workers in the coal, iron-ore and steel industries. Coal mining was the main supplier of labour, with 127 092 workers or 74.4 per cent of the total of foreign workers (Table 14.7). The bulk of foreign workers comprised Italians and Poles. Some 54 per cent of the Italians employed in the mining and steel industries worked outside Italy. They constituted more than 44 per cent of the total of foreign workers employed within the ECSC. Meanwhile the Poles constituted 22.5 per cent, and the 'others', 17 per cent. Within the ECSC, Belgium and France were the main

Table 14.7   Foreign and native workers employed in coal mining and steel industry of the ECSC, 31 December 1955

Six member states	Foreign workers		Native workers	Total	% of foreigners
	Number	%			
Coal mining	127 092	74.4	928 112	1 055 204	12
Iron ore mining	9 651	5.7	41 027	50 678	19
Steel plants	34 044	19.9	393 413	427 457	8
Total	170 787	100.0	1 362 552	1 533 339	11

Source: Table based on figures provided by Communauté Européenne du Charbon et de l'Acier, *Obstacles à la mobilité des travailleurs et problèmes sociaux de réadaptation, préface de Paul Pinet, membre de la Haute Autorité*, Luxembourg, Communauté Européenne du Charbon et de l'Acier, s.d. [1958], annexe III, tableau 1.

receiving countries (Tables 14.8 and 14.9). In Belgium, 70 per cent of the foreign workers were Italians and 9 per cent were Poles. In France, Poles represented 45 per cent of the foreigners employed in coal mining and Italians only 13 per cent. More than 11 100 people (23 per cent) came from other countries: 7013 from French territories outside Europe, and 4097 from several countries outside the ECSC. In Belgium, the bulk of foreign workers employed in coal mining were Italians, Poles and 'others' mainly from European countries outside the ECSC (Tables 14.9 and 14.10). Belgium had very few iron-ore mines in the 1950s and the number of non-Belgian workers employed in those mines were negligible. In France, however, the iron ore mines of the Lorraine attracted more than 9000 foreigners, mainly Italians

Table 14.8   Distribution by nationality of foreign workers employed in coal and iron ore mining, and steel industry within the ECSC, 1955

Workers from	Working in another country of the ECSC						
	Germany	Belgium	France	Sarre	Italy	Lux.	NL
Germany	—	1 950	4 138	2 786	1	75	835
Belgium	69	—	3 831	3	0	1 073	297
France	64	1 823	—	165	0	358	8
Saarland	235	2	4 476	—		3	2
Italy	415	52 953	20 820	97	—	1 040	208
Luxembourg	19	165	632	14	0	—	1
Netherlands	1 305	3 145	36	5	0	7	—
Poland	577	6 951	29 812	50	1	79	866
Others	3 361	8 230	15 689	141	19	279	1 676
Total	6 045	75 219	79 434	3 261	21	2 914	3 893

Source: See Table 14.7.

Table 14.9   Distribution by nationality of foreign workers employed in coal mining within the ECSC, 1955

Countries	(1) Total of foreign workers	(2) Total of workers	(3) % of foreign workers (1)/(2)	(4) Italians	(5) Italians among foreign workers (%) (4)/(1)	(6) Poles	(7) Poles among foreign workers (%) (6)/(1)
Germany	5 081	522 386	10	359	7	503	9.9
Belgium	67 446	157 585	43	47 445	70	6 237	9.2
France	48 163	243 828	20	6 448	13	22 042	45.6
Saar	2 567	63 368	4	22	0	32	1.2
Italy	0	7 210	0	—	—	—	0
Netherlands	3 835	60 827	6	208	5	838	21.9
Total	127 092	1 055 204	12	54 482	43	29 652	23.3

Source: See Table 14.7.

and Poles, while the iron-ore mines of Luxembourg represented another opportunity for Italians (Table 14.11).[18]

1956: the Marcinelle Disaster

Italian miners in Belgium were the largest group of ECSC workers employed by another member state. From 1947 to 1955, the *Fédéchar* recruited around 143 000 foreign workers, the majority of whom were from Italy. Between 1948 and 1957, the number of Italians in underground mines fluctuated between

Table 14.10   Distribution of foreign coal miners working in France and Belgium, 1955

Country of origin	France		Belgium	
	Number	%	Number	%
Germany	3 352	7.0	1 924	2.9
Belgium	733	1.5	—	—
France	—	—	1 338	2.0
Sarre	4 434	9.2	1	0.0
Italy	6 448	13.4	47 445	70.4
Luxembourg	24	0.0	48	0.0
Netherlands	20	0.0	2 992	4.4
Poland	22 042	45.8	6 237	9.2
Other	11 110	23.1	7 461	11.1
Total	48 163	100.0	67 446	100.0

Source: See Table 14.7.

Table 14.11   Distribution by nationality of foreign workers employed in iron ore mining within the ECSC, 1955

Countries	(1) Total of foreign workers	(2) Total of workers	(3) % of foreign workers (1)/(2)	(4) Italians	(5) Italians among foreign workers (%) (4)/(1)	(6) Poles	(7) Poles among foreign workers (%) (6)/(1)
Germany	105	19 074	0.6	0	0	5	0
France	9 026	25 214	35.8	4 475	50	2 817	31
Luxembourg	520	2 471	21.0	294	57	21	4
Total	9 651	46 759	20.6	4749	49	2 843	29

Source: See Table 14.7.

33 000 and 47 500. However, many of them were temporary workers and the labour turnover in Belgian coal mines was very high. The movement into and out of Belgian coal mines represented something like one-third of the underground miners in 1955–6, twice the rate in France or Germany.[19] Moreover, 20 per cent of the miners frequently moved from one coal mine to another.[20]

More than 45 500 Italians were working in Belgian collieries when a new disaster occurred at the *Charbonnage du Rieu du Coeur* in Quaregnon on 8 February 1956. Seven Italians died. The Italian government immediately decided to suspend the emigration. It also sent an officer to draw up a 'black list' of Belgian collieries that failed to offer enough guarantees in terms of health and safety. At the request of the Italian government, the Belgo-Italian Committee (*Commission mixte italo-belge*), established by the agreement of March 1954, met at Rome in April 1956. The Italians refused to send new migrants unless the Belgian Federation took the requisite safety measures without any delay.

Meanwhile, Belgian trade unions supplied the Italian representatives of the Belgo-Italian Committee with a detailed list of claims from Italian coal miners in Belgium. Health and safety were the highest priority. However the whole meeting collapsed after only two days of intense discussions. The Belgian members abruptly declared themselves unable to come to a decision on Italian claims (and therefore claims of the Belgian trade unions), while the Italian government maintained its opposition to further migration.[21]

Such a position was unacceptable within the ECSC. According to the Treaty of Paris (Articles 46, 47 and 55), the ECSC had the mission to stimulate research and innovation in mining safety. Of course, it was not entitled to impose new regulations but it could encourage the member states to revise their legislation in favour of new safety devices. So the High

Authority offered its good offices by reopening the meeting of the Belgo-Italian Committee at Luxembourg in July 1956. The negotiations had just begun when a new tragedy unfolded. On 8 August 1956, the disaster of the *Charbonnage du Bois-du-Cazier* at Marcinelle killed 269 miners, among them 136 Italians. It had a tremendous impact throughout Europe. Once again, the Italian government forbade emigration to Belgium and asked for serious explanations and compensation. After painful negotiations, a new Belgo-Italian agreement was signed on 11 December 1957, but it was too late. Marcinelle had already marked the end of the flow of Italians to Belgian collieries.

Fédéchar then had to look elsewhere for foreign workers. Bilateral agreements were signed in a hurry with Spain (November 1956) and Greece (August 1957), while refugees from Hungary were 'welcomed' in underground mines. In the following years, the coal crisis seriously threatened the Belgian coal industry. An era of closure and restructuring coincided with the control of the ECSC. From 1957 until the closure of the last coal pit, the number of Italians employed in Belgian coal mines fell from 44 000 in 1957 to 2500 in 1975. Italians were replaced by Spanish and Greek miners and, in the 1960s–70s, workers from Morocco and Turkey. The proportion of Italians among foreign workers in underground mines decreased, from 71 per cent in 1955, to 37 per cent in 1965, and 20 per cent in 1975.[22]

The Marcinelle disaster was a watershed in the social history of Europe. In Belgium, the crisis united foreign workers and Belgians in defence of their common interests. Italians entered trade unions – especially the Christian trade union – and professional organizations like the *conseils d'entreprise*, while the *Centrale des Mineurs* pushed hard for the abolition of all kinds of social discrimination (in particular, in terms of retirement pensions) between Belgian and foreign miners.[23] As far as the ECSC was concerned, the Marcinelle disaster was a unique opportunity to take tangible initiatives in the social field. The Community organized a conference on safety in the coal mining industry at the end of 1956. The meeting brought together representatives of governments, employers and trade unions from the six member states, while experts were appointed to produce a series of recommendations on safety conditions. A European 'miner's statute' was even suggested, but finally rejected by the employers. However, the conference paved the way for important reforms in working conditions for coal miners in Europe. It insisted upon the training of new miners, improvements in safety devices and rescue facilities, and so on. Moreover, the conference recommended that governments should set up a Permanent Mines Safety Commission at a European level. On 9 May 1957, the Council of Ministers of the ECSC endorsed the proposal. The Commission on Mines Safety started to work in September 1957. 'By 31 December 1960, despite the considerable problems remaining to be disposed of and the difficulties of adapting or remodelling the safety regulations, most of the Conference's recommendations had been either incorporated into national legislation and regulations or simply applied in practice', the ECSC declared in its 10th Report.

Between 1961 and 1965, the Commission formulated 61 recommendations, 85 per cent of which were adopted by Belgium, France, Germany and Italy. Alongside the High Authority, the Commission elaborated a system of gathering comparable statistical data on mining accidents and injuries. The powers of the Commission on Mines Safety were strengthened after a new disaster, in February 1962, at Völklingen in the Saar. In other words, Marcinelle had forced Europe to enter the battlefield of working conditions in coal mines and paved the way for a broader European policy in industrial health and safety.[24]

The Treaty of Rome and the Free Movement of Labour

The Marcinelle disaster took place during the negotiations over the establishment of the European Economic Community, that took place from June 1955 until March 1957. The concept of the free movement of labour evolved from a freedom enjoyed by a limited group of skilled workers in coal mining and the steel industry to a general freedom for all workers within a general common market. The 'Report of the Heads of Delegations to the Foreign Ministers at the Messina Conference, 21 April 1956' (the Spaak Report) was the starting point for relaunching the European project. It stressed the importance of the free movement of labour and the necessity 'to remove prejudice against it and to be aware of the very important expansion factor which an influx of labour can present to a country'. The free movement of workers was considered an important factor for economic growth and as an instrument in the process of economic integration.[25]

The Treaty of Rome, signed on 25 March 1957, endorsed the free movement of labour within a framework which emphasized the mobility of all the factors of production. Article 48 stated that 'the free movement of workers shall be ensured within the Community not later than at the date of the expiry of the transitional period'. Thus was the policy for the free movement of labour extended to all workers and not only to those in the coal and steel industries. It involved 'the abolition of any discrimination based on nationality between workers of the Member States as regards employment, remuneration and other working conditions'. It included 'the right, subject to limitations justified by reasons of public order, public safety and public health, to accept offers of employment actually made; to move about freely for this purpose within the territory of Member States; to stay in any Member State in order to carry on an employment in conformity with the legislative and administrative provisions governing the employment of the workers of that State; and to live [. . .] in the territory of a Member State after having been employed there'.[26] During the transitional period a series of regulations established incrementally the free labour market that came into existence on 1 July 1968 over the whole territory of six member states of the Community.[27]

However, the free movement of labour had a limited impact upon the recruitment of foreign workers for coal mining. There were several reasons for this. Firstly, it concerned skilled workers, while unskilled workers continued to be recruited on a bilateral basis. Secondly, the Italian 'economic miracle' reduced the number of candidates looking to emigrate to Belgian coal mines. Finally, the structural decline of coal mining in Belgium, France and Germany implied a reduced and less attractive labour market. But, at the same time, underground coal miners remained in demand and a significant proportion of underground workers were made up of non-Belgian labour.[28] From the mid-1950s there was a constant decline in the total numbers of underground workers in Belgium (Figure 14.1). Nevertheless, more than 60 per cent of miners working in Belgium in the mid-1960s were still recruited from outside the country, although the numbers of Italian workers fell constantly. From the late 1960s, the proportion of Belgians increased. Yet it was not until the era of massive pit closures in the late 1970s and early 1980s that Belgians comprised the majority of underground workers.

From 'Guest Workers' to Community Members

The Marcinelle disaster and the various European and trades union initiatives *vis-à-vis* foreign workers slowly modified the way foreign labour was perceived both by the state and by local communities. Of the foreign workers who came to Belgium or France during the late 1940s and 1950s, many either returned home or moved elsewhere, but an important proportion remained and was progressively assimilated into the local population of the coalfields. In the 1960s, the structural decline of the coal industry meant that more and more collieries closed forever. Year after year, the coalfields saw their populations diminish, sometimes to a point that the demographic stability and economic viability of the regions were seriously threatened by the departure of young workers. Coal communities were in danger of becoming retirement homes for an increasingly ageing population. Reports appeared detailing the extent of the demographic crisis unfolding in Wallonia,[29] and against such a context, the foreign workers began to be seen in a much more positive light. Italian families came to be seen as vital members of the community rather than simply temporary workers. Studies launched by the High Authority on the question of redeployment and modernization of the industrial areas insisted upon the potential of this foreign population to

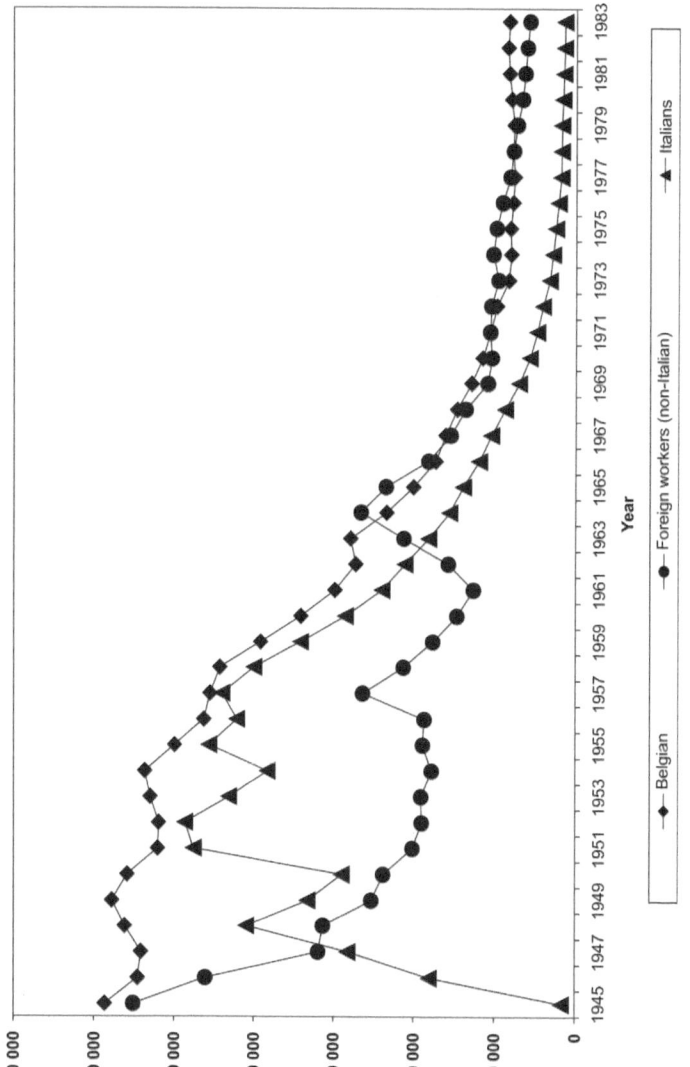

Source: based on the figures given by Harald Bergman and Jean-Louis Delaet, 'Memento statistique', in *Italiens de Wallonie*, Archives de Wallonie (Charleroi, 1996), p.239.

Figure 14.1 Underground workers in Belgian coal mining

act as brakes that could halt the demographic decline.[30] The challenge was to keep and maintain the foreign workers and their families in the industrial regions by introducing social reforms and social improvements in terms of housing, working conditions and the creation of new jobs.[31]

This new perception was clearly expressed in 1962, by the authors of a study of the situation in the Hainaut coalfield. Many of the foreign workers who came into the region were not simply temporary workers; lots remained to make their homes in Belgium, albeit in the face of much prejudice and hostility from the host society. The first measure that had to be taken therefore was to struggle against such negative attitudes and encourage a more tolerant and welcoming culture that would work to keep the immigrants within the coalfield communities. As Italian families came to be seen as an indispensible factor in the long process of industrial redeployment and regional modernization,[32] so the role of the European Community, alongside national and regional authorities, has been to foster these multi-ethnic communities.[33]

Conclusion

Belgian coal mining had attracted thousands of Italians and other foreign workers since the end of the Second World War. The ebbs and flows of 'guest workers' were regulated by the demand for coal. So often ill-trained and unskilled, foreign workers were more exposed to the dangers of underground mining than their skilled, Belgian counterparts. Accidents and disasters occurred with depressing regularity during the early years of the ECSC. However, the regulations in the Treaty of Paris said nothing about the massive influx of unskilled workers and, consequently, the issue remained in the hands of national authorities. On 8 August 1956, the disaster at Marcinelle sounded the death knell for massive migrations of Italians to Belgian coal mines. From that moment on, the character of the migration of unskilled labour changed. Italians were progressively replaced by workers coming from Spain, Portugal, Greece, Morocco and Turkey. And, crucially, the main legacy of the Marcinelle disaster was the international conference on mines safety, organized by the Council of Ministers of the ECSC, that led to the establishment of the Permanent Mines Safety Commission in 1957.

As far as the free movement of labour is concerned, neither the Treaty of Paris (1951) nor the Treaty of Rome (1957) dealt directly with mass migration of unskilled workers. However, in the long run, the European Community played a significant role in terms of industrial health and safety and in altering the ways in which foreign workers were perceived. New, multi-ethnic communities in the industrial areas have developed. Increasing demand for labour migrants after the Second World War helped bring this situation about. When the process of industrial decline began in the late 1950s, the importance of the immigrant workers' contribution to these local communities was more clearly revealed. Since the late 1950s and early 1960s, the ECSC and the European Economic Community have developed social policies designed to support the former immigrants and encourage them to stay. From the 1970s onwards, the free movement of labour has been a key issue in the process of European integration. As we have seen, such a process has deep roots in the coalfield communities of Belgium.

Notes

1 Communauté Européenne du Charbon et de l'Acier, *Obstacles à la mobilité des travailleurs et problèmes sociaux de réadaptation, préface de Paul Pinet membre de la Haute Autorité*, Luxembourg, Communauté Européenne du Charbon et de l'Acier, s.d. [1958], annexe III, tableau 1.

2 For more on the theory of economic integration and labour markets, see George N. Yannopoulos, 'Economic integration and labour movements', in G.R. Denton (ed.), *Economic Integration in Europe* (London, 1969), pp.220–45.

3 Alan S. Milward, *The European Rescue of the Nation-State* (London, 1994), pp.52–5.

4 Anne Morelli, 'L'appel à la main-d'œuvre italienne pour les charbonnages et sa mise en charge à son arrivée en Belgique dans l'immédiat après-guerre', *Revue Belge d'Histoire Contemporaine*, **XIX**(1–2) (1988), 83–130.

5 Jean-Louis Delaet, 'Les Belges ne veulent plus descendre. Recours à la main-d'œuvre italienne de 1922 à 1946', in Archives de Wallonie, *Italiens de Wallonie* (Charleroi, 1996), pp.15–29; W. Eder, 'Ruch wychodzczy ludnosci Polskiej do Belgii w okresie miedzywojennym (1919–1938) [Movement of Polish migration to Belgium, 1919–1938], *Przeglad Polonijny*, **4** (1981), 79–95. This study is confirmed by R. Clemens, G. Vosse-Smal and P. Minon, *L'Assimilation culturelle des Immigrants en Belgique. Italiens et Polonais dans la région liégeoise* (Liège, 1953).

6 A. Martens, *Les immigrés. Flux et reflux d'une main-d'œuvre d'appoint. La politique belge de l'immigration de 1945 à 1970* (Louvain, 1976), pp.41, 47–8; L.C. Hunter and G.L. Reid, *European Economic Integration and the Movement of Labour* (Ontario, 1970), p.32.

7 Milward, *European Rescue of the Nation-State*, p.51; Martens, *Les immigrés*, pp.67–70.

8 Franck Caestecker, 'Alien policy in Belgium, 1830–1940: the creation of guestworkers, refugees and illegal immigrants', European University Institute, PhD, Florence (1994).

9 Morelli, 'L'appel à la main-d'oeuvre italienne', 83–130; Michel Dumoulin, 'L'émergence du facteur Europe dans la politique à l'immigration de la Belgique à l'égard des Italiens au début des années cinquantes', in M. Dumoulin (ed.), *Mouvements et politiques migratoires en Europe depuis 1945: le cas italien* (Louvain-la-Neuve, 1989), pp.53–64; Claudio Calvaruso, 'La politica italiana in materia di emigrazione dal 1945 al 1960', in M. Dumoulin (ed.), *Mouvements et politiques*, pp.33–6; M. Dumoulin, 'Pour une histoire de l'immigration italienne en

Belgique, 1945–1956', in R. Aubert (ed.), *L'immigration italienne en Belgique. Histoire, langue, identité* (Louvain-la-Neuve, 1985), pp.27–52; Michel Dumoulin, 'Les mineurs italiens en Belgique (1945–1957). Des relations bilatérales à la dimension européenne', *Relations Internationales*, **54** (1968), 205–16. France too signed a similar agreement with Italy on 30 November 1946 (Pierre Guillen, 'L'immigration italienne en France après 1945, enjeu dans les relations franco-italiennes', in M. Dumoulin (ed.), *Mouvements et politiques*, pp.37–51).

10 Albert Delperee, 'L'organisation des mouvements de main-d'œuvre dans la Communauté Européenne du Charbon et de l'Acier', Centro Italiano di Studi Giuridici, *Actes officiels du Congrès international d'études sur la Communauté Européenne du Charbon et de l'Acier, Milan-Stresa, 31 mai-9 juin 1957*, vol. 7, *L'orientation sociale de la Communauté* (Milan, 1958), pp.323–78, especially 343–7.

11 Alan S. Milward, *European Rescue of the Nation-State*, pp.51–2.

12 Jean-Louis Delaet, 'Cinquante mille Italiens. La main-d'œuvre italienne dans les charbonnages de 1946 à 1958', *Italiens de Wallonie*, p.136.

13 William Diebold, *The Schuman Plan: A Study in Economic Cooperation 1950–1959* (New York, 1959), pp.443–4.

14 A. Martens, *Les immigrés*, pp.75–9, 90–93. The complex problems related to foreign workers and trade unions were the key point of an important conference at Brussels in 1952. See R. Petre, 'Le syndicat et le problème des travailleurs étrangers en Belgique', in *L'industrie charbonnière belge dans ses aspects sociaux, Conférence du B.I.T. et de l'Office Européen des Nations Unies, Bruxelles, 10–20 December 1952* (Brussels, 1953).

15 British Iron and Steel Federation and High Authority of the European Coal and Steel Community, *Treaty Establishing the European Coal and Steel Community*, translated by the British Iron and Steel Federation London. British Iron and Steel Federation (1951); Diebold, *Schuman Plan*, pp.442–3.

16 Carlo Lega, 'La liberté de circulation des travailleurs dans les pays de la Communauté Européenne du Charbon et de l'Acier', in Centro Italiano di Studi Giuridici, *Actes officiels du Congrès international d'études sur la Communauté Européenne du Charbon et de l'Acier, Milan-Stresa, 31 mai-9 juin 1957*, vol. 7 (Milan, 1958), pp.417–30; Albert Delperee, 'L'organisation des mouvements de main-d'œuvre dans la Communauté Européenne du Charbon et de l'Acier', in Centro Italiano di Studi Giuridici, *Actes officiels*, pp.323–78.

17 Doreen Collins, *The European Communities: The Social Policy of the First Phase*, vol. 1: *The European Coal and Steel Community 1951–70* (London, 1975), pp.65–71.

18 Figures provided by the Communauté Européenne du Charbon et de l'Acier, *Obstacles à la mobilité des travailleurs et problèmes sociaux de réadaptation, préface de Paul Pinet, membre de la Haute Autorité*, Luxembourg, Communauté Européenne du Charbon et de l'Acier, s.d. [1958], annexe III, tableau 1.

19 Diebold, *Schuman Plan*, p.445.

20 R. Leboutte, 'Mobilité spatiale de la main-d'œuvre dans les bassins industriels au 19e siècle – L'apport des livrets d'ouvriers', in Y. Landry, J.A. Dickinson, S. Pasleau and C. Desama (eds), *Les chemins de la migration en Belgique et au Québec, XVIIe–XXe siècles* (Louvain-la-Neuve, 1995), pp.155–64.

21 Martens, *Les immigrés*, p.106.

22 Delaet, 'Cinquante mille Italiens', pp.136–9; Felice Dassetto and Michel Dumoulin (eds), *8 août 1956. Marcinelle* (Louvain-la-Neuve, 1986).

23 Congrès de la Centrale de Francs Mineurs, September 1956; Congrès de la Centrale des Mineurs, October 1958. See Martens, *Les immigrés*, pp.114–15.

24 Collins, *The European Communities*, pp.82–9, and quotation from ECSC 10th Report, p.84; Diebold, *Schuman Plan*, pp.433, 444, 455.

25 Report of the Heads of Delegations to the Foreign Ministers at the Messina Conference, 21 April 1956, translated by J.P. Farrar (Brussels, Council Secretariat, 1956); Simone Goedings, 'Labour market developments, national migration policies and the integration of Western Europe, 1948–1968', in R. Leboutte (ed.), *Migrations and Migrants in Historical Perspective: Permanencies and Innovations* (Brussels, 2000), pp.311–30; F. Romero, 'Migration as an issue in European interdependence and integration: the case of Italy', in A.S. Milward (ed.), *The Frontier of National Sovereignty: History and Theory, 1945–1992* (London, 1992), p.52; F. Romero, *Emigrazione e integrazione europea, 1945–1973* (Rome, 1991).

26 European Communities, *Treaty Establishing the European Economic Community, Rome, 1957* (Luxembourg, 1958); D. Collins, *The European Communities*, vol. 2: *The European Economic Community 1958–72* (London, 1975), pp.99–116.

27 Regulation 15, August 1961, only authorized, rather than ensured the right of, workers of the Community to hold a job in another member state if no native workers were available in the national labour market. Regulation 38/64/EEC provided the measures to set up the free movement of labour, which was realized in July 1968 (Kenneth A. Dahlberg, 'The EEC Commission and the politics of the free movement of labour', *Journal of Common Market Studies*, **6** (1967–8), 310–33).

28 L.C. Hunter and G.L. Reid, *European Economic Integration*, pp.27–8.

29 Alfred Sauvy, *Le rapport Sauvy sur le problème de l'économie et de la population en Wallonie* (Liège, 1962).

30 M. Drechsel, 'Introduction à l'étude des problèmes de la reconversion du Centre et du Borinage', *Les régions du Borinage et du Centre à l'heure de la Reconversion*, XXIXe Semaine Sociale Universitaire, novembre 1961 (Brussells, 1962), pp.19–34.

31 Communauté Européenne du Charbon et de l'Acier, Haute Autorité, *Rapport spécial de la Haute Autorité à l'Assemblée Parlementaire Européenne concernant la question charbonnière (31 janvier au 15 mai 1959)* (Luxembourg, 1959).

32 Communauté Européenne du Charbon et de l'Acier, Haute Autorité, *Etude du développement économique des Régions de Charleroi, du Centre et du Borinage* (Luxembourg, 1962), pp.23–4, 106.

33 Communauté Européenne du Charbon et de l'Acier, Haute Autorité, *Etude sur la zone de Carbonia. Les conséquences sociales de la crise minière dans le bassin du Sulcis (Sardaigne)* (Luxembourg, 1966), p.18.

CHAPTER FIFTEEN

A Moral Economy, an Isolated Mass and Paternalized Migrants: Transvaal Colliery Strikes, 1925–49

Peter Alexander

At least for the period under consideration, 1925–49, black colliery workers were, in all likelihood, the most strike-prone component of South Africa's working class. To what extent should we explain this militancy in terms of dynamics that have been used to understand similar phenomena in the United States and Britain, and to what degree was it a product of purely local factors? In the account that follows, the first part considers some background data on coal and colliers in the Transvaal and Orange Free State (the two provinces responsible for the bulk of South Africa's coal production). The second analyses the colliery strikes that occurred in these provinces during the quarter-century to 1949 (that is, from just after the historic 1922 general strike to just after the pro-apartheid, National Party victory in the 1948 general election). The third and final part draws upon theories of 'moral economy' and 'isolated mass' as part of a discussion of these strikes. We conclude by suggesting that there may be merit in investigating policing as a factor affecting levels of militancy.

Background

By 1949, South Africa was producing nearly 28 million tons of coal, nearly double the output of 25 years before, but still less than middle-ranking countries like France, Japan and India.[1] Of this total, 81.1 per cent came from the Transvaal and Orange Free State, which were generally regarded as having a single coal industry (referred to here, as in contemporary literature, as that of 'Transvaal').[2] Nearly half of all South African coal came from the district around the town of Witbank, 90 miles east of Johannesburg, where there were huge, thick bituminous seams, close to the surface and containing few natural obstacles to relatively safe and efficient mining.[3] Production in this area took off in 1906, following the completion of a rail link to the gold mines of the Witwatersrand, close to Johannesburg. Most output went to local railways, gold mines and power stations, though from 1934 some went to the newly established steel industry, and, during the Second World War, there was increased demand for exports.[4]

The industry was marked by centralization of capital, internal cohesion and a high degree of integration with the Witwatersrand gold mines. In 1924, while only 16 per cent of collieries in Britain were employing 1000 or more workers, at least 20 per cent of the Witbank collieries were this size.[5] Similarly, in 1921, whereas Transvaal and Alabama (which, in many respects, had a comparable industry) both had 11 collieries producing more than 250 000 tons, the 11 in the former region were responsible for 73.7 per cent of total output, whilst those in the latter contributed only 28.4 per cent. Moreover, by 1920, 82 per cent of Transvaal coal was machine-cut, compared to 45 per cent in Alabama (and 60 per cent for the United States as a whole).[6] Cohesion, and reduced competition between the colliery owners, came about through the establishment of a price-fixing cartel in 1907 and the inauguration of an employers' association, the Collieries' Committee (CC), in 1914. The link to gold mining existed through purchases and ownership, but also through the Witwatersrand Native Labour Association, usually referred to as Wenela, which was controlled by the gold industry's Chamber of Mines. In 1900, with support from the British and Portuguese governments, Wenela had secured the sole right to recruit Mozambicans for work in South Africa and, as a consequence, it was able to provide cheap labour for both the collieries and the gold mines.[7]

In the collieries, as in the gold mines, the racial cleavage was qualitatively different from the deep division that existed even in Alabama. Rather than one, racially divided community, Transvaal collieries had two distinct communities. All the whites had higher-status employment than all the blacks, and the average white miner or mechanic was paid about nine times as much as the average African worker. Whilst most white workers lived with their families, typically in a five-roomed house (usually with some additional accommodation for a servant), Africans generally 'oscillated' between work, where most stayed in single-sex barracks, and a rural home. Language differences ensured that communication across the racial divide was minimal, food and clothes were completely different, and churches, schools and social and sporting activities provided further separation. Underground, a miner, always a white man, supervised a team of 40 to 100 'boys', that is, black men, and overall whites were outnumbered by about 20 to one. White workers never provided any solidarity for black struggles, and on the one occasion that they participated in a strike, in 1922, blacks continued to work.[8] The account presented here is, therefore, only about black workers. By 1949, there were some 32 742 of them employed in Transvaal collieries.[9]

Among these black workers, there were some significant differences. Figures for Chamber of Mines collieries, which were employing roughly nine out of ten Africans in the industry, show the proportion of Mozambicans fluctuating between 53.4 per cent in 1925, 43.3 per cent in 1936, 61.2 per cent in 1944 and 52.9 per cent in 1949. After Mozambique, the next largest sources of labour were the Transvaal (in all likelihood, these were mostly workers from Witbank's Sotho-speaking hinterland) and Basotholand. Combining figures for these two territories shows participation peaking in 1941, when it represented 42.1 per cent of total black labour.[10] Single-sex barracks were segregated according to nationality, and there was some inter-ethnic conflict (though this should not be exaggerated). However, not all Africans stayed in this accommodation. In 1926, the government had insisted that collieries provide housing for all black employees (with no more than 15 per cent in married quarters), but during the Second World War, when higher production was coupled with building controls, many workers moved into huts located on neighbouring farms. In this setting, with men often 'marrying' women of a different nationality, and some even adopting local names, there was some blurring of ethnic difference.[11]

The first major colliery-related collective action was in 1913, when incoming Mozambican recruits staged 'nearly an absolute mutiny' after being told they would be sent to collieries rather than gold mines. Following an investigation, Wenela forced the CC to accept reforms that included guaranteed minimum rates of pay, a 57-hour working week and a 313-day contract, food and accommodation in accordance with legislative require-ments and full access for Wenela's inspectors. Behind Wenela stood the Chamber of Mines, which, though it wanted cheap coal, was not prepared to allow distaste for the collieries to jeopardize recruitment in general. Between August and November 1918, a wave of eight strikes in the Witbank district led to a clarification and buttressing of the 1914 agreement. These strikes were typified by marches to the Wenela compound in Witbank and by sympathetic responses from the authorities. The following year, in a stoppage at Tweefontein Colliery, workers called for ten shillings per day, justifying this with comparison to white workers. This was an offensive demand – even perhaps a revolutionary one – and this time the strike was met with considerable repression, thereby circumscribing limits to what might be regarded as legitimate behaviour.[12]

It should be clear that our colliery workers were neither independent miners nor archetypal proletarians.[13] At the risk of creating yet another oversimplification, they might, perhaps, be regarded as *paternalized migrants*. The emphasis on 'migrant' is significant, not only because migrancy conditioned these workers' racial segregation and immediate living condi-tions, but also because, to a greater or lesser degree, they retained social security linked to rural subsistence. 'Paternalized' is taken to mean that they were regarded as 'boys', which implied not only that they could be physically punished for any wrongdoing, but also that they could expect, for instance, a reasonable amount of food and limits to mistreatment. These expectations, institutionalized by the 1914 agreement, had their roots in the combination of migrancy and the hegemony of gold-mining capital. We might contrast our communities of paternalized migrants with those of the South Wales and West Virginia colliers compared by Roger Fagge.[14] Whilst the Transvaal workers certainly lacked the social capital and civic spaces of the South Wales miners, they probably escaped some of the worst excesses of West Virginian-style 'feudalism', and they had a more reliable, if smaller, income than the colliery workers of these other two regions.

Strikes

For the 25 years under consideration, we have a record of 57 strikes in the Transvaal coal industry (and doubtless more will come to light as research continues). For historians of US and UK mining history this must seem like a small number, even if one were to make allowance for the relatively small size of the industry, and even, perhaps, if we were to add that these were non-unionized, non-citizens engaged in illegal acts. However, for South Africa, it meant that coal mining was the most strike-prone major industry. Narrowing our focus to the years 1939–48, a decade for which we also have detailed data for other industries, in Transvaal coal mining there were, on average, about 57 strikers per thousand black workers per year. This was higher than for any other groups of black workers. The most interesting contrast is with the closely-related gold industry, which, even though it employed, on average, about 344 000 workers during these years, experienced, to my knowledge, only nine stoppages involving black workers. Notwithstanding the fact that these disputes included an industry-wide strike in 1946, the number of strikers per thousand workers averaged only about 22 per year.[15] The chronological distribution of the 57 strikes is presented in Table 15.1. There was a cluster of seven strikes in the years 1926–7, and a major concentration in 1944–5, when

Table 15.1   Summary of Transvaal coal strikes, 1925–49

Period	No. of strikes	No. of strikers	No. of strike days
1925–29	10	5 192	8 692
1930–34	5	1 040	1 440
1935–39	6	1 605	1 605
1940–44	16	8 950	11 150
1945–49	20	8 805	14 218
Total	57	25 592	37 105

Table 15.2   **Complaints raised in Transvaal coal strikes,
1925–49**

Complaint	Times raised
Assault	25
Food-related	20
Wage-related	13
Hours	13
Dismissal of person in authority	10
Task-related	9
Marking of ticket	6
Inaction of person in authority	6
Washing water-related	6
Discrimination	5
Other	24
Total	137

there were 14 strikes. This pattern reflected the national sequence of black militancy, but why was it more dramatic?

Table 15.2 shows the 137 complaints raised by strikers. All of these were essentially defensive in character. There was not a single dispute akin to the 1919 Tweefontein strike, or the 1920 and 1946 strikes on the gold mines, all of which raised a demand – the same one in each case – that challenged the status quo. The proportion of wage complaints was relatively small, and all such disputes involved infringements of established rates rather than any attempt to secure a new norm. In addition to stoppages that were specifically 'wage-related', those that were 'task-related' or involved 'marking of a ticket' also had implications for pay. These reflected the fact that, for about half of all African workers (those engaged on cutting, drilling and loading), bonus payments and/or hours worked depended on the completion of a daily 'task', as evidenced by the accurate marking of a 'ticket'.

Assaults were the most common issue mentioned by strikers. It is likely that, as Dunbar Moodie noted for the gold mines, a certain level of force was 'taken for granted' and disputes would only arise if this went beyond the normal slaps and kicks. However, in contrast to the gold mines, where, according to Moodie, assaults persisted until the 1970s, the collieries may have experienced a shift in attitudes some 20 years before. Whilst assaults were mentioned in 60 per cent of strikes between 1925 and 1939, they only figured in 47 per cent of those occurring between 1940 and 1944, and 30 per cent of those that happened between 1945 and 1949, and, according to one ex-collier, by 1951 'people were no more beating workers'.[16] Whilst demands for dismissal of somebody in authority could be related to 'inaction' (that is, a failure to respond to a complaint), they might also be linked to an abuse of power, such as an assault or favouritism. The second most common complaints were those associated with quantity and quality of the 'free' food that workers received. Seven of these 20 'food-related' disputes were in the years 1945–6, when similar complaints contributed to the 1946 strike. Drought and bad harvests had brought about shortages, but these were made worse by forced reductions in imports and by the failure of the government to introduce a generalized scheme of rationing.[17]

Notwithstanding a lack of formal education and the absence of union organization, Transvaal colliery workers had an extensive 'repertoire of contention'. Four main tactics are discernible. The first of these was the simple protest meeting. Workers would assemble, as a mass, in some prominent position on the colliery, and would simply wait there, in an orderly manner, until the local commissioner of the Department of Native Affairs, plus other officials, arrived to consider their grievances. The second main tactic was a march to the nearest Wenela compound, which in most cases was the one in Witbank. Nineteen of the strikes started in this way, compared with 24 that started with a protest meeting. The routine was for the marchers to be persuaded to return to their colliery where, having demonstrated their unity of purpose, they would be given a good hearing by the commissioner and others. By marching in a large group and involving outside officials, workers protected themselves from victimization by colliery managers.[18] As Wenela recruits, Mozambican miners could expect a sympathetic ear at the compound and, as one Wenela inspector noted, they regarded it as their 'Father's House'.[19] As a quintessentially Mozambican tactic, it was less common on the gold mines, where other nationalities predominated. There was only one instance of a Sotho-led march by colliery workers and, on that occasion, and only that occasion, it terminated at the Witbank native commissioner's office rather than the Wenela compound.

The third tactic was a hunger strike. Five strikes started in this way, the first in 1944 and the last in 1946. This was the period when there were widespread food shortages, and the main reason for these actions was probably – as with the 1945 hunger strike at the gold-producing Crown Mines – to dramatize discontent about inadequacy of meals.[20] Finally there was violence, with four strikes starting this way and five others involving some use of violence, either against individuals or against property. Thus, as in the gold mines, violent acts were not commonly associated with strikes,[21] and when they did occur there was a specific purpose. This included an attempt to release Sotho youths from gaol, to apprehend a white miner who had badly assaulted a black worker and had then fled to the safety of the compound manager's office, to intimidate strike breakers and the police (evidence for this is only circumstantial) and to cause expensive damage (though again we cannot be certain about motivation). It is possible that the Sotho, or at least the Basotholand Sotho, were more prone to violence than Mozambicans. To quote one of Lauren Segal's interviewees: 'The Basothos are the ones who like to fight. Basothos are very aggressive. Jo! Just pick a small thing and they make it a big fight.' Segal suggests that this may have been linked to the fact that the Sotho tended to be fitter and stronger than Mozambicans (and often worked as loaders, the most physically demanding job, though one of the best paid).[22]

We can be confident that Mozambicans led at least 26 of the 57 strikes, yet there is clear evidence of Sotho leadership in only three cases. Given the numerical domination of Mozambicans (especially, perhaps, among the older workers), this pattern is not surprising. However the logistics of organizing a strike would also have had implications for its ethnic characteristics. There is some evidence that, on any particular colliery, the different national groupings consisted of networks of miners from the same rural areas, and, according to Elijah Shangwe, who worked at Minaar Colliery in the 1940s, Mozambicans would even arrange for someone from their home district to take over their position when their contract expired.[23] Mobilization linked to rural connections and potential retributions reduced the chance that a participant would disclose plans to a management representative. Once the decision had been taken to hold a protest, messengers would be sent to the various rooms to communicate the outcome. Moodie describes this process for the gold mines, but it seems to have been similar in the collieries.[24] Thus, in 1939 at Navigation Colliery, 'three East Coast Natives were discovered by the Mine Police going round the Compound rooms inciting East Coast Natives [that is, Mozambicans] not to go back to work in the morning'. Once there was a large body of strikers it would then be possible to prevent others from starting work, and this had the potential to produce ethnic tensions. For instance, at Tweefontein in 1944, South African 'natives' wanted to work, or so it was claimed by the commissioner, but, so he said, they feared reprisals from the Mozambicans.

Most strikes were ended through the intervention of the local native commissioner, the government representative charged with this task. To be effective, he had to appear even-handed, and usually began by listening to the workers' complaints. He would then go through the issues that had been raised and determine which of them were justified, acting, in effect, as an arbitrator (even though War Measure 145, promulgated in December 1942, made it almost impossible for other groups of black workers to secure arbitration). In a fairly typical example, the 1926 strike at Navigation Colliery, he decided (a) there had been undermarking of tickets, and this would have to be rectified; (b) since nobody came forward to substantiate allegations of assault against the mine captain, that matter would fall; and (c) he would ask the management to provide more *mareu* (a liquid porridge). There was a strong element of paternalism in this process and, after resolving a 1942 strike at Clydesdale Colliery, the commissioner commented, 'Like a lot of naughty children [the workers] cheered and went on their way back to the Colliery dancing and singing.'[25] This method of dispute resolution was quick and simple, which partially explains why the strikes were so short. The average strike lasted for less than two days, and the longest only continued for five days.

However, the perceived merits of a case were not the sole factor in resolving a dispute. For instance, in a 1944 strike at Landau Colliery, the commissioner responded to a workers' demand for the sacking of an *induna* (that is, a black headman) by arguing that 'it was not their place or right to select their own Induna and that his appointment rested entirely in the hands of the Mine Management'. 'But,' continued the commissioner, 'this explanation was received with such hostility ... I feared a general riot and suggested to the General Manager that it would be [advisable] in the circumstances to accede to their request.' After some protesting, the general manager complied.[26] An old Greenside colliery worker, Albert Simelane, interpreted the situation thus: 'The Native Commissioner will talk to them nicely ... he'll never shout. When people are many and when they are cross, when there's a strike, they can kill you.'[27] War Measure 145 reinforced an existing ban on strikes, but even after its promulgation only a minority of strikes led to detentions (seven out of the 27 strikes that occurred between 1943 and 1947). In cases where there were arrests this was always related to a breech of an accepted norm, that is, the use of violence, either against people or against property.

So, between 1925 and 1949, coal mining was the most militant major industry in South Africa. However, all its strikes were defensive in character, taking place around issues such as assaults and inadequate food. In these disputes, strikers gained strength from ethnic networks and from innovative tactics, such as marches to the local offices of Wenela, and conflicts were generally resolved through the intervention of a third party, normally the local native commissioner. Given that these strikes were illegal, and that for much of the period South Africa was at war, there were relatively few arrests.

Discussion

In attempting to make sense of these strikes, explaining, in particular, the relatively higher level of militancy among black colliers than among other South African workers, the idea of 'moral economy' provides a valuable starting point. This concept, initially deployed by E.P. Thompson, was applied by Moodie to understand disputes involving South Africa's black gold miners.[28] The latter author's description of this phenomenon is apposite: 'Workers expected food of a certain minimal quality, as well as wages comparable to those at other mines, a limit on the amount of personal assault underground, fair adjudication of personal disputes, equal treatment for each "tribal" category of workers, and a measure of latitude in allowing workers private lives of their own.'[29] The same expectations also existed in the collieries, where, as in the gold mines, they provided the main focus for strikes in this period (which was broadly similar to Moodie's). As with Thompson's eighteenth-century rioters, workers were defending what they regarded as established norms, rather than challenging the status quo, and, as with those 'plebs', our colliers did so by appealing to the guardians of these norms. Indeed we would know little of the strikes if it were not for the fact that they generally involved the intervention of the principal guardian, the local Commissioner of Native Affairs (or his representative).[30] Tactics, too, reflected the essentially 'moral economy' character of the action.

Significantly, the eighteenth-century rioters justified themselves in terms of 'rules' that had origins in an earlier period, and in their case these had been codified in the Book of Orders prior to 1630.[31] In like manner, our colliery strikers expected their masters to fulfil obligations that, in the main, dated back to the agreement of 1914 (and, I suspect, Moodie's moral economy of the gold mines had similar origins, though this is not an observation he makes). That is, they were appealing to the institutionalization of a balance of class forces that prevailed at the time of that agreement. When the native commissioner sided, as he sometimes did, with striking workers, he was representing dominant capitalist interests, particularly the mining houses, which required uninterrupted supplies of labour and a reasonable level of public order. Thus *moral economy was rooted in political economy*.

This assessment can be taken further. In attempting to comprehend why there was an upsurge of militancy in the 1940s, the moral economy approach can assist, but it only contributes a partial understanding. It is surely no coincidence that colliery unrest occurred at about the same time as a peak of militancy in the South African working class as a whole. As argued elsewhere, the key features of this wider discontent were wartime inflation, a rapid expansion in job opportunities and improved union organization.[32] Even if coal miners lacked the means to demand increased rates of pay, their frustrations were magnified by a decline in living standards and, while they did not flock into a union, they cannot have been left untouched by the new mood of confidence that developed during the war. Indeed there are occasional glimpses of a new assertiveness, as when, after the 1944 Tweefontein strike, a boss boy told a white miner: 'We have got rid of the Compound Manager and the Mine Captain and now we are going to get rid of the Mine Manager.'[33]

It is also worth recalling the following comment made by Thompson in his famous polemic supporting the 1972 strike by British coal miners: 'the miners,' he claimed, 'have always had difficulty in comprehending the simplest propositions as to market-regulation of wages, and have clung tenaciously to unscientific notions such as "justice" and "fair play". Hence every major wage conflict has turned into an argument about the "system" as a whole'.[34] Our South African strikes were also about 'justice' and 'fair play', yet did not turn into an argument about the 'system'. Maybe this was partly because South Africa's paternalists did not credit black workers with the intelligence to comprehend arguments about market forces. More significantly, perhaps, there was no reason why small disputes in a rising industry (unlike Britain's coal industry in the 1970s) needed to become a threat to the 'system'. Once again, it is important to situate moral economy within political economy.

With regard to the present chapter, the main limitation of 'moral economy' is that it does not help us explain the contrast between the different levels of militancy in the collieries and gold mines. In a seminal article, Clark Kerr and Abraham Siegel argued that the most militant workers, including miners, were those that constituted an 'isolated mass'.[35] This thesis is much disputed, but it continues to have resonance, albeit in disparate form.[36] Notwithstanding national and other differences, Transvaal's black colliery workers did have some of the homogeneity that one might expect of such a mass, with racial oppression, in particular, contributing to group cohesion. Moreover collieries did create 'occupational communities', albeit ones that were less stable than those in, for instance, the USA and the UK.[37] For the Transvaal communities, the colliery not only employed the workers, it was also responsible for housing and food, and in very limited ways for health care and social activities as well. Thus, compared with an increasing number of South Africa's industrial workers, colliery workers had a greater number of causes of complaint that they could lay at the door of their employer (and it is interesting to note how many of their grievances concerned issues of reproduction).

Whilst black gold miners were no less a 'mass' than their colliery cousins, they were less *isolated*. This was important, in particular, because it meant that the collieries were far from large concentrations of police. In the case of sustained rioting, police were brought in from Pretoria, but the city was 70 miles away, and it could take a day to mobilize and transport sufficient men to the disturbance. Crucially, perhaps, as a native sub-commissioner observed in 1926: 'The Police force here is so small as to be practically helpless if several hundred natives are determined to march into Witbank.'[38] Within the local balance of forces, this simple geographical reality gave weight to the workers' claims. I am aware of only one march by gold miners to the Johannesburg Wenela compound during the 1940s; it took place during the 1946 strike and was brutally attacked by the police.[39] A second, though indirect, effect of greater isolation was that, in this period, the colliery workers were never unionized. This was in contrast to the gold mines, where the African Mine Workers' Union, which had 25 000 members by 1944, used its influence to discourage strike action.[40] (Whilst the union did lead a massive strike in 1946, by that stage it was in a much weaker position than during the war, and the outcome was a demoralizing defeat.)

In their comprehensive analysis of British colliery strikes, Roy Church and Quintin Outram also concur with aspects of the 'isolated mass' thesis, but they place greatest weight on colliery size; that is, other things being equal, larger collieries tended to be more strike-prone.[41] However, in a comment on the analysis presented here, Gwede Mantashe, the general secretary of South Africa's National Union of Mine Workers, and a former colliery worker, suggested that the greater militancy of the collieries, compared to the gold mines, might have been because they were smaller; this, he said, had been the case in the 1980s, especially in the 1987 strike (the largest in South African history). In contrast to the collieries, where the average mine employed about 700 workers in 1924, the gold mines were huge, employing, on average, roughly ten times this number by 1936, so one can sympathize with the proposition that it was more difficult to mobilize gold miners for the kind of effective action that could halt production across a whole mine.[42] Despite the apparent contradiction between the British and South African arguments, it is possible that, while enterprises of the average colliery size – and Britain's were not much smaller than South Africa's – created a better basis for strong organization than in the average factory, which was much smaller in both countries, the enormity of South Africa's gold mines created new problems.[43] This notion of an optimum size for mobilization, which deserves further attention, was common currency among industrial militants in West London when this author was an activist in that area in the late 1970s.[44]

Other differences between the collieries and gold mines might also have been significant. Between 1939 and 1945, while there was a slight decline in the number of black workers employed in South African gold mines, the number employed in Chamber of Mines collieries increased by well over 50 per cent.[45] It is likely, therefore, that colliery workers tended to be younger than those on the gold mines, but further research would be required to judge both the credibility of this assumption and its significance. It is clear, though, that ethnic ties played a significant role in mobilization, and the fact that Mozambicans dominated the collieries in a way that no nationality prevailed in the gold mines, may also have had an impact on overall levels of strike action.

To summarize, in attempting to understand why colliery workers were the most militant black workers in South Africa between 1925 and 1949, a number of factors have been advanced, all of which require further exploration. These include, in particular, a firm foundation for moral economy disputes, substantial distance from large concentrations of police, a strong ethnic basis for strike mobilizations and the development of innovative tactics (notably the protest march). While some support was provided for Moodie's analysis of the workings of moral economy, it was argued that this needs to be linked to an historicized account of political economy. In terms of international comparison, our assessment offers some sustenance for the 'isolated mass' thesis, and proposes that there might be value in further investigating the ability of police to control striking mine workers.

Whilst Transvaal colliery workers' militancy was rooted in the efficacy of their practices, once the state was prepared to use force, as it did from the mid-1940s, this was found wanting. This is not to argue that, had the colliery workers had unions, as in the gold mines, they would have fared any better. As the future would demonstrate, it would require another era, full of new contradictions and new forms of resistance, before colliery workers would be able to advance their interests by means of successful trade-union mobilization.

Acknowledgments

A substantially longer version of this article was presented to the Comparative Coalfield Communities Conference and the RAU Sociology Seminar, and I am grateful for comments received on both occasions. The original paper, only lightly revised, has been published in *Social Dynamics*, **29**(1) (2003), and readers requiring full archival references should consult this version. I am grateful to South Africa's National Research Foundation for supporting the present phase of my research.

Notes

1   Office of Census and Statistics, *Official Yearbook of the Union of South Africa*, 25-1949 (Government Printer, Pretoria, 1950), p.975. It was not until the mid-1970s, following the signing of a massive contract with Japanese buyers, that exports, and production in general, began a steep upward curve.
2   Office of Census and Statistics, *Official Yearbook*, 25, p.966. For the history of coal mining in Natal, the province which produced the remainder of South Africa's coal, see Ruth Edgecombe and Bill Guest, 'The Natal coal industry in the South African economy 1910–1985', *South African Journal of Economic History*, **25**(1)(1987).
3   See Christopher K. Soutter, 'Labour migration and stabilization on the Transvaal coal mines, 1920–1990', MA, Queen's University, Kingston, 1995, 44.
4   See John Lang, *Power Base: Coal Mining in the Life of South Africa* (Johannesburg, 1995), especially pp.104–25, and Office of Census and Statistics, *Official Yearbook*, 25, p.973.
5   Roy Church and Quentin Outram, *Strikes and Solidarity: Coalfield Conflict in Britain 1889–1966* (Cambridge, 2002), p.19; Peter Alexander, 'Oscillating migrants, "detribalised" families and militancy: Mozambicans on Witbank collieries, 1918–1927', *Journal of Southern African Studies*, **27**(3)(2001), 516.
6   Peter Alexander, 'Race, class loyalty and the structure of capitalism: coal miners in Alabama and the Transvaal 1918–22', *Journal of Southern African Studies* **30**(1)(2004) 117, 121.

7   Peter Alexander, 'Coal, control and class experience in South Africa's Rand revolt of 1922', *Comparative Studies of South Asia, Africa and the Middle East*, **19**(1)(1999), 33.

8   Alexander, 'Race, class loyalty and the structure of capitalism'.

9   Office of Census and Statistics, *Official Yearbook*, 25, p.974.

10 Chamber of Mines Statistics Department, 'Black workers to COM affiliated collieries in the Transvaal', unpublished data, reproduced in Soutter, 'Labour migration', 91.

11 Peter Alexander, 'Patterns in protest: a preliminary analysis of strikes by black miners in the Transvaal, 1925–50', presented to conference 'Towards a comparative history of coalfield societies', University of Glamorgan, 2002, 4–9. See also Lauren Segal, 'Mines, migrants and women: strike action and labour unrest on the Witbank collieries from 1940-1950', BA Honours, University of the Witwatersrand, 1989, 140.

12 Alexander, 'Oscillating migrants', 508–23.

13 See R. Harrison (ed.), *Independent Collier: the Coal Miner as Archetypal Proletarian Reconsidered* (Hassocks, 1978).

14 Roger Fagge, *Power, Culture and Conflict in the Coalfields: West Virginia and South Wales, 1900–22* (Manchester, 1996), especially p.64.

15 Calculations draw on data presented in Peter Alexander, 'Industrial conflict, race and the South African state, 1939–1948', PhD, London University, 1994, 72–5 and Appendix 1, and Bureau of Census and Statistics, *Union Statistics*, G-5. Since evidence presented in my doctoral dissertation grossly underestimated the number of colliery strikes, assessments for the gold mines might also be questioned. However, Dunbar Moodie, the leading authority on the labour history of the gold mines, also recognizes the higher level of strike action on the collieries (email 5 February 2002). Indeed, he drew my attention to the fact, and I am grateful to him for this.

16 T. Dunbar Moodie, *Going for Gold: Men, Mines and Migration* (Johannesburg, 1994), p.54; Daniel Langa, interviewed by author and Nini Xulu, Witbank Location, 16 December 1997. In 1954, a Native Affairs official would report: 'Assaults: The position in the main satisfactory and on many collieries an improvement has been brought about.' See National Archives, NTS box 2339 file 1029/280, Labour Officer, Brakpan to DNL, 'Visit of collieries in Witbank area', 24 August1954.

17 Peter Alexander, *Workers, War and the Origins of Apartheid: Labour and Politics in South Africa 1939–48* (Oxford, 2000), pp.101–3; Moodie, *Going for Gold*, pp.224–5.

18 See Moodie, *Going for Gold*, p.91.

19 Alexander, 'Oscillating migrants', 519.

20 Moodie, *Going for Gold*, pp.224–5.

21 Ibid., pp.91–2.

22 Segal, 'Mines, migrants and women', 92–3, quoting Joseph Mathda.

23 Segal, 'Mines, migrants and women', particularly 85.

24 Moodie, *Going for Gold*, p.91.

25 National Archives, KWB box 217 file N1/7/2. H. Frost, Native Commissioner, Witbank, to Director of Native Labour, 29 September 1942.

26   National Archives, KWB box 217 file N1/7/2, Native Commissioner, Witbank, to Director of Native Labour, 1 May 1944.

27   Quoted in Segal, 'Mines, migrants and women', 63.

28   E.P. Thompson, 'The moral economy of the English crowd in the eighteenth century', in E.P. Thompson, *Customs in Common* (London, 1991), p.224; Moodie, *Going for Gold*, p.86. For some sceptical comments on Moodie's use of moral economy, see Alexander, *Workers, War*, pp.104–5, and for Thompson's own concerns about circumscribing the use of the term, see his 'Moral economy reviewed', in *Customs in Common*.

29   Moodie, *Going for Gold*, p.86.

30   In terms of the contribution they make to this account, the most important scribes were these commissioners and sometimes their assistants. According to government policy it was they who had the primary responsibility for responding to disputes involving 'natives' employed in gold mines and collieries. A close second in significance were senior local police officers, who normally played some part in such stoppages. Then, sometimes, strikes involved Wenela's compound managers, and occasionally their reports have also been used. Finally, strikes often drew in the Curator of Portuguese Natives, an appointee of the Mozambique government, or one of his inspectors. These were the officials most trusted by Mozambican labourers, so it is unfortunate that, as yet, I have not been able to locate their files. Of course, as always, basing an interpretation on second-hand description can multiply the possibilities for bias and misunderstanding. In this instance, though, the reports were, mostly at least, unvarnished summaries presented to a more senior official, commonly the Director of Native Labour, in the case of Native Affairs, or the deputy commissioner responsible for the Transvaal, with respect to the police. They were presented in the form of letters that, in the style of the period, invariably began with something like 'I have the honour to report ...'.

31   Thompson, 'Moral economy of the English crowd', p.224.

32   Alexander, *Workers, War*.

33   National Archives, KWB 217 N1/7/2, Native Commissioner, Witbank, 'Report on strike at Waterpan compound: Tweefontein on 16/5/44.'

34   E.P. Thompson, 'A special case', in *Writing by Candlelight* (London, 1980), p.66.

35   C. Kerr and A. Siegel, 'The interindustry propensity to strike: an international comparison', in A. Kornhauser, R. Dubin and A.M. Ross (eds), *Industrial Conflict* (London, 1954).

36   For a recent discussion, see Church and Outram, *Strikes and Solidarity*, especially pp.139–43.

37   For 'occupational community', see Robert Blauner, 'Work satisfaction and industrial trends in modern society', in W. Galenson and S.M. Lipset (eds), *Labor and Trade Unionism* (New York, 1960).

38   National Archives, NTS box 2077 file 177/280, E.K. Whitehead, Native Sub-commissioner, Witbank, to Director of Native Labour, 13 October 1926.

39   Alexander, *Workers, War*, p.103. There were also two attempts by colliery workers to march to the Johannesburg compound, both from collieries south of Johannesburg, and the police blocked both of these.

40   Alexander, *Workers, War*, especially p.82.

41   Church and Outram, *Strikes and Solidarity*, pp.149–50.

42   Alexander, 'Oscillating migrants'; Alexander, *Workers, War*, p.22.
43   Alexander, *Workers, War*, p.22; Church and Outram, *Strikes and Solidarity*, pp.19, 165.
44   However, in a general analysis of strikes in South Africa, 1939–48, I found no basis for this argument; see Alexander, *Workers, War*, p.130.
45   *Union Statistics for Fifty Years*, G-5; Chamber of Mines, Statistics Department, 'Black Workers to COM Affiliated Collieries in the Transvaal'.

CHAPTER SIXTEEN

Trade Union Development in the Ruhr and South Wales, 1890–1914

Leighton James

During the nineteenth and early twentieth centuries, South Wales and the Ruhr were two major powerhouses of the industrial revolution. Their coal fed the growing metal industries, enabled the rail networks to spread and facilitated the increased use of steam power in shipping. However the development of the industry in the two regions was characterized by different patterns. South Wales experienced an extended period of industrialization. From the late eighteenth century, coal mines supplied the blast furnaces in the burgeoning iron district centred around Merthyr Tydfil. From the mid-nineteenth century, the focus moved towards coal exports. In contrast, industrialization occurred later and more rapidly in Germany. In the Ruhr, the industry had been under state control before the liberalization of the economy in the 1860s. Moreover, there was a greater degree of horizontal concentration than was the case in South Wales. Yet, despite these structural differences, the coal industry in both areas shared one key feature: its labour intensity. Although advances in drainage and ventilation technology encouraged the growth of the industry, the actual winning of coal remained largely unchanged before the First World War. Coal getting was still based on the physical strength and skill of the miner. Thus, if the employers were to increase output, they had also to expand their workforce. By 1913 there were some 233 134 miners in South Wales and 394 569 in the Ruhr.[1]

This chapter is concerned with the efforts made to organize these workers. Despite similarities in work conditions and processes, a brief glance at the history of trade unionism in the two coalfields reveals striking differences between them in terms of miners' organization. In the Ruhr, the first lasting trade union organization, the *Alter Verband*, was formed in 1890 following a mass strike movement the previous year. The South Wales Miners' Federation (SWMF), established in 1898, was a comparative latecomer to the British industrial scene. However, by the outbreak of the First World War, the SWMF organized over half of the workforce. By contrast, the trade union movement in the Ruhr had split into several competing bodies. The Christian-social element within the *Alter Verband* broke away because of fears of social democratic influence. In 1894, they established a counter-organization, the *Gewerkverein christlicher Bergarbeiter* (Trade Association of Christian Miners). The picture was further complicated in 1902 by the formation of a specifically Polish organization, the *Zjednoczenie Zawdowe Polskie* (ZZP), by immigrant workers who felt that the German unions looked after their own first. These three unions together could only organize a third of the workers. Therefore, during the period under investigation, the development of miners' organization in the two regions was generally characterized by two different trends. In South Wales this was one of homogenization, in the Ruhr one of fragmentation. This begs the question: why in South Wales was a single trade union ultimately able to dominate workers' organization by 1914, while in the Ruhr several miners' unions competed for the allegiance of the colliers?

Structure, Lifeworld and Identity

Those who have commented on trade union development in South Wales and the Ruhr tend to explain these differences either in terms of the pattern of industrialization or of the different degree of religious and ethnic diversity in the mining communities.[2] These are, of course, important explanatory factors in any investigation of miners' trade unionism. The mining valleys of South Wales were virtually monoindustrial in character. Other industries, such as iron and tinplate, were present, but the former, at least, was in decline by the time the coalfield began to be exploited extensively. As regards religion, Nonconformity dominated the small pit villages[3] and the chapels made an important contribution to the social and cultural life of the mining communities.[4] Finally, although South Wales was not entirely free from religious and ethnic tension,[5] the origin of the workforce was relatively homogeneous, with the vast majority of workers coming from the rural counties of Wales and the west of England.[6]

The mining communities of the Ruhr, on the other hand, were far more diverse. Although pit villages similar to the South Wales pattern did exist in the region, there was generally a greater degree of urbanization.[7] In towns like Bochum, Dortmund and Essen, miners rubbed shoulders with other occupational groups such as metalworkers.[8] Indeed many enterprises in the area included both mines and steelworks. Substantial populations of both Protestants and Catholics also coexisted in the region, a feature unusual for Germany. In addition, a substantial portion of the mining workforce was made up of Polish-speaking immigrants originating from East Prussia. By 1904, almost a third of the mining workforce came from this region.[9] There was also a sizeable group of Masurian immigrants. Religious and ethnic cleavages bisected each other, for, while the Poles were predominantly Catholic, the Masurians tended to be Protestant. Unlike the assimilation of

English in-migrants into Welsh culture, and their consequent adaptation of it, the integration of the Polish immigrants into the industrial communities of the Ruhr proved far more of a problem and the formation of a distinctive Polish sub-culture has been identified.[10] The region also possessed vibrant Catholic, Protestant and Social Democratic cultures. Each was made up of a dense network of associations, clubs, parties and institutions and the various trade unions themselves were integral parts of these milieux.[11]

However, while acknowledging the importance of these structural explanations, it is apparent that they do not provide the whole picture. For example, the Pennsylvanian coalfield was just as socially diverse as the Ruhr, yet only one trade union (the United Mineworkers of America) held sway. This suggests that reading organizational structures off pre-existing milieux can be highly problematic; one runs the risk of oversimplification and crude sociological determinism. More recent work on the two coalfields has utilized Habermas's theory of the 'lifeworld'. Habermas argued that 'the cultural tradition shared by a community is constitutive of the lifeworld which the individual member finds already interpreted. This intersubjectively shared lifeworld forms the background for communicative action'.[12] This communicative action is the key to the lifeworld. For Habermas it is through this communicative action in the lifeworld that identities are formed, cultural traditions transmitted and understandings reached. Since the actors within the lifeworld engage in communicative action to reach understandings they would seem to have some independence from sociological factors. Thus the link between the milieux and the actors becomes more than a causal relationship. Examination of how the trade union-movements engaged in communicative action as organizations offers a fruitful avenue through which to approach the question of trade union development. This approach accepts the importance of the material context within which organizations were formed, but also emphasizes the role trade-union activists played in their creation and maintenance. It is suggested here that it can help to explain how the miners of South Wales had a far more unified trade union than their counterparts in the Ruhr.[13]

The present work seeks to approach the communicative action of the trade unions through an examination of their discourse and the means by which they sought to construct a discursive identity. This notion of the constructed identity recognizes the unions as active in creating an identity, namely, that of the 'ideal' miner to which all others could aspire. Such identities were used as a banner for recruitment and as an ideological weapon. This concept owes a debt to Patrick Joyce, who has argued that the 'ideology and practice of the unions played a central role in creating popular conceptions of the social order, particularly class perceptions. Union discourse was more than the reflection of the workers' worlds. It actively shaped it'.[14] However, as noted above, although this chapter is primarily concerned with the trade union discourses, it does not go as far as Joyce does in disassociating it from materialism. More recent work has tried to find a medium between structure and discourse.[15] In a similar vein, here discourse is seen as an effort to understand and interpret the 'real' world. As Dai Smith has noted, 'when leaders ... speak they are often as concerned with making sense of what they feel around them as they are with leading as such'.[16] Of course, individual attempts at 'making sense' raise the possibility of myriad, sometimes conflicting, sometimes complementary, interpretations. Thus fragmentation or homogenization in the trade union movement were not simply reflections of social divisions or the lack of them. Rather there was a more complex relationship between social structures and organizations. In the case of South Wales, part of the SWMF's success in recruiting miners to its ranks lay in the flexible identity it created, while the identities created by the unions of the Ruhr were more doctrinaire, serving to reinforce social divisions that already existed in the mining communities.

Discourses of Unionism

In an effort to justify the latest wage demand, much of trade union activists' language in both South Wales and the Ruhr was heavily factual, and comprised a mass of figures and statistics concerned with wages, prices and the cost of living. It was in disputes over these figures that the underlying economic concerns of the trade unions were most evident. However, in seeking to add flesh to the bare bones of figures, trade-union activists couched their figures in more metaphorical language ostensibly aimed at establishing the miners as a homogeneous group. In examining the discursive practices of the trade unions it is possible to distinguish a fundamental difference between the trade unions in the Ruhr and in South Wales. Within the discourse of the SWMF the most important motif was that of unity of the miners. This unity was placed before any religious or ethnic differences. Admittedly there were often bitter struggles between various ideological groups within the organization, but these internecine conflicts were contained within the existing structure of the SWMF. Differences between trade-union activists in South Wales tended to revolve around policy direction, rather than the more fundamental question of the 'essential' identity of the miner himself.

In the Ruhr, the miner's identity was problematized. Through their discursive practices the three main unions created organizational identities that were predicated on a specific conception of the miner. For example, the ZZP, as a Polish organization, constructed its identity upon the basis of ethnicity.[17] The *Gewerkverein* used Christianity in much the same way. In industrial relations it expounded an ideology of class collaboration, seeking to work with the employers in the spirit of Christian brotherhood. The *Alter*

Verband was, in turn, influenced by Marxist thought, and regarded the miners as part of the class war, which they believed was the inevitable result of the capitalist system.

These different tendencies were expressed through the literature produced by the union activists. It is perhaps pertinent to note here that there was a much greater volume of such literature in the Ruhr, something that was symptomatic of the competition between the various trade unions. Although a full survey of trade union literature is not possible here, several pieces, two from the SWMF and one each from the *Alter Verband* and *Gewerkverein*, can be highlighted as particularly paradigmatic.[18]

The Miners' Next Step, penned by members of the Unofficial Reform Committee (URC) and published early in 1912, has been regarded as a classic document of syndicalist thought. Seeing the document as being essentially about the unity of the miners might at first seem surprising. The background of the document was after all the Cambrian Combine dispute of 1910–11, during which the strikers refused to grant the SWMF executive plenary powers in negotiations and rejected terms agreed to by the Miners' Federation of Great Britain (MFGB). Indeed the document contained a scathing indictment of the policy of an older generation of leaders such as the moderate William Abraham (better known by his bardic name of Mabon). The pamphlet argued that the previous conciliatory policy 'of identity of interest between employers and ourselves be abolished, and a policy of open hostility installed'.[19] In fact, the union was to be restructured to make it into a more effective fighting force, able to carry the struggle to the employers. The leaders themselves also came in for heavy criticism. Conciliation, it was argued, tended to take power away from the men and concentrate it in the hands of the leaders, resulting in autocracy.

However it is here that the image of the miners as a united group of workers becomes most apparent. The pamphlet constructed a dichotomous relationship between the leaders and the men, rather than between different groups of miners. The leaders, although necessary in the past, had become, by virtue of the social prestige their position accrued, opposed to the interests of the rank and file. '*The leader then had an interest – a vested interest – in stopping progress.* They have therefore in some things an antagonism of interests with the rank and file. The condition of things in South Wales has reached a point when this difference of interest, this antagonism has become manifest.'[20] There was an implicit assumption behind this that the interests of 'the men' were identical and unified. Indeed, throughout the pamphlet, there was a juxtaposition of 'the leaders' and 'the men', the former frustrating the interests of the latter.

The authors' examination of the 'good' and the 'bad' side of leadership further emphasized the issue of unity. The leader was portrayed as an obstacle to the solidarity of the miners:

Sheep cannot be said to have solidarity. In obedience to a shepherd, they will go up or down, backwards or forwards as they are driven by him and his dogs. But they have no solidarity, for that means unity and loyalty. Unity and loyalty, not to an individual, or the policy of an individual, but to an interest and a policy which is understood and worked for by all.

Since leadership prevented the emergence of solidarity among the men, it also hindered the development of large-scale, national initiatives.[21] The pamphlet therefore presupposed the existence of unity among the miners. Its proposed changes to the SWMF constitution were aimed at allowing this existing unity its full expression. Centralization would break down the district autonomy of the lodges, while the issue of minimum wages would provide a common issue on which to fight. These changes would ultimately reinforce the solidarity of the miners since 'the enjoyment of benefits, derived from association, makes an atmosphere in which non-unionism cannot live'.[22]

The URC campaign to realize their goal of restructuring the SWMF naturally met with resistance from the older generation of leaders. Much of the struggle between the different schools of thought within the movement was conducted through the press in the form of letters and editorials. This war of words could often become extremely acrimonious.[23] However, despite these differences, the emphasis on the unity of the South Wales miners remained the paramount consideration of all activists. It was in this vein that Vernon Hartshorn wrote an article entitled 'Federation War'. Hartshorn himself occupied an ideological position somewhere between the syndicalists of the URC and the old guard of trade-union leaders.[24] He had been present at several early meetings of the URC, but was not among the authors of *The Miners' Next Step*.[25] Following the syndicalists' failed attempt to restructure the SWMF, Hartshorn noted that there was considerable criticism of those who were members of both the SWMF and the syndicalist Industrial Democracy League.[26] He feared that there was a movement afoot to expel 'some of the most prominent of these Industrialists from the official positions which they now hold within the Federation'.

Hartshorn's reaction was to appeal for calm. He recognized that the acrimony of the struggle was more concerned with personalities than with policies and that both sides were responsible for letting debate degenerate into personal invective. That different schools of thought regarding the policy of the Federation existed was only natural and right:

Within a vast organisation like the Miners' Federation, the clash and struggle of rival schools of theorists will always be at work. This is unavoidable, for it is the law of progress. The struggle for existence, for the survival of the fittest, is as pronounced on the plane of ideas as it is among species in the animal world. Freedom of personal expression and an unfettered right of propaganda are absolutely necessary to the fullest development of democratic organisations like the Miners' Federation.

However, Federation members had to exercise toleration 'if we are to preserve that unity which is the beginning and end of our strength'. Personal recrimination was to be avoided because it was 'not only morally unjustifiable, but it is very bad policy because it tends to disruption in the ranks of the workers, and we need all the unity we can command for pushing on the constructive programme which is immediately before us'. Unlike the *Miners' Next Step*, there was no dichotomous relationship between leaders and led in Hartshorn's piece. Differences of opinion there were, but the goal of raising the standards of the miners remained common to all. Yet, like the *Next Step*, the unity of the miners remained the first consideration. In effect, Hartshorn was urging the activists to remember that this unity should be placed before any 'barren war between personalities'. The official was to remember that, while he might have views of his own, he was also an administrator, 'whose duty it is to administer the declared policy of the Federation as a whole. The legislative power is in the hands of the rank and file, whose decisions the official is bound to observe'.[27] Ultimately the URC activists shared this view, for when their attempt to reshape the SWMF failed, they remained within the organization rather than establishing a counter-association.

The theme of unity that acted as a bridge between different schools of thought in the miners' trade-union movement in South Wales was lacking in the Ruhr. This is not to say that the various activists were unconcerned with the issue of unity. All sought to encourage solidarity among the miners. However, as noted, the very concept of the miner was more problematic in the Ruhr than in South Wales. This was most clearly seen in the relationship between religion and trade unionism. While the *Gewerkverein* was a self-declared Christian organization,[28] the *Alter Verband*, owing to its perceived links to the Social Democratic Party (SPD), was seen as an anti-religious body. From the mid-1890s, Otto Hue, the leading ideologue of the *Alter Verband*, sought to distance the organization from the SPD.[29] In his pamphlet, *Neutrale oder parteiische Gewerkschaften* ('Neutral or Party Political Trade Unions?'), Hue argued that strictly neutral trade unions would, by excluding political and religious issues, allow inter-union cooperation and strengthen the position of the working class. The pamphlet constructed several dichotomous relationships between the neutral union and, variously, the *Gewerkverein*, religion (as applied to social problems) and the employers. Hue recognized that there were schools of thought in the SPD that regarded the trade-union movement as a mere recruiting ground for the party and saw politics as the only important area of activity.[30] This approach was rejected. Instead emphasis was placed on the practical gains that could be made by trade unions. In fact the bread-and-butter demands of the *Alter Verband* and *Gewerkverein* were remarkably similar.[31] By focusing upon these practical goals, Hue was attempting to find some basis for a common identity of the miners stripped of politico-religious shibboleths.

> First I consider the neutral trade union an absolute necessity should the most important socio-political workers' associations win influence to improve the professional conditions of their members. This I believe is the central point of the trade union question; everything else is of incidental importance. Whoever wants to gain decisive economic influence for their trade union must keep it open to all work colleagues and tolerate neither bashful nor barefaced party politics in the professional association; religious questions are really not to be discussed in the union. Therefore, all those trade unions, who by statute exclude members from this or that political party, who commit their members to a religious or anti-religious creed, are not to be seen as really neutral.[32]

In this final sentence, Hue was effectively labelling the *Gewerkverein* as a non-neutral trade union, thereby implying that it weakened the whole trade union movement. He directly criticized its statutes, questioning how the union could be free only when it 'it is free from Social Democracy', and argued that, as far as the priests were concerned, its sole purpose was to combat the SPD.[33]

Although Hue was seeking to play down anti-religious sentiments within the *Alter Verband*, he failed to understand the extent to which religion was the ideological basis upon which *Gewerkverein* identity and policy was predicated. First, Hue attacked the notion that religion could offer any solution to social questions and pointed to the conflict between the various denominations: 'The apparent homogeneous Christian social policy is internally so fractious as one can only imagine. There is really not one, but two, three, four and who knows how many "Christianities".[34] In part, this preoccupation with religion reflected the confessional division of the Ruhr coalfield. This was in stark contrast to the SWMF activists, who made no mention of religion in their literature. Even though Hue was advocating a non-religious rather than an anti-religious identity, he still created a dichotomous relationship between the Churches and the trade unions. To an extent, this dichotomy was forced upon Hue by the organizational rivalry that existed between the two trade unions. He needed to show that religious questions could be excluded, but also that religion did not provide an effective ideological tool with which to improve the lot of the working class. In his attempt to delegitimize the efficacy of the Churches regarding social problems he portrayed them as an obstacle to trade unionism, their focus upon the afterlife justifying a quietist approach.

> I claim once more: If the orthodox Church authorities of both great confessions want to follow their social views consistently, then they must be opponents of every

trade union movement. In almost every religious edict it can be read that 'the earth is not our actual residence' ... In addition to this comes the beatitude of poverty in the oft-quoted parable of poor Lazarus. The more wretchedly lives the person on earth, the sooner he has the prospect of Heaven, if he is otherwise devout.[35]

Second, Hue's discursive construction of class was based on an openly Marxist reading of industrial society:

> The trade union movement is not the product of systematic agitation but owes its formation to the thinking proletarian's feeling of the conflict between capital and labour. This conflict forces the labourer to resist as soon as he recognizes the necessity of economic organization of the working class. Every real worker's organization is an alliance of the working class, the organization of the class struggle.[36]

Hue expressed his ideological convictions even more clearly when he argued that the neutral trade unionists were the 'true Marxists'.[37] The identity being created by the *Alter Verband* was predicated on the idea that the miners represented a class whose economic interests were in opposition to those of the employing class. Because the relationship between the two classes was essentially a conflicting one, 'the neutral trade union movement can never lead to an attenuation of the fighting character of the workers' associations'.[38] Recognition of the shared interest between workers and the conflict with employers formed the basis for unity. The exclusion of religion and politics would allow the true expression of what he called 'worker's politics' (*Arbeiterpolitik*).[39] This third dichotomous relationship within the discourse has some parallels in South Wales, where the unity stressed in the *Miners' Next Step* and by Hartshorn was ultimately directed at winning concessions from employers.

Heinrich Imbusch was Hue's counterpart in the *Gewerkverein* movement. He was influential in constructing his organization's identity through its organ, *Die Bergknappe*, and various pamphlets.[40] Imbusch's *Ist eine Verschmelzung der Bergarbeiterorganisationen möglich* ('Is an Amalgamation of the Miners' Organizations Possible?') was designed primarily to counter suggestions that the two unions should amalgamate following cooperation during a strike in 1905. As such it offers a particularly good example of the *Gewerkverein* discourse. The crux of the argument, as in Hue's pamphlet, was the issue of religion. However, rather than seeking to exclude it, Imbusch placed it at the centre of *Gewerkverein* identity. Unsurprisingly he emphatically rejected accusations that the *Gewerkverein* split the workers and served only the interests of the employers and Zentrum Party. Although admitting that it was regrettable that the movement had been divided, he argued that the union existed to service a particular requirement of the workers. In exploring the reasons for the split, Imbusch, like Hue, established a dichotomy between the movements based on their world view (*Weltanschauung*). For Imbusch the SPD stood

> on the basis of materialism, for which everything is only matter; they renounce hereby the existence of God, of an ever-lasting human soul, of an afterlife and of retribution in the hereafter. According to their materialistic view of history there is nothing unchangeable; everything is comprehended in a continuous, never-ending process of development ... this fundamental belief of Social Democracy stands diametrically opposed to the fundamental beliefs of Christianity.

In this understanding, the SPD was seen as being involved in a struggle against Christianity.[41] Imbusch conflated the SPD and the Free Trade Unions, assessing both on the basis of their materialism.[42] He rejected the determinism of Marxist thought, which acted as an ideological basis for the *Alter Verband*.

Turning to the issue of amalgamation specifically, Imbusch applied the same arguments against the *Alter Verband*:

> The relationship of the Gewerkverein to the Alter Verband is different from the relationship of the entire Christian to the entire social-democratic movement ... The Christian workers can and must not be allowed to become traitors to their religious and political convictions and affiliate themselves to the materialistic anti-Christian standpoint of the Alter Verband; they want to help the working class to gain rights on the basis of their Christian viewpoint, because they are convinced that every economic policy that runs counter to the ideals of Christianity and directly fights this can ultimately only bring the working class harm. The Christian workers are conscious that in following their material interests they have no opponents in Christianity; on the contrary, their demands, recognition of human dignity and the equality of the worker, just wage, and so on, find their first principal substantiation through Christianity, with its high value of humanity, its ideal view of work, its perceptions of justice and of love.[43]

Thus, for Imbusch, religious thought not only offered a point around which to rally the miners, but also offered a means for realizing their demands. The contrast between himself and Hue on this point could not be clearer. Nevertheless, Imbusch freely admitted that the demands of the two trade unions were similar. Indeed they might have provided a stage for united action had they not been refracted through the two unions' inflexible ideological positions, which neither side seemed able to transcend. Instead their rivalry was consciously played out between the two identities each had created. 'The fierce, bitter struggle between the two movements corresponds not so much to disagreement over foreseeable economic disagreements, as the difference of fundamental ideas.'[44]

The *Gewerkverein*'s discursive construction of the miners as a class also differed markedly from that employed by the Alter Verband activists. That the union saw the workers as a distinct group with their own interests is beyond doubt. However, rather than seeing the workers as locked in inevitable conflict with the employers owing to the workings of the capitalist system, it sought to create a more cooperative relationship between the two groups. The language both sides employed directly represented this cooperation/conflict dichotomy. Although both referred to the workers (*Arbeiter*), Imbusch repeatedly used the term *Arbeiterstand*, while Hue used *Arbeiterklasse*. While the latter can be translated as 'working class', the former has the overtones of an 'estate' (*Stand*), suggesting a more reciprocal relationship between it and other groups in society. In policy terms, this led Imbusch to condemn the class-war rhetoric of the *Alter Verband*. He deplored what he saw as the growing militancy both within and without that movement: 'Again and again Social Democracy points out that the present state, the present order of society, is not at all capable of fulfilling the demands of the workers.' For Imbusch such attitudes were bound up with the ultimate aim of overthrowing the existing order. This was in contrast to 'the formation of the Christian trade unions, who want to improve the situation of the workers and fight for the equality of the *Arbeiterstand*, within the existing state, within the existing order of society'.[45] In his 'othering' of the *Alter Verband*, Imbusch was presenting it as a danger not only to the miners' faith, but also to society at large. At the same time Imbusch was placing the *Gewerkverein* solidly behind the existing social order, effectively adding nationalism to his discursive construct. It was for these reasons that the *Gewerkverein* considered any amalgamation of the two unions impossible and that the two unions were to remain in conflict until the outbreak of the First World War.

Conclusion

This chapter has attempted to highlight the role that discourse and organizational identities played in the development of the trade-union movements in South Wales and the Ruhr before the First World War. This is not to deny the role sociological factors played in the shaping of these organizations, but merely to suggest that the relationship between the material context and language is more complex than has been accounted for. That the SWMF was more successful than the *Alter Verband* or the *Gewerkverein* in organizing miners was due to the constant emphasis of solidarity within their discourse, even when there was bitter division among leaders over policy. The identity of the miner constructed by the SWMF was comparatively unproblematic and was largely devoid of religious or political overtones. The more cohesive communities contributed to this, while the union simultaneously reinforced their homogeneity. All activists ultimately aimed at improving the lot of the miners. It was simply a question of how to do so.

The similarity of the demands made by individual workforces during the 1905 strike suggests that there existed a common base upon which a more unified movement might have been possible in the Ruhr. As both sides admitted, their practical goals were virtually the same. However, unlike the SWMF, the *Gewerkverein* and the *Alter Verband* were unable to transcend their sectionalist, doctrinaire identities, based on Christianity and Marxism, respectively. In large measure, these discourses were a reflection of the diverse communities in the Ruhr, but they also contained a dynamic of their own, feeding back into social divisions and serving to reinforce them. All three unions, then, were not only a product of their societies, but played an active role in shaping them for their own members.

Notes

1 Figures compiled from Chris Williams, *Capitalism, Community and Conflict: The South Wales Coalfield, 1898–1947* (Cardiff, 1998), p.89, and S.H.F. Hickey, *Workers in Imperial Germany: The Miners of the Ruhr* (Oxford, 1985), p.13. For accounts of the development of the coal industry, see E.D. Lewis, *The Rhondda Valleys: A Study in Industrial Development, 1800 to the Present Day* (London, 1958) and Klaus Tenfelde, *Sozialgeschichte der Bergarbeiterschaft an der Ruhr im 19. Jahrhundert* (Bonn/Bad Godesberg, 1977).

2 Werner Berg, *Wirtschaft und Gesellschaft in Deutschland und Großbritannien im Übergang zum 'organisierten Kapitalismus': Unternehmer, Angestellte, Arbeiter und Staat im Steinkohlenbergbau des Ruhrgebietes und von Südwales, 1850–1914* (Berlin, 1984); Roy Church and Quentin Outram, *Strikes and Solidarity: Coalfield Conflict in Britain, 1889–1966* (Cambridge, 1998), pp.244–5.

3 See Michael Lieven, *Sengehnnydd: The Universal Pit Village, 1890–1930* (Llandysul, 1994) and the relevant sections of David Gilbert's *Class, Community and Conflict: Social Change in Two British Coalfields, 1850–1926* (Oxford, 1992).

4 Cyril E. Gwyther, 'Sidelights on religion and politics in the Rhondda valley, 1906–26', *Llafur*, **3**(1)(1980), 30–31.

5 See, for example, Neil Evans, 'Through the prism of ethnic violence: riots and racial attacks in Wales, 1826–2002', in Charlotte Williams, Neil Evans and Paul O'Leary (eds), *A Tolerant Nation? Exploring Ethnic Diversity in Wales* (Cardiff, 2003).

6 Brinley Thomas, 'The migration of labour into the Glamorganshire coalfield, 1861–1911', W.E. Machinton (ed.), *Industrial South Wales 1750–1914: Essays in Welsh Economic History* (London, 1969), pp.41–8.

7 See Wolfgang Köllmann, 'The process of urbanization in Germany at the height of the industrialization period', *Journal of Contemporary History*, **4**(3)(1969), 59–77.

8 See David F. Crew, *Town in the Ruhr: A Social History of Bochum, 1860–1914* (New York, 1979).

9 Berg, *Wirtschaft*, p.241.

10 See Christoph Kleßmann, *Polnische Bergarbeiter im Ruhrgebiet 1870–1945* (Göttingen, 1978), pp.83–92.

11 Wolfgang Jäger, *Bergarbeitermilieus und Parteien im Ruhrgebiet: Zum Wahlverhalten des katholischen Bergarbeitermilieus bis 1933* (Munich, 1996); Stefan Goch, *Sozialdemokratische Arbeiterbewegung und Arbeiterkultur im Ruhrgebiet. Eine Untersuchung am Beispiel Gelsenkirchen, 1848–1975* (Düsseldorf, 1990); Michael Schäfer, 'Das Milieu der katholischen Arbeiter im Ruhrgebiet (1890–1914)', in Dagmar Kift (ed.), *Kirmes-Kneipe-Kino. Arbeiterkultur im Ruhrgebiet zwischen Kommerz und Kontrolle (1850–1914)* (Paderborn, 1992).

12 From Jürgen Habermas, *The Theory of Communicative Action*, vol. 2, *Lifeworld and System: A Critique of Functionalist Reason* (Boston, 1987), quoted in William Outhwaite (ed.), *The Habermas Reader* (Cambridge, 1996), p.133, original emphasis.

13 Stefan Berger, 'Working-class culture and the labour movement in the South Wales and the Ruhr coalfields: a comparison', *Llafur*, **8**(2)(2001), 5–40.

14 Patrick Joyce, *Visions of the People: Industrial England and the Question of Class, 1848–1914* (Cambridge, 1991), p.137.

15 Neville Kirk and John Belchem (eds), *Languages of Labour* (Aldershot, 1997); Jon Lawrence, *Speaking for the People: Party, Language and Popular Politics in England 1867–1914* (Cambridge, 1998); James Vernon, *Politics and the People: A Study in English Political Culture, 1815–1867* (Cambridge, 1993). For an overview, see Andy Croll, 'The impact of postmodernism on British social history', *Mitteilungsblatt*, des Insitutes für die Geschichte der sozialen Bewegungen, **27**(2002), 137–52.

16 David Smith, 'Leaders and led', in K.S. Hopkins (ed.), *Rhondda Past and Future* (Ferndale, 1974), p.43.

17 The ZZP combined both nationalism and class consciousness to provide a particularly militant stance, John J. Kulczycki, *The Foreign Worker and the German Labor Movement* (Oxford, 1994), 263.

18 A specific study of the Polish identity constructed by the ZZP awaits its own Polish-speaking historian.

19 *The Miners' Next Step* (Tonypandy, 1912), p.29.

20 Ibid., p.15; original emphasis.

21 Ibid., p.20.

22 Ibid., p.27.

23 For example, see *The Rhondda Socialist*, 8 June 1912 and 31 August 1912, for a sharp exchange between David Evan and Noah Ablett.

24 Hartshorn was miners' agent for Maesteg from 1905. See Peter Stead, 'Vernon Hartshorn: miners' agent and cabinet minister', *Glamorgan Historian*, **6**(1969), 84–93.

25 David Egan, 'The Unofficial Reform Committee and *The Miners' Next Step*: documents from the W.H. Mainwaring Papers with an introduction and notes', *Llafur*, **2**(3)(1978), 64–80.

26 For the Industrial Democracy League, see Joseph White, 'Syndicalism in a mature industrial setting: the case of Britain', in Marcel von der Linden and Wyne Thorpe (eds), *Revolutionary Syndicalism: An International Perspective* (Aldershot, 1990), p.109. See the *South Wales Worker*, 2 August 1913, for a serialization of the League's aims and objectives written by W.F. Hay.

27 *South Wales Worker*, 21 June 1913.

28 For the *Gewerkverein* statutes, see Heinrich Imbusch, *Arbeitsverhältnis und Arbeiterorganisationen im deutschen Bergbau. Eine geschichtliche Darstellung* (Essen-Ruhr, 1908), pp.716–20.

29 Hue's biographer, Nikolaus Osterroth, noted that such was Hue's presence and role that many found it hard to believe that he was not leader of the *Alter Verband*, but merely editor of its organ, *Die Bergarbeiterzeitung*. See Nikolaus Osterroth, *Otto Hue: Sein Leben und Wirken* (Bochum, 1922), p.13.

30 Otto Hue, *Neutrale oder parteiische Gewerkschaften? Ein Beitrag zur Gewerkschaftsfrage zugleich eine Geschichte der deutschen Bergarbeiterbewegung* (1900), p.112.

31 Ibid., p.83.

32 Ibid., p.155.

33 Ibid., pp.133–4.

34 Ibid., p.110.

35 Ibid., p.114.

36 Ibid., p.9.

37 Ibid., pp.147–8.

38 Ibid., p.150.

39 Ibid., p.142.

40 See Michael Schäfer, *Heinrich Imbusch, Christlicher Gewerkschaftler und Widerstandskämpfer* (Munich, 1990), pp.34–6.

41 Heinrich Imbusch, *Ist eine Verschmelzung der Bergarbeiterorganisationen möglich? Kritische Betrachtungen zur Frage der Verschmelzung der beiden großen Bergarbeiterverbände* (Essen, 1906), p.6.

42 Ibid., pp.16–17.

43 Ibid., p.61.

44 Ibid., p.62.

45 Ibid., p.63.

CHAPTER SEVENTEEN

Coalfield Leaders, Trade Unionism and Communist Politics: Exploring Arthur Horner and Abe Moffat

John McIlroy and Alan Campbell

Arthur Horner (1894–1968) and Abe Moffat (1896–1975) are sometimes bracketed together as successful architects and builders of twentieth-century trade unionism in the British coalfields. Reference to their careers studs the literature but neither has attracted proper analytical study. Mention is frequently made of their membership of the British Communist Party (CP) but the influence of this central allegiance and its interplay with their role as trade union leaders has received inadequate attention.[1] Communism is neglected in the best known and long influential histories of the miners by Robin Page Arnot, himself a party member. Here CP strategy and the distinctive role of its supporters are marginalized and merged: party activists are typically and simplistically constituted as just union activists, good guys who unproblematically represent the real interests of the miners.[2] The finest, although skeletal, account, of Horner's career until 1945 is threaded through Hywel Francis and David Smith's narrative of the South Wales Miners' Federation (SWMF), embedded in an expansive union history and constrained, as far as the CP dimension is concerned, by the inadequate sources available in the 1970s.[3] Stephen McBride, in a terse outline, depicts Horner as general secretary of the National Union of Mineworkers determined to prosecute the transition to socialism but thwarted by the unprepossessing context of postwar Britain. It is an argument which neglects Horner's earlier evolution, his relations with CP politics and, by 1945, his restricted radicalism.[4]

More recently, Nina Fishman has asserted that Horner, Moffat and other CP mining leaders were conscious adherents of a calculated strategy on the part of the CP leaders, a political design which ensured that from 1932 into the Cold War years there was 'a practical coincidence of views and outlook between Party and non-Party mining activists'.[5] A successful battle against CP sectarians ensured that Horner and Moffat operated on the basis of a 'trade union loyalism' and 'revolutionary pragmatism'. It was a strategy which coherently subordinated the party to the union and secured political and industrial unity with their labourist colleagues. The claim is that

> the underlying motivation for refusing to enter into political conflict was union loyalism. The deep commitment to placing the union first had marked the CP's approach to economic struggle since the early 1930s. CP activists espoused rank and file militancy, but when the economic struggle forced them to choose between the two ideological imperatives, Communist mining trade unionists were firmly guided by Horner, Abe Moffat from Scotland, and Harry Pollitt and Johnnie Campbell at the Communist Party centre towards union loyalism.[6]

Remarkably, this view has little to say about the fundamental and enduring problem which trade unionism constitutes for Communists or the pressures of economism and bureaucratization which have explained, for many observers from Lenin and Michels, how activists were drawn away from radical politics and party discipline. Pinning down what was a real if demarcated distancing between the CP and its trade unionists, Fishman's assertion replaces process, evolution and drift with conscious decision: it substitutes for consideration of the practical difficulties and consequent erosion of democratic centralism an explicit project on the part of the party leadership simply to disregard it in favour of economism. It neglects the directive Russian dimension in the determination of CP policy, its very real significance in the political psyche of coalfield leaders such as Horner and Moffat and the different attitudes they took to domestic and international issues in their trade union practice. The broad sweep and dearth of detailed evidence driving such labels as 'union loyalism' and 'revolutionary pragmatism' (it is difficult to see Horner and Moffat as in any significant sense 'revolutionaries' by the 1940s, while their pragmatism was always circumscribed) blur important differences between different periods. Correctly discerning areas of agreement and cooperation between CP activists and non-party activists, this approach diminishes significant areas of conflict. It makes one wonder why a range of miners' leaders – Alf Davies, Lawrence Daly, Bert Wynn, Arthur Scargill – left the CP, and precisely why Labour supporters such as Will Lawther, Jim Bowman and Will Arthur moved from collaborating with it to antagonism towards it.

Biography, still underutilized in the history of mining unionism, can provide another way of looking at these problems. Life histories can provide both longitude and a more intense means of exploring and illuminating the formation and evolution of historical actors, their intricate relationships with others and the complexity of their interaction with industrial and political structures. Collective biography can throw into relief the relative importance of different, sometimes competing, allegiances and suggest both what was distinctive about the protagonists and what they had in common with cothinkers and opponents. Here we are restricted to a rough sketch. It may still provide some basis for challenging, dissolving and reassembling existing but inadequate conceptualizations.

Learning Leadership in Scotland and Wales

Only two years separated Horner and Moffat but the differences in their political development suggest the complexities of generational analysis. Both came from religious backgrounds: Horner was trained as a preacher in the Churches of Christ, Moffat was brought up amongst the Plymouth Brethren. Initial socialization may have encouraged the qualities of dedication, service, sacrifice, a sense of mission and faith in the Idea which they brought, in 1920 and 1922, respectively, to the secular faith of communism. While both were associated with Little Moscows, Lumphinnans was part of the fabric of Moffat's existence from birth. Horner, raised in the relative diversity of Merthyr Tydfil, came to Mardy already a socialist activist. Here is an important difference. By 1919, when Moffat was finding his feet as a working miner after fighting in France, Horner was serving six months' hard labour for opposition to the war, already hardened by experience in the Miners' Reform Movement, the South Wales Socialist Society and the Irish Republican Army. By 1927, victimized after the national lockout, Moffat remained a relatively inexperienced rank-and-file Communist in Fife. It was only the following year that, together with his younger brother Alec, who then overshadowed him in the party, he gained his first office as a checkweighman. Horner, in contrast, was by then an experienced leader who had been a checkweigher, a member of the CP Executive, Political Bureau and industrial organizer, animator of the Minority Movement and the party's leading activist in the Miners' Federation of Great Britain (MFGB).[7]

In comparison, Moffat entered the Comintern's Third Period and its world of ultra-leftism and social fascism a political ingénu. He had, however, already experienced sectarianism. In the fissiparous Scottish coalfields, he had favoured purism and fragmentation over unity by supporting the breakaway Fife Reform Union in 1922 in defiance of party policy. He now became a partisan of the new CP breakaway, the United Mineworkers of Scotland (UMS) and an enthusiast for its policy of stimulating strikes. After a stint as UMS organizer in Central Fife, he was appointed General Secretary in September 1931 in what proved a vain attempt to rescue a failing union and a disastrous policy. It was only then, as the leader of a CP auxiliary organization and on the basis of active ultra-leftism, that he achieved prominence in the party. He was in his mid-thirties when he was first elected to the CP Central Committee in 1932, visiting Russia for the first time three years later.[8]

Where Moffat was a beneficiary of the Third Period, Horner was a prime victim. Like a sizeable section of the CP leadership, he could not perceive the relevance or viability in the mining unions of seeking to win the rank and file away from the 'social fascist apparatus' and if possible into a revolutionary United Mineworkers of Great Britain. He recoiled at the prospect of fighting inside the Federation and then, if defeated, ignoring its democracy and providing independent leadership outside it and against it. Surveying the divisions among Scotland's miners, Moffat backed new ruptures. After 20 years fighting the fragmentations of geology, geography, occupation and politics in Wales, Horner had no stomach for a new schism.[9] After the Leeds Congress of December 1929 which affirmed the new line, he was sent to Moscow to work in the Red International of Labour Unions. He was permitted to return only after public recantation of his mistakes. He soon found himself the victim of Stalinist heresy hunting orchestrated by the CP leader Harry Pollitt. Horner was hammered by Pollitt and his supporters for the cardinal sins of 'trade union legalism' and 'lack of faith in independent leadership'. He had demonstrated a half-hearted attitude towards an abortive attempt to continue a strike which had been called off in January 1931 by a SWMF conference and thereafter had criticized the role which party leaders had played in the dispute. Refusing to recant, he was vilified in the party press, packed off to Moscow once more, tried by Stalinist votaries and convicted of transgression of Russian doctrine. It was only after further demeaning self-flagellation in the pages of the *Daily Worker* that he attained a measure of rehabilitation.[10]

Thus far we can read very little at all of trade union loyalism from the activity of the CP in the unions. Horner's rebellion and its outcome affirmed his ultimate subordination to the party, otherwise he would have left it for a union career, not bent at the knee to the Kremlin *curia* and publicly renounced his beliefs. We can only speak of union loyalism on Horner's part at this stage in the limited sense that he believed that the SWMF constituted the optimal arena for revolutionary activity and that effective work precluded open flouting of its democracy. Moffat, for his part, had demonstrated an almost total absence of trade union loyalism. It is unquestionable that both of them – and the party leaders – had learnt or affirmed lessons during the bizarre Third Period. Both Horner and Moffat had come to grips with the problems of organizing. Both grasped the indispensability of starting from miners' problems and both were developing as experts in collective agreements and union rules and as accomplished advocates who dealt in facts, not rhetoric. Moffat, for example, was training himself in safety legislation and would soon establish a reputation for effective prosecution of miners' cases.[11]

But the primary, indispensable agent of change, and this must be emphasized, was not the lived experience of British Communists: it was the perceived interests of the Russian bureaucracy sifted through the Comintern with the results relayed to its affiliates. It was Moscow's decision to move from ultra-leftism to Popular Front rightism which primarily shaped the next phase, not the subordinate predilections of Pollitt or Horner for a change in trade union strategy. The imperatives of the new line – building cross-class, anti-fascist alliances, enhanced realism in the face of the threat to Russia from the new order in Germany, entering more organically into the mainstream of civil society and the labour movement – facilitated greater recognition of the actual consciousness of workers and greater acceptance of the limitations of collective bargaining and the constraints of union democracy.

But the move was not from opposition to social fascist bureaucracy to loyalty to the union apparatus. The change was from the position that a pro-capitalist bureaucracy could not be impelled to act in workers' interests to the Popular Front axiom that that this was in fact possible. Moreover it could be largely achieved through official channels rather than through 'independent' structures like the Minority Movement. To accept this was not to embrace union loyalism. The CP still saw its interests as in conflict with those of a union leadership oscillating between labour and capital. It still organized disciplined party fractions in unions and it still resorted to rank-and-file organization when necessary to force union leaders to fight. However, in the context of the perennial pressures of economism and incorporation, growing union membership, enhanced bargaining power and expanded collective bargaining, as well as more developed bureaucracy, this new approach produced powerful temptations for communists. As the 1930s progressed, the unions were much changed from the 1920s. For the party leaders, who prioritized party loyalism, it must again be stressed that trade union loyalism remained a snare to which unsteeled activists could succumb, not a strategic imperative of the CP.[12]

As the party moved to Popular Front politics, Moffat encountered new problems. The UMS was liquidated in December 1935 in accordance with the decisions of the Seventh World Congress of the Comintern. Moffat explained, 'we are demonstrating to CI that we are carrying out decisions not only in words but in actual practice'.[13] However, the National Union of Scottish Mine Workers (NUSMW) refused to readmit the UMS officials on the grounds that they were not working miners. It was only in late 1938, after a spell as a CP organizer, that Moffat found work underground and gained re-entry to the former social fascist organization. In 1940, he was elected to the NUSMW Executive. But his newfound enthusiasm for unity in the coalfields was fractured by the overriding necessities of the Hitler–Stalin pact. Far from sharing an identity of outlook with 'the old gang' of the NUSMW and MFGB whom he had spent much of the previous decade vilifying, he opposed the war which they loyally supported. He campaigned for the MFGB to use its power, including strikes, to make what was bound to be a humiliating peace with Hitler. Reverting to his earlier self, he opposed the 'Fascist methods' which he believed Chamberlain and Churchill were utilizing to incorporate the unions and discipline the working class.

On the occasion of Hitler's invasion of Russia, Moffat somersaulted with all the aplomb of a performing seal and rediscovered his discarded zeal for fighting German Fascism. Helped by this *volte-face*, by his undoubted abilities and by unprecedented turnover of officials, he was elected president of the NUSMW in November 1942 at the age of 46. Thereafter he excelled the right wing, with whom he now belatedly shared a temporary identity of outlook, albeit anchored in Russian rather than British interests, in his demands for extra output and opposition to strikes. He secured executive powers to expel those participating in industrial action and excoriated strike leaders as 'Fascist'. He worked assiduously for nationalization: consonant with CP policy, he never argued for a greater element of workers' control but accepted in its entirety the Morrisonian, state-capitalist National Coal Board (NCB). And he was second to none in his support for the new National Union of Mineworkers (NUM), becoming the first president of the Scottish Area (NUMSA).[14]

If Moffat only demonstrated a limited form of trade union loyalism, and one activated by purely Russian considerations from 1941, Horner's development was more contradictory and distinctive. His recollected invocation of reading Clausewitz *On War* in prison in 1932 as indicative of a sea change in his conception of trade unionism may be read as Gramscian code for a turn not only from the Third Period but from the politics he had embraced earlier, a turn from revolution towards strategic reformism.[15] Horner's career until 1940 cut with the grain of the Popular Front, but in the SWMF he pursued a personal project cast from calculative labourism, one which owed little to Pollitt or the party. Rather the enduring stigmata of the Third Period fostered an independence which had the party leaders in King Street and in South Wales railing against his new assertion of autonomy. To take one example, Horner defied CP instructions to advocate a coalfield-wide strike against company unionism in October 1935, as a critical party report made clear:

> With regard to the question of discipline in not carrying out the line put forward by Comrades Cox and Pollitt. Comrade Horner poured contempt on the very idea that a decision of Comrades Cox and Pollitt represented the line of the Party ... He also claimed that Comrade Pollitt could not take a decision on his own and also criticized Comrade Pollitt that he did not get in touch with the District Party leadership ... he claimed that those responsible to take a decision in a situation where there is no time for consultation with the leading Party organs are the fraction leaders in the Trade Unions, and not the Party leaders elected by Party Congresses. He also went on to attack the Party, claiming it had done nothing to fight against the Company union in Parc and Dare.[16]

These problems endured; but in practice Pollitt and Campbell settled for second best.

Horner's appointment as the agent for the anthracite district in 1933 gave him a new and influential bastion in the SWMF, one which he was not prepared to endanger. As president of 'the Fed' from 1936, the structural limitations of his position generated new influences, new responsibilities, new possibilities and new constraints. If a year earlier he had cited his role as 'fraction leader' to defy Pollitt, he now invoked his position as union president. His earlier thinking crystallized and he worked to modernize and unify the union rather than utilize his union practice to make revolutionaries. The compromises Horner made to win back members of the South Wales breakaway union and the concessions he made to the employers to establish the SWMF as a responsible bargaining agent, pledged to adhere to agreements and control strikes, drew criticism from the left, including CP members. While he made no attempt to manoeuvre through the union the CP's opposition to the war and conciliation of Hitler, he did not, on the evidence, criticize, still less actively oppose, party policy.[17]

Rather he opposed the consequences of the war. In September 1940, for example, Horner asserted that Churchill's Order 1305 and 'outlawing of strikes' was 'bringing the unions into line with the Nazi labour front'. In response he called for the creation of a new opposition in industry, based on the model of the shop stewards' movement of 1914–18.[18] This is interesting in view of his subsequent statement that, without the knowledge of the SWMF or the CP, he served on Churchill's Invasion Committee.[19] As with Moffat, it was only in June 1941, when the German invasion of Russia required a crusade against Fascism unnecessary when Hitler and Stalin invaded Poland or when Hitler conquered France and threatened Britain, that Horner changed course along with his party, emerging as a fully-fledged proponent of productionism and scourge of strikers. Where other CP union leaders, such as Dai Dan Evans, wavered or, like Dai Llewellyn, broke, Horner remained implacable on these issues. Like Moffat, he was a fervent supporter of nationalization and its limits, and of the new national union: in 1946, he became the first general secretary of the NUM.[20]

Leading Britain's Miners

In the postwar era the industrial politics of both Horner and Moffat were circumscribed by the positions they held and the ideas which they brought to these positions: first, by loyalty to the NCB, acceptance of the limits of nationalization and the need to protect what was perceived as a major achievement from the attention of the Conservative governments after 1951; second, by the perceived need to nurture and extend the fragile unity of the

NUM in a situation where coalfield fissures, of occupation, remuneration or denomination, religious and political, ensured that even area cohesion was brittle. The leadership of both men, for better or worse, looked backwards to the prewar position of private ownership, disunity and defeat. The attitude of caution, of 'never again' and the mantra of 'unity', while understandable in relation to the lives not only of Horner and Moffat but of thousands of miners, fostered restraint and moderation in the context of partnership with the NCB. But it was a static, conservative, troubled unity which facilitated relative decline in wages, pit closures and contraction of employment. Bargaining power, in at times favourable conditions of coal shortage, was suppressed to protect a status quo only questionably endangered. Sectional action, resilient despite virulent denunciation by our protagonists, was seen not as a precursor of more general mobilization but as a challenge to an inert, often imagined and, in terms of wages and jobs, minimally effective unity.[21]

But the dialectic between the values and party principles which they brought to defined roles in the union bureaucracy and the pressures which the demands of those roles exercised on their practice require further exploration. And we must pay regard to precision. Horner and Moffat had had different experiences, they were in different positions and they were subject to different structural pressures. Horner, relatively isolated at the head office in London as the civil servant of the executive, bristling with facts, figures and clauses, lacked autonomy and a regional base, although, given his earlier views, it is one-sided to see him as simply 'a prisoner of the executive'. Moffat, an increasingly authoritative patriarch in his own domain, was constrained as a member of the Communist minority on the NUM executive but possessed more independent leverage. Nonetheless, both remained party spokesmen on mining matters, both were members of the CP executive and, as the Cold War developed, the shadow of Stalin fell on the NUM. After the ruptures of 1927–34, the CP had mended its fences with non-party activists. Now growing suspicion of the CP on the part of the president, Will Lawther, vice-president, Jim Bowman, and regional leaders such as Will Arthur and Sam Watson re-emerged. It was bound up with their new establishment role in a nationalized industry, loyalty to Labour and Attlee's incomes policy and fear of expanding Stalinism. Their antipathy was reciprocated. There were very real political conflicts.[22]

Horner's final watershed and further estrangement from the CP came in 1948, not in 1932, and it was not entirely voluntary. The rap across the knuckles which he received from the NUM executive in May 1947 for speaking for the CP on NUM issues was reinforced the following year by a firm warning circulated across the coalfields. The executive repudiated his support for a French mining strike led by Stalinist-influenced unions. Horner's isolation was symbolized by Moffat's signature on the executive report on the affair, although Moffat stated his disagreement with its political comments,[23] and by the lack of support Horner received from the South Wales Area. Many saw this as an important terminus: 'From then on Comrade H took a back seat and Mr H moved in.'[24] As the CP reverted to attacking both capitalist nationalization and the moderation of the NUM leadership, Horner maintained its industrial politics of 1941–7. For his subordination of party to union he was criticized by CP activists and 'treated like a recalcitrant schoolboy' by the party's industrial organizer, Peter Kerrigan.[25]

Moffat maintained a higher profile. The Scottish Area defended Horner and, as the Cold War set in, in conflict with the NUM establishment, spearheaded opposition to wage restraint and took a harder line on collective bargaining issues. But, as with sectional strikes within his area, Moffat drew the line at independent area action. His only response to colliery closures was legalistic: he argued each case through procedures which could not reverse NCB decisions. By 1959, 90 Scottish pits had closed since Vesting Day. When miners struck in protest, he pressured them back to work. When a stay-down strike at the Devon Colliery in Clackmannanshire stimulated solidarity action from 25 000 men at 43 collieries, he urged the men to leave the pit, accepted the offer of the NCB chairman to meet a deputation if the strikes were called off, secured a return to work and subsequently roundly condemned the unofficial action. As the industry contracted, all he could offer, like Horner, was demonstrations and lobbies of Parliament. The NCB's suspicion that Moffat was soft on sectional stoppages proved unfounded. If he differed from his Labour Party colleagues on the national executive over increased wage differentials and NCB structures, his sentiments were never acted on. Like Horner, he never questioned the ethos of partnership and the prioritizing of unity. He always remained a staunch opponent of militancy and membership mobilization and perceived his role as balancing different interests and delivering limited, incremental improvements to his members in an orderly, regulated fashion.[26]

For Moffat, 1954 was a landmark: the tensions between the CP and their opponents in the NUM culminated in his standing for the presidency in succession to Lawther against Ernest Jones. This violated what some saw as a newly minted convention that the two top jobs should be shared between left and right. The election, which Moffat lost by a creditable 162 396 to 348 391 votes, affirmed the very real antagonisms between CP leaders in the NUM and their opponents. The events of 1956 fanned the flames, put the CP on the defensive and led to a weakening of factional organization. Thereafter, Moffat retreated behind Hadrian's Wall. But he remained a party man. He not only consolidated an industrial unity only forged in the 1940s; he made the union the bedrock of CP influence in Scotland. Communists dominated its administrative staff. Moffat's personal secretaries, Renée Barnes and later Honor Tweedie, were party members, while his daughter Ella played an important organizing role. Instruments of cohesion such as the educational scheme and the youth committee were under his direct influence and used to recruit and groom party members, while the posts of union agents in Ayrshire, Lanarkshire, Fife and the Lothians were increasingly occupied by CP members. The six-strong committee of the journal, the *Scottish Miner*, consisted of Moffat and five fellow party members and criticism was ruthlessly suppressed. Moffat remained, through and through, a CP loyalist. Aware of his responsibilities, he spoke on its platforms, sponsored its policies and, if he failed to act on its policy over closures and industrial action, he facilitated building party branches in the pits.[27]

Horner, Moffat and Stalinist Labourism

Both our protagonists, despite disjunctures in their practice and constraints of structure which they accepted both by seeking office and by internalizing its limits, persisted with their politics and their party to the end. The political practice and ideology of Stalinism and its incarnation in Russia remained, as their autobiographies attest, central to their view of the world. Moffat would not have demurred from Horner's considered judgment:

> history will pay tribute to the many great achievements of Stalin during the most difficult period of the Soviet Union in peace and war, and will record that the triumphs of Socialism, which he saw carried through, is [*sic*] the hope for the people all over the world. Stalin made mistakes ... the crime of re-arming the Nazi generals and spreading nuclear arms all over the Western alliance, including Western Germany, will be recorded by history as a greater evil.[28]

Horner myopically believed not only that the Russian economy had overtaken that of Britain and America, but that 'The Soviet Union and the other countries building Socialism have developed their own sort of democracy, a democracy not based on two or more parties representing different class interests facing each other, but a democracy based on the elimination of class.'[29] Moffat, in turn, asserted that Russia was more democratic than Britain: 'there is a dictatorship here but dictatorship of what? Dictatorship of the ruling class. Whereas in the Soviet Union, while there is a dictatorship, it is a dictatorship of the working class, a dictatorship of the people'.[30] Moffat made no bones about his attempts to carry his Stalinism into the union and he continued to propound the line on party platforms.[31] At the CP's 1949 Congress, for example, he moved a resolution: 'To strengthen the understanding of the vanguard role of the Socialist Soviet Union as the powerful bulwark and champion of peace, democracy and socialism in the common cause of the working class and peoples of the whole world.'[32] In disputes within the Soviet bloc, he knew which side he was on, pronouncing for the party: 'it was quite plain that Tito and his gang were placing the entire economy of Yugoslavia in the hands of British and American capital'.[33] Needless to say, 'Like Trotskyism from which it derives, Titoism seeks to organize disruption in the international working class movement.'[34]

Such views, far from coinciding with those of their Labour Party colleagues, brought Horner and Moffat into political conflict with them. Hungary demonstrated the limits of trade union loyalism, the depth of their allegiance to Stalinism and the extent to which they would fight to defend its depredations. The NUM executive condemned the Russian invasion and called for the withdrawal of troops. It wrote to the Soviet miners' union deploring the incursion as murderous, morally wrong and inexcusable. Pressure was brought to bear particularly by the president, Ernest Jones, to ensure that Horner, particularly in view of his prominence, made a statement disowning the Russians in accordance with the position of the union. Horner refused to do so, blaming only American imperialism.[35]

Outnumbered at national level, Moffat assiduously manoeuvred to ensure that the Scottish Area implemented the CP line. In the face of demands from branches for condemnation of the Russian tanks, he opined that, unlike the British in the contemporary invasion of Egypt, they had been invited in. At a Special Conference, the Scottish executive tabled an anodyne, evasive motion lamenting 'the tragic loss of life', pledging support for 'socialism' and 'withdrawal of all foreign troops from all foreign countries' and demanding that the TUC organize relief for Egyptian and Hungarian victims of war. Moving the statement, Moffat unashamedly defended the tanks. He asserted that, while (in the aftermath of Khrushchev's speech to the 20th Congress of the Russian party) there were 'weaknesses' in the Soviet Union, 'it was now generally accepted' that Stalin had been 'correct' to invade Poland and Finland. A 'similar working class approach' should be taken to the invasion of Hungary. Despite the president's best efforts, an amendment to the motion was carried explicitly condemning 'Soviet military aggression'.[36] Moffat did not leave matters there: a 'financial vote' of the branches endorsed the unamended motion with strong support from the CP bastions in Fife. Despite strikes at two collieries and a demonstration outside his office, he weathered the storm, ruling out of order resolutions from branches proposing a ban on CP members holding union office and blessing suppression of criticism by party dissidents in the *Scottish Miner*.[37]

Horner retired in 1959, his last years shadowed by John Barleycorn. Moffat quit the scene two years later. Even this sketch demonstrates that their easy categorization (and Horner is the best case) as 'union loyalists' may simplify or distort the role of the CP and what its activists had in common with those of different allegiance, while playing down their Stalinism. Such labelling can elide into the whitewashing which asserts that the Horners and the Moffats were, just like the Lawthers or Joneses, 'ordinary socialists' and similar pronouncements that the Russian dimension was a subsidiary element in Communist lives.[38] It is fitting to emphasize what communist coalfield leaders had in common with their colleagues, but it is incumbent on the conscientious historian also to emphasize the differences between them. If we evoke their labourism, we must evoke their Stalinism.

Allowing for differences between individuals, accepting the need for more research and proffering this as a broad framework for further detailed analysis and comparison of individual careers, we can tentatively characterize the practice of Horner, Moffat and other CP miners from the early 1930s into the 1960s as the practice of Stalinist labourism. We use the term 'Stalinist' to denote endorsement of the policies and practice of Stalin and the rulers of Russia who were seen as the guarantors of 'the gains of October'; their belief that the primary, overriding interests of the British working class lay in the defence of 'socialism' in Russia and the Soviet bloc; their understanding that the rulers of Russia were the best judges of what was best for the world's workers; and their trust that, suitably guided, the CP would usher in the Russian system, suitably refined, in Britain. We use 'labourism' to denote the practice of gradualism, immediately dependent on collective bargaining and incremental political reformism sited centrally in and on the state, which in Lenin's terms failed to transcend economism, trade union consciousness and illusions in the bourgeois state.

Stalinist labourism was not a conscious strategy of CP leaders. It developed gradually and organically and aspects of it were resisted. It was influenced by processes within trade unionism, by union growth and collective bargaining, by the pressures of position, specialization and bureaucratization – and by processes within the CP: not only the difficulties of implementing any thorough-going democratic centralism in the face of union pressures and the prospective loss of cadre, but also the progress from *Class Against Class* and *For a Soviet Britain* to *The British Road to Socialism*. Stalinist labourism had four main strands. First, its proponents developed a primary, specialized identity as union activists, attempting to operate the institutions of trade unionism and accepting the reformist constraints they impose. This may be contrasted with 'the tribune of the people', equally active in other arenas, and the revolutionary cadre utilizing the union as a forum for propaganda and mobilization. Final decisions on implementation of party policy were reserved to protagonists and there was a tendency to conventional reformist practice. Second, there was a commitment to accepting and extending the social community dimension of mining trade unionism rather than, as in the Third Period, seeking to supersede it. CP mining activists sought to develop the social, educational and welfare role of the union, particularly in the Welsh and Scottish labour movements.

Third, what distinguished Stalinist labourism from mainstream labourism was its enduring aspiration to organize politically inside workplace and union, closing the divide between the industrial and the political. But this must be qualified. The right wing in the NUM also organized, if more fitfully and less elaborately. The role of CP activists in building party pit branches was at best sporadic, and for high-ranking officials the project was often rhetorical: the labourist divide was sustained. Fourth, if many CP full-time officers separated their party's industrial politics from their own practice, they articulated the CP's international policies, forthrightly declaring their allegiance to Stalinism. This was not incidental or episodic, as the minutes of the NUM and its predecessors in the 1930s, 1940s and 1950s demonstrate. And when the chips were down, as we have seen, even those whose industrial practice was essentially labourist could be trusted to hurry to the aid of the perverted 'already existing socialism'.

Postscript: the Decline of Stalinist Labourism

But even before Horner and Moffat had passed from the coalfields, things were changing like the seasons. Stalinist labourism was evanescent and crumbling. The avatar of change emerged from its heartland. Lawrence Daly was born in Fife in 1925. Raised as a Catholic, he started work at fourteen at Glencraig Colliery near Lumphinnans and in 1940, inspired 'with something akin to religious fervour', he joined the Young Communist League.[39] He was a protégé of Moffat, groomed early for leadership: he became chair of the NUMSA youth committee, secretary of the Glencraig branch and a member of the Scottish Committee of the CP. He lived in Ballingry, where Moffat had been a Communist councillor.[40]

He was an incisive thinker, outstanding orator and amateur of poetry, folksong and whisky. Politically Khrushchev's speech to the 20th Congress crystallized his longstanding doubts about the unquestioning acceptance of the policies of the Russian state by the leadership of his union and his party. Writing to the local press and to CP general secretary, John Gollan, he argued that what was central to the CP's existence, what condemned it in the eyes of the vast majority of British workers and what must be terminated forthwith, was its enduring endorsement of Soviet policy. British workers would never accept a Stalinist party. They were well aware that there was no democracy, no freedom, no human rights in Russia, whatever Horner or Moffat said. They believed, with reason, that, whatever party leaders claimed, if the CP came to power a similar state of affairs would pertain in Britain. The 'disgusting and disgraceful somersault' over Stalin, far from heralding a new dawn, simply confirmed that the CP leadership would swallow whole whatever Moscow dictated and was 'as reprehensible as was its previous unquestioning acceptance of every aspect of Soviet Policy while Stalin was in power'. Daly called for the dissolution of the CP and, when he encountered a brick wall in King Street, turning his back irrevocably on Stalinism, resigned from the party in June 1956.[41]

In the crisis over Hungary, he affirmed the need for a new, anti-Stalinist left in the pages of the *Reasoner* and in the deliberations of the union. Distinguishing himself from right-wing critics of Stalinism, he campaigned against Moffat, who fought back ferociously, denying Daly access to the *Scottish Miner* on the grounds that 'our paper is a Trade Union paper and is not for the purpose of discussing internal differences within political parties or religious bodies'.[42] In 1957, Daly formed the short-lived Fife Socialist League, enrolling a number of miners among its membership. Given the CP's entrenched position, he found it heavy going in the union, but he speedily became a thorn in the side of the leadership.[43]

And he won in the end. In 1962, soon after Moffat retired, Daly was elected to the Scottish executive and became full-time agent for Fife the following year, in the face of sustained CP opposition. He defeated the ineffectual Communist candidate, Guy Stobbs, in the election for NUMSA general secretary in 1965. Wiser counsels in the shape of Michael McGahey prevailed and the CP came to terms with him. By then discontent with the inheritance of Moffat and Horner was advanced. Patience with loyalty to the NCB and moderation in all things was reaching breaking point. Daly led the union towards a new era. He became a member of the NUM executive, played a part in building the national left which Horner and Moffat had disregarded and, in 1968, was elected to Horner's old position as general secretary of the NUM.[44] If labourism was resilient, it was becoming a more radical, militant labourism. Stalinism never recovered from the sapping blows of 1956; it remained a pallid remnant of the fervent faith which had infused Horner and Moffat in their heyday. History marched on: in a few short years, Daly would constitute yet another of its coalfield casualties.

Acknowledgment

The original paper was prepared for the Conference on the Comparative History of Coalfield Societies at the University of Glamorgan in April 2002, but was not delivered owing to illness. We acknowledge the support of ESRC Award R000237924 and British Academy Award SG 33730.

Notes

1   For discussion of the literature on Horner, see John McIlroy and Alan Campbell, 'The heresy of Arthur Horner', *Llafur: Journal of Welsh Labour History*, **8**(2)(2001), 105–18. For the literature on Moffat, see Alan Campbell and John McIlroy, 'Moffat, Abe, 1896–1975', in Keith Gildart and David Howell (eds), *Dictionary of Labour Biography*, vol. 12 (London, 2005).

2   See, for example, Robin Page Arnot, *The Miners in Crisis and in War* (London, 1961); *The Miners: One Union, One Industry* (London, 1979); *A History of the Scottish Miners* (London, 1955).

3   Hywel Francis and David Smith, *The Fed: A History of the South Wales Miners in the Twentieth Century* (London, 1980).

4   Stephen McBride, 'Trade union leadership in capitalist society: between context and disposition', *Studies in History and Politics*, **5**(1986), 10–12.

5   Nina Fishman, *The British Communist Party and the Trade Unions, 1933–45* (Aldershot, 1995), p.330. See, for example, pp.333–6, where 'trade union loyalism' is declared an initiative stemming from the party, rather than a problem stemming from trade unionism which the leadership had to manage and curb, and 'revolutionary pragmatism' is perceived as deriving from religious non-conformism, presumably leaving matters to individual conscience. Not only is democratic centralism thus liquidated by the stroke of a pen, but Stalinism is related to Protestantism rather than, perhaps more fittingly, to Catholicism.

6   Nina Fishman, 'Coal: owned and managed on behalf of the people', in Jim Fyrth (ed.), *Labour's High Noon: The Government and the Economy, 1945–51* (London, 1993), p.73. See also Nina Fishman, 'Heroes and anti-heroes: Communists in the coalfields', in Alan Campbell, Nina Fishman and David Howell (eds), *Miners, Unions and Politics, 1910–47* (Aldershot, 1996), pp.93–117.

7   For Horner's early career, see Arthur Horner, *Incorrigible Rebel* (London, 1960), pp.9–90; John Saville, 'Horner, Arthur, 1894–1968', in Joyce Bellamy and John Saville (eds), *Dictionary of Labour Biography*, vol. 5 (London, 1979), pp.112–18; Michael Casey and Peter Ackers, 'The enigma of the young Arthur Horner: from Churches of Christ preacher to Communist militant, 1894–1920', *Labour History Review*, **66**(1)(2001), 3–23. For Moffat, see Abe Moffat, *My Life With The Miners* (London, 1965), pp.1–48; Campbell and McIlroy, 'Moffat' pp.182–3.

8   Campbell and McIlroy, 'Moffat' pp.184–5.

9   For the Scottish and Welsh coalfields in this period, see Alan Campbell, *The Scottish Miners, 1874–1939*, two vols (Aldershot, 2000); Stefan Berger, 'Working-class culture and the labour movement in the South Wales and Ruhr coalfields, 1850–2000', *Llafur: Journal of Welsh Labour History*, **8**(2)(2001), 5–40; Chris Williams, *Capitalism, Community and Conflict: The South Wales Coalfield, 1898–1947* (Cardiff, 1998).

10   McIlroy and Campbell, 'Heresy' pp.116–7.

11   Campbell and McIlroy, 'Moffat' pp.184–5.

12   For critical discussion of the move to the Popular Front and its impact on CP policy in the unions, see John McIlroy and Alan Campbell, 'Histories of the British Communist Party: a user's guide', *Labour History Review*, **68**(1)(2003), 33–59. The literature on union bureaucracy and its pressures is immense and inconclusive. For valuable introductions, see Richard Hyman, *Marxism and the*

Sociology of Trade Unionism (London, 1971); John Kelly, *Trade Unions and Socialist Politics* (London, 1988), pp.147–83.

13 Russian State Archive of Social and Political History (RGASPI), 495/100/1006, Moffat to CP Secretariat, 26 September 1935.

14 Campbell and McIlroy, 'Moffat', p.186; John McIlroy and Alan Campbell, 'Beyond Betteshanger: Order 1305 in the Scottish coalfields during the Second World War, part 1: politics, prosecutions and protest', *Historical Studies in Industrial Relations*, **15** (Spring 2003), 27–72; John McIlroy and Alan Campbell, 'Beyond Betteshanger: Order 1305 in the Scottish coalfields during the Second World War, part 2: the Cardowan story', *Historical Studies in Industrial Relations*, **16** (Autumn 2003), 39–80.

15 John McIlroy and Alan Campbell, 'Communist trade union leaders, 1947–1991: the case of the South Wales Miners', paper presented to International Conference on Comparative Communist Biography, Manchester, 6–8 April 2001.

16 RGASPI, 495/100/1004, Statement on the Events in South Wales in the Recent Mining Struggles [1935]. For CP criticism of Horner's lack of involvement in the South Wales party, see University of Swansea Archives, Glyn Evans Papers, G1, Report on the South Wales District, March 1935, p.4.

17 For Horner as SWMF President after 1936, see Francis and Smith, *Fed*, pp.308–424.

18 *Daily Worker*, 4 September 1940.

19 Horner, *Incorrigible Rebel*, p.163.

20 Francis and Smith, *Fed*, pp.395–419.

21 McIlroy and Campbell, 'Communist trade union leaders'.

22 For his opposition to Lawther, Bowman, Watson and Jones, see Moffat, *Life*, pp.265–96. For Horner's differences with the right wing, see *Incorrigible Rebel*, pp.184–92.

23 Moffat, *Lite*, pp.268–70.

24 Michael Stewart, 'Comrade Arthur bows out', *News Chronicle*, 26 January 1959.

25 South Wales Miners' Library, Swansea, AUD 467, interview with Vol Tofts.

26 Campbell and McIlroy, 'Moffat' pp.187–8.

27 Ibid.

28 Horner, *Incorrigible Rebel*, p.218.

29 Ibid., p.217.

30 Moffat, *Life*, p.317.

31 Ibid., p.313.

32 *Daily Worker*, 28 November 1949.

33 Ibid.

34 Ibid.

35 National Union of Mineworkers, Executive Committee Minutes, 1957, 82–3; *Daily Worker*, 19 November 1956.

36 NUMSA, Executive Committee (EC) Minutes and Minutes of Special Conference, 12 November 1956.

37 NUMSA, EC, 3 December 1956, 1 April 1957, Annual Conference (AC), 5–7 June 1957.

38 Kevin Morgan, 'Parts of people', in John McIlroy, Kevin Morgan and Alan Campbell (eds), *Party People, Communist Lives* (London, 2001), p.26; and see Richard Croucher, review of McIlroy *et al.*, *Party People*, in *Historical Studies in Industrial Relations*, **14** (2002), 180.

39   Modern Record Centre, University of Warwick (MRC), MSS 302/5/8, Lawrence Daly Papers.

40   Moffat, *Life*, p.31.

41   Interview with Willie Clark, 4 May 2000; *The Reasoner*, September 1956; MRC, MSS 302/3/2, John Gollan to Daly, 16 May 1956, Daly to 'The Editor', 20 June 1956.

42   MRC, MSS 302/2/1, John Wood to Daly, 1 April 1957.

43   MRC, MSS 302/3/11, Fife Socialist League Minutes; Clark interview; NUMSA, EC, 1 April 1957, AC, 5–7 June 1957.

44   Vic Allen, *The Militancy of British Miners* (Shipley, 1981), pp.124–36.

Index